Mathematical Methods in Engineering

Nuno Miguel Fonseca Ferreira • José António Tenreiro Machado
Editors

Mathematical Methods in Engineering

Springer

Editors
Nuno Miguel Fonseca Ferreira
Coimbra Institute of Engineering
Coimbra
Portugal

José António Tenreiro Machado
Dept of Electrical Engineering
Institute of Engineering of Porto
Porto
Portugal

ISBN 978-94-007-7182-6 ISBN 978-94-007-7183-3 (eBook)
DOI 10.1007/978-94-007-7183-3
Springer Dordrecht Heidelberg New York London

Library of Congress Control Number: 2014938246

Printed on acid-free paper

Springer is part of Springer Science+Business Media (www.springer.com)

Preface

This book addresses, in a single volume, some of the contributions that are carefully selected according to the reports of referees, presented at the International Symposium, MME10 Mathematical Methods in Engineering, held in Polytechnic Institute of Coimbra- Engineering Institute of Coimbra (IPC/ISEC), Portugal, October 21–24, 2010.

The Symposium provided a setting for discussing recent developments issues about theoretical and applied areas of mathematics and engineering. The conference was intended to be an international forum where an effective exchange of knowledge and experience amongst researchers active could take place.

The members of the organizing committee were Micael Couceiro and Nuno Ferreira.

We would like to thank all the referees and other colleagues who helped in preparing this book for publication. Our thanks are also due to all participants for their contributions to the symposium and to this book.

Our special thanks are due to Nathalie Jacobs and Cynthia Feenstra from Springer, for their continuous help and work in connection with this book.

Contents

Mathematical Modeling for Software-in-the-Loop Prototyping of Automated Manufacturing Systems

Claudio Bonivento, Matteo Cacciari, Andrea Paoli and Matteo Sartini

Abstract Nowadays automated manufacturing systems are designed as the complex interconnection of components belonging to different engineering domains. Actually high performances are required in order to satisfy market needs and standards. In this framework the validation via simulation plays a crucial role as it allows to verify the system during the design phase. Software-in-the-loop architectures represent a good practice to take into account also technological side-effects that represent a classical cause of long time-to-market or, in the worst case, to project failure. In this paper we present a mathematical simulator to be used within a software-in-the-loop prototyping system.

Keywords Multi-domain simulator · Mechatronics · Rapid prototyping · Validation by simulation

1 Introduction

Recently in most of industrial processes an ever increasing degree of automation has been observed. This is motivated by new request of systems with high performances in terms of quality of products and services, productivity, efficiency and low costs in the design, realization and maintenance. This trend in the growth of complex automation systems is rapidly spreading over Automated Manufacturing Systems (AMS). Nowadays automation is based on the integration between different areas: automatica and mathematical control theory, mechanics, electrical devices and electronics and

C. Bonivento (✉) · M. Cacciari · A. Paoli · M. Sartini
CASY—Center for Research on Complex Automated Systems, University of Bologna, Viale Carlo Pepoli 3/2, Bologna 40136, Italy
e-mail: claudio.bonivento@unibo.it

M. Cacciari
e-mail: matteo.cacciari@unibo.it

A. Paoli
e-mail: andrea.paoli@unibo.it

M. Sartini
e-mail: matteo.sartini@unibo.it

N. M. Fonseca Ferreira, J. A. Tenreiro Machado (eds.), *Mathematical Methods in Engineering*, DOI 10.1007/978-94-007-7183-3_1,

computer engineering (see [1–3]), and this makes automation a tough task. Time-to-market is a crucial issue in developing any industrial product and, consequently, the request of reducing development time in realizing automated industrial plants is getting more and more tight. Toward this purpose, many tools have been developed to speed-up the testing phase of control logic in the development life-cycle. One way to improve development time is to develop hardware and software in parallel. Usually, this approach involves separate hardware and software professional teams development that perform their work simultaneously and independently. As soon as an hardware prototype and a substantial portion of the embedded code become available, such hardware and software are combined in a system integration phase and the testing task begins. Too frequently, serious problems arise during this system integration process and this typically causes significant hardware reconfiguration or software workarounds which worst the time to market. To develop software independently from hardware while avoiding integration problems an important key factor is the availability of simulation tools. In fact integration problems are due to side effects deriving from technological aspects which cannot be considered if software is developed independently from hardware. Besides this, the simulation is not only important to reduce software development time, but also to analyze system properties such as, fault diagnosis, fault reconfiguration and safety that cannot be tested on the field (see [4, 5]). From this brief discussion it turns out that a good simulation tool is such if is based on a properly detailed mathematical model of the system, and it is capable to capture all technological aspects linked to the field device (mechanical implementation, communications etc.).

Several software tools exist to model and simulate separately all different aspects. In particular we refer to physical domain simulator of the system dynamics, logic control CACSD tools, electronic and mechanics CAD. So, the integration problem still remain open. An approach to test the control software algorithm while considering technological aspects is the Software-In-the-Loop (SIL) simulation. In this approach the control unit interacts with a mathematical simulator of the plant evaluating both logic correctness and the side effects coming from the implementation (see [6–8]).

The aim of this work is to present a SIL technological architecture for rapid prototyping that embeds a multi-domain mathematical simulator implemented on a PC, a logic control running on a PLC and a communication infrastructure between the simulator and the controller.

This paper is organized as follows. In Sect. 2 we present a description of our proposed architecture and, in Sect. 3, the mathematical model of the system. In Sect. 4 we show the application to a micro FMS.

2 A SIL Technological Architecture for Rapid Prototyping

In order to present the needs and requirements for the proposed SIL architecture, we present a simple FMS that has motivated in this work. This system will be used as a testbed along the the work.

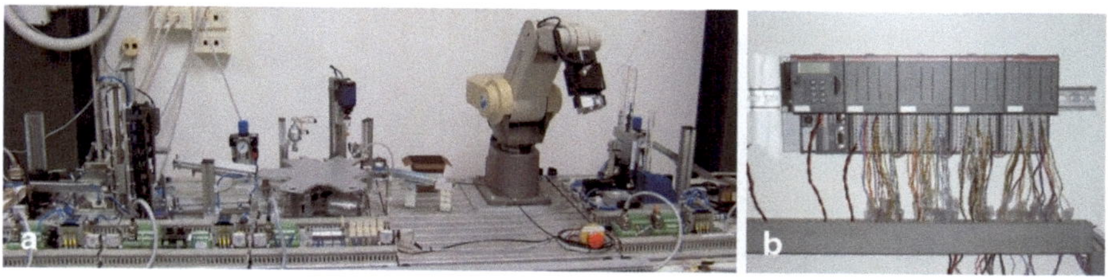

Fig. 1 Testbed hardware. **a** Flexible manufacturing system. **b** Control hardware

The testbed is a miniaturized flexible manufacturing system (FMS) produced by FESTO-DIDACTIC (see Fig. 1a); the plant is devote to produce short-stroke cylinders each of them composed by a basic body, a piston, a spring and a cap. In particular the system starts from raw pieces which are worked to realize the bodies and assemblies them with the other parts to obtain the desired cylinder. Thanks to the use of different basic bodies it is possible to realize different diameter cylinders. In the following cylinders' bodies will be referred as workpieces. The FMS is composed by four stations (see Fig. 1a): the first station is the distribution station, where the workpiece is picked from the raw materials warehouse and moved to the second station, the testing station. In testing station the workpiece is measured and its color and height is identified. According to this measurements the workpiece is discarded or moved to the processing station; in this station the workpiece is tested to verify if it can be worked or not. If the workpiece positively passes the test, it is drilled and then moved to the last station, the assembly station, where workpieces are assembled by a robotic manipulator to realize the cylinder. For a complete description of the system the reader is referred to [9]. The control of the FMS is implemented on a ABB PLC belonging to AC500 family equipped with CPU PM581-ETH with four input/output modules DC523 (see Fig. 1b)

The FESTO-FMS is equipped with sensors and actuators related with pneumatic and electric technology and is driven by control logic (event-driven) implemented on the real time computational architecture (PLC). The interaction of all these different fields makes the simulation of the system a complex task.

An important remark should be done on the communication between the control logic and the field; The state of the plant is read by the set of sensors and communicated to the PLC. On the basis of "picture" that freeze the state of the plant, the control logic compute the actual control action that is communicated to the set of actuators. This scenario must be reproduced also in the SIL architecture: the mathematical model should also simulate the logic sensor readings that will be send to the PLC in the form of a data vector. In the same way, the control action computed by the control logic (all boolean values) is sent back to the simulator that must accept it as input of the drivers of the simulated physical components. The communication is performed via ethernet using OPC communication. The main advantage of OPC is that is realize data exchange between different hardware nodes guaranteeing interoperability and flexibility (see [10–12]). In our architecture both the simulator and the control logic

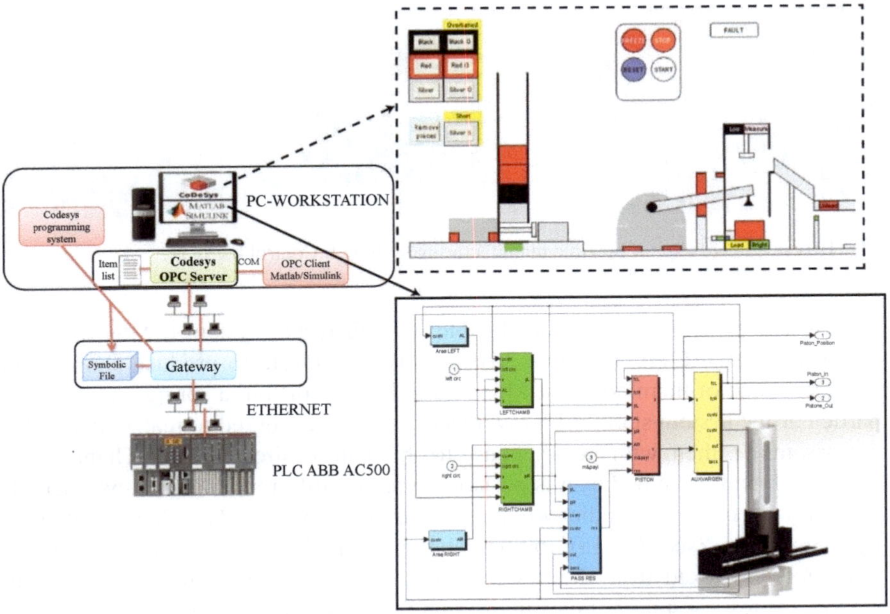

Fig. 2 SIL technological architecture

interface with an OPC client (the first is installed on the PC, the latter on the PLC), while an OPc server (installed in the PC) realize the communication between the two clients. More in details the OPC server manage symbolic files containing variables from time to time read or written by PC or PLC. At the same time is guarantied the data synchronization using time-stamps of data. it is important to stress at this stage the OPC do not realize a real time communication. This extension belongs to our ongoing research and can be realized physically (using a real time network) or virtually (managing time at both simulator and controller sides).

The task of the simulator implemented on the PC is to simulated components dynamics and prepare input data vector for PLC. We implemented the mathematical model of the components with Matlab/Simulink. Matlab/Simulink supports different toolboxes to help the designer to model systems, using a multi-domain approach. Using Matlab/Simulink we have an easy integration between different components and it is possible to build open model, changing in easy way the level of accuracy of the mathematical model. Moreover Matlab/Simulink embeds the OPC client that is used for communicated. The control logic has been developed with Codesys, a logic CACSD tool, the same that also run in the PLC. Codesys can also be used to design a logic simulator when all components are modeled as discrete integrators. Having Codesys also installed in the PC where the simulator runs (besides Matlab), it is possible, from time to time to decide which simulator should be used, changing the level of details of the simulator see (Fig. 2). Finally, since Codesys embed an OPC client, also the communication can be performed without the needs of any other tool.

3 Mathematical Modeling of Automation Systems Components

In this section we are going to show the mathematical models and within the simulator. For the sake of brevity here we present only the model of a pneumatic component. The intrinsic difficulty of modeling a system of this type is the strong interaction between mechanical and pneumatic physical domains. In a pneumatic system the potential energy generated by compressed fluid is converted to mechanical energy using valves and actuators. For this reason we can speak of multi-domain modeling.

Before seeing our implementation it is usefull to introduce some pneumatic basics. It is well known that, like others physical domains, a pneumatic circuit can be associated to an equivalent electric circuit (see [13]). The pressure (p) can be assimilated to a voltage and the mass flow rate (G), different from the volume flow rate because of air compressibility to the electric current. As in electrical circuits, also in pneumatic domain can be introduced the concepts of resistance (R), inductance (L_p) and capacitance (C_p) of a component.

Resistance is defined as the pressure derivative respect to mass flow. Analytically deducing this value is very difficult because is variable with the mass flow rate. Instead of using a complex formula like $G = f(\Delta p)/R(\Delta p)$, it is preferable to use two experimentally deducible constant parameters (sonic conductance C and critical pressure ratio b) that are characteristic of each pneumatic resistive component.

Exploiting these constants and assuming polytropic transformations, the value of G can be found as:

$$G = \rho_a \cdot Q = \rho_a \cdot C \cdot p_1 \cdot \sqrt{1 - \left(\frac{\frac{p_2}{p_1} - b}{1 - b} \right)^2} \cdot \sqrt{\frac{293}{T}} \qquad \textit{subsonic case} \tag{1}$$

$$G = \rho_a \cdot Q = \rho_a \cdot C \cdot p_1 \cdot \sqrt{\frac{293}{T}} \qquad \textit{sonic case} \tag{2}$$

where p_1 and p_2 are respectively the upstream and downstream pressure of any components, ρ_a is the air density and Q is the volume flow rate. According to the rate between downstream and upstream pressure ($b = p_2/p_1$) we can be in subsonic case ($b < \theta$) or in sonic case ($b \geq \theta$) where θ is a constant function of the gas specific heat ratio. In few words, due to the Venturi effect, when upstream pressure is larger than θ times the downstream pressure ($b = 0.5283$), the flow is independent from the downstream pressure (in this case we say that the flow is chocked or that the flow has sonic velocity).

Pneumatic capacitance is typical of components with non negligible volume and, assuming a perfect gas, it is linked with the pressure by the equation:

$$\frac{dp}{dt} = \frac{G_{in} - G_{out}}{C_p} \tag{3}$$

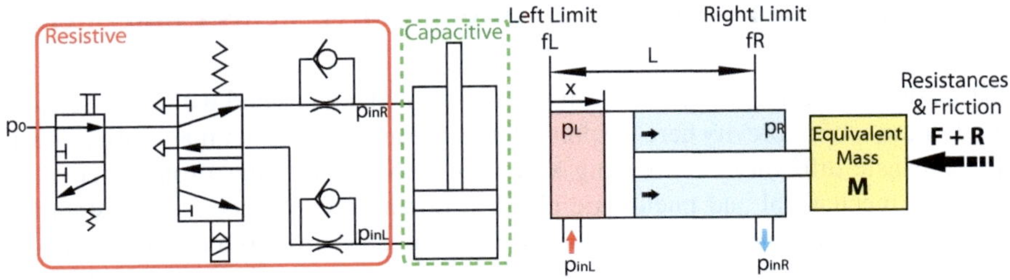

Fig. 3 Pneumatic system

Finally the pneumatic inductance is linked with the pressure by the equation:

$$p_1 - p_2 = L_p \cdot \frac{dQ}{dt}$$

Every pneumatic component has a resistance, a capacitance and an inductance. We consider a pneumatic component as resistive if the resistive effect is dominant respect to the others. Besides it is important to notice that usually pneumatic circuits are represented as simple RC circuits, in fact inertial phenomena are usually negligible with respect to resistive and capacitive ones. In rough words the dynamical behavior of a pneumatic circuit is equivalent to filling/emptying capacitive elements (volumes) through resistive elements at the inputs and outputs.

In the following we will study the mathematical model of the system depicted in Fig. 3. The model can be split in two parts: a resistive part composed by a set of resistive elements and a capacitive part including the pneumatic cylinder.

The pressure source p_0 feeds the first part composed by distribution valves, unidirectional flow control valves, pneumatic fits and pipes. All these components are resistive elements characterized by a sonic conductance C and a critical pressure ratio b constant. As in an electric circuits, the resistive element in series can be substituted with an equivalent resistive element with an equivalent C_{eq} and an equivalent b_{eq}. The relations are:

$$C_{eq} = \sqrt[3]{\frac{1}{\sum \frac{1}{C_j^3}}} \qquad b_{eq} = 1 - C_{eq}^2 \cdot \left(\sum \frac{1 - b_j}{C_j^2} \right)$$

Concerning the second part of the scheme, the pneumatic cylinder can be seen as two chambers element with a sliding common wall determining the piston movements. The dynamical behavior of the piston in a double effect cylinder can be described by:

$$p_L \cdot A_L - p_R \cdot A_R - (F + R) = M \cdot \ddot{x} \tag{4}$$

where p_L and p_R are the chambers pressures, A_L and A_R are the areas of the piston face subject to the pressure, M is the piston equivalent mass (the mass of the moving parts) and F and R are respectively the force due to the friction and to the payload.

Since the volume of the chambers is variable, the pneumatic capacitances of the pneumatic cylinder are variable and can be computed as:

$$C_{pL} = (A_L \cdot x + V_{0s}) \cdot \frac{\rho_i}{n \cdot p_{atm}} \cdot \left(\frac{p_L}{p_{atm}}\right)^{\frac{1-n}{n}} + \rho_i \cdot \left(\frac{p_L}{p_{atm}}\right)^{\frac{1}{n}} \cdot A_L \cdot \dot{x} \quad (5)$$

$$C_{pR} = (A_L \cdot [L - x] + V_{0d}) \cdot \frac{\rho_i}{n \cdot p_{atm}} \cdot \left(\frac{p_R}{p_{atm}}\right)^{\frac{1-n}{n}} - \rho_i \cdot \left(\frac{p_R}{p_{atm}}\right)^{\frac{1}{n}} \cdot A_R \cdot \dot{x} \quad (6)$$

where C_{pL} and C_{pR} are respectively the left and right chamber capacitances, L is the length of the cylinder, p_{atm} is the atmospheric pressure, n is the polytropic index ($n = 1$) and ρ_i is the air density at the initial condition.

Knowing C_{eq}, b_{eq} and the pressure at the extremities of the duct (p_{inL} or p_{inR} and p_L or p_R), remembering Eqs. (1) and (2), we can find the mass flow rate entering or outgoing the chambers:

$$G_{in} = \rho_a \cdot C_e \cdot pL \cdot \sqrt{1 - \left(\frac{\frac{p_L}{p_{inL}} - b}{1 - b}\right)^2} \cdot \sqrt{\frac{293}{T}} \quad \textit{Subsonic entering flow} \quad (7)$$

$$G_{in} = \rho_a \cdot C \cdot p_{inL} \cdot \sqrt{\frac{293}{T}} \quad \textit{Sonic entering flow} \quad (8)$$

For the outgoing flow the equations are the same. The only difference is that, since $p_L > p_{inL}$, the two pressures must be swapped.

Using Eqs. (1), (2), (5) and (6), and exploiting the relation (3), we can find the p_L and p_R evolution.

Having p_L, p_R, the piston areas (A_L and A_R), the payload and the friction force, it is possible to find the piston position (x in Fig. 3). This model has been implemented using Matlab/Simulink (see Fig. 4) to simulate a controlled device of the FMS described in Sect. 2.

4 Application to the Experimental Setup

In order to show the application of the SIL architecture to the FESTO-FMS (see Fig. 5), we present here the model of the cylinder warehouse whose aim is to distributed workpieces to testing station using a singular pneumatic cylinder. The Cylinder is fed by PLC using digital signal to force the movement and a different digital signal to enable the air supply. The device is equipped with two sensor that read the two limit of its stroke. These two digital signal are send to the PLC.

It is possible to define the parameters C and b of the resistive elements and the characteristics of the pneumatic cylinder like: piston diameter, rod diameter (right

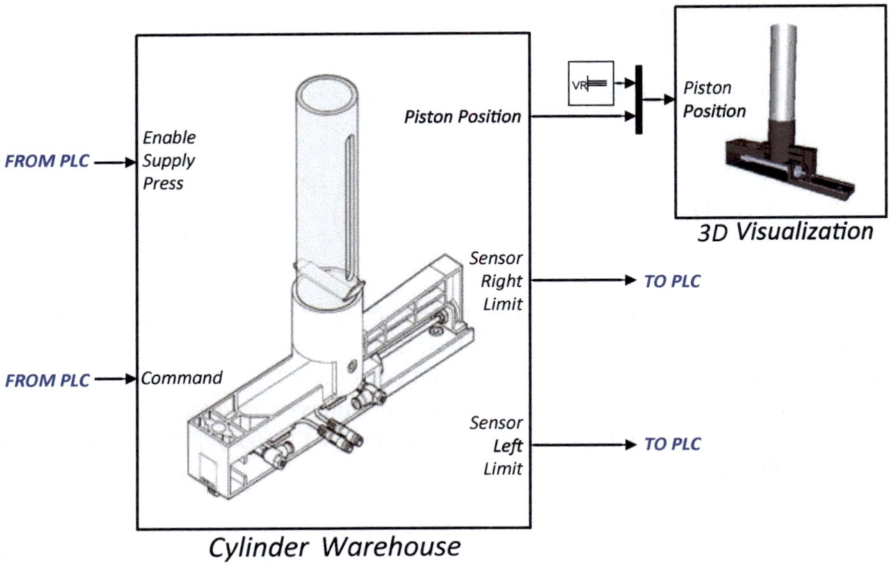

Fig. 4 Simulink scheme

and left), the friction parameters, the extraction delay, the payload during the entering and the outgoing of the piston and the equivalent mass of the piston (calculated using CAD) etc.

The presented Simulink block is directly connected to the PLC and to the 3D visualization. The Cylinder Warehouse receives as inputs from the PLC the signals that enable the supply pressure and that command the distribution valves (hence the direction of the piston move). On the other hand the PLC receives from Simulink the value of the limit switch sensors. These are virtual sensors created inside the block that give 1 if the piston reach the stroke limit or 0 otherwise. The 3D visualization has been realized importing a CAD model of the system into the Matlab toolbox Virtual Reality Toolbox. This toolbox allows the interaction with the 3D object using signals coming from Simulink. In this case the signal of interest is the piston position x.

To tune the model we have used producer (where possible) data sheet, but data like piston seal friction coefficient or the regulation of the flow valves are unknown. Obtaining these data requires experimental identification of parameters with expensive and not easy to find instruments. Due to this problem and due to the high number of parameters using a grey box identification approach, we preferred tuning the parameters to have a comparable response with the response obtained with Festo ProPneu simulator. Since the parameters are linked each other, the tuning is not easy. For example the mass of the components has been calculated using CAD tools, the friction has been achieved multiplying the weight of the moving components for the friction coefficient of the material and the resistance associated to the flow valves depends on

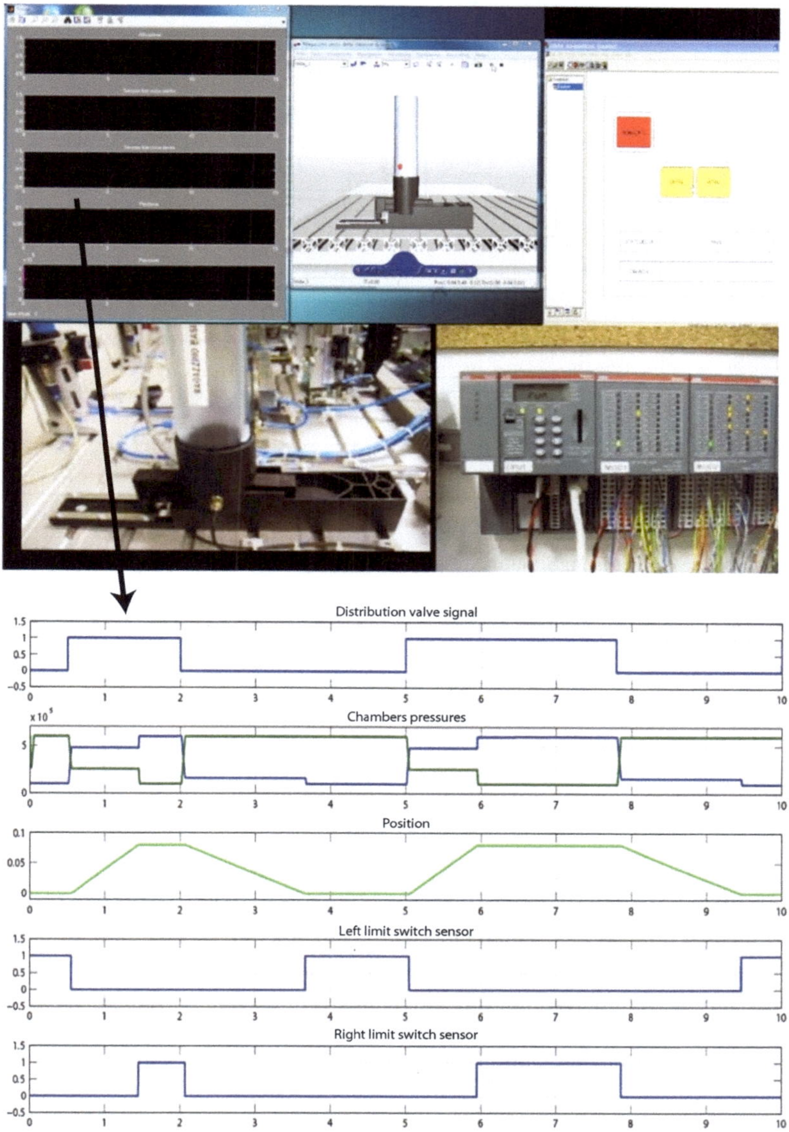

Fig. 5 Application to testbed: user front-end (*above*) and simulation results (*below*)

the mechanical resistance due to the friction. The tested control logic is very simple and roughly realize a cyclic run of the device along its stroke.

In Fig. 6 the comparison of position, velocity and chambers pressure are presented. The position and the velocity are very similar to the results obtained with ProPneu, although the mass and the loads are very little (0.06 Kg and 0.2/0.3 N).

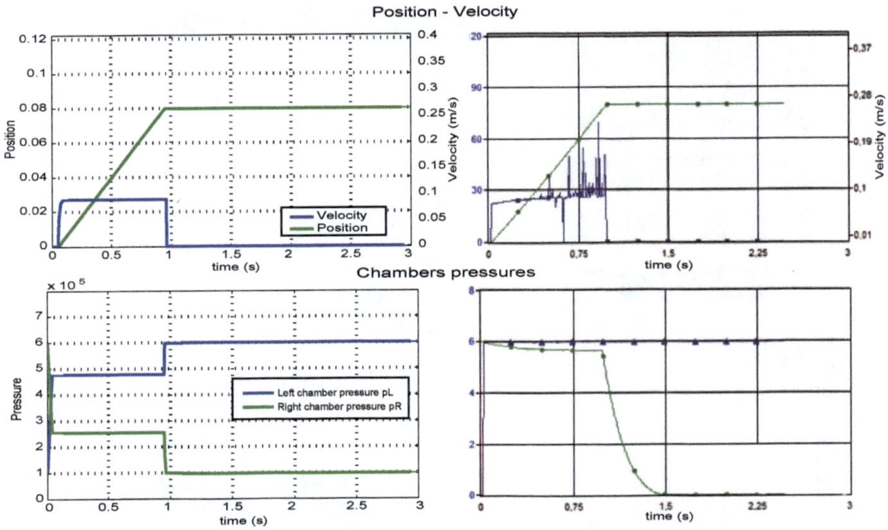

Fig. 6 Response comparison: Simulink response on the *left* and ProPneu response on the *right*

The chamber pressures have a similar behavior in the initial part but the piston starts moving before the pressure reach the 6 bar. This behavior is due to friction parameters difficult to estimate.

5 Conclusions

In this paper we have presented a software-in-the-loop prototyping architecture for automated manufacturing system based on a mathematical simulator. This architecture embeds a multi-domain mathematical simulator implemented on a PC, a logic controller running on a PLC, and a communication infrastructure properly linking together simulator and controller. Both the key roles of mathematical modeling and governor architecture are emphasized as key component of an integrated solution necessary for engineering application such as the FESTO-FMS presented and discussed.

References

1. Thramboulidis KC (2005) Model integrated mechatronics—towards a new paradigm in the development of manufacturing systems. IEEE Trans Industr Inform 1:54–61
2. Bonfè M, Fantuzzi C (2004) Application of object-oriented modeling tools to design the logic control system of a packaging machine. IEEE international conference on control application industrial informatics, pp 506–574

3. Faldella E, Paoli A, Tilli A, Sartini M, Guidi D (2009) Architectural design patterns for logic control of manufacturing systems: the generalized device. XXII international symposium on information, communication and automation technologies
4. Benosman M (2009) A survey of some recent results on nonlinear fault tolerant control. Math Probl Eng 1:1–26
5. Blanke M, Kinnaert M, Lunze J, Staroswiecki M (2006) Diagnosis and fault-tolerant control. Springer, Heidelberg
6. Demers S, Gopalakrishnan P, Kant L (2007) A generic solution to software-in-the-loop. Military communications conference
7. Virzonis D, Jukna T, Ramunas D (2004) Design of the embedded software using flexible hardware-in-the-loop simulation scheme. 9th IEEE Mediterranean electrotechnical conference
8. Vyatkin V, Hanish HH, Pang C, Yang C (2009) Closed-loop modeling in future automation system engineering and validation. Trans Syst Man Cybern 39:17–28
9. FESTO-Didactic (2003) Distribution, testing, processing, assembly station manual
10. OPC Overview OPC foundation. https://opcfoundation.org/
11. Mahnke W, Leitner S, Damm M (2009) OPC unified architecture. Springer, New York
12. Santos RA, Normey-Rico JE, Gomez AM, Acebes Arconada LF, Moraga C (2005) OPC based distributed real time simulation of complex continuous processes. Simul Model Pract Theory 13:525–549
13. Merrit HE (1967) Hydraulic control systems. Wiley, New York

A New Parallel Matrix Multiplication Algorithm for Wormhole-Routed All-Port 2D/3D Torus Networks

Cesur Baransel, Kayhan Imre and Harun Artuner

Abstract A new matrix multiplication algorithm is proposed for massively parallel supercomputers with 2D/3D, all-port torus interconnection networks. The proposed algorithm is based on the traditional row-by-column multiplication matrix product model and employs a special routing pattern for better scalability. It compares favorably to the variants of Cannon's and DNS algorithms since it allows matrices of the same size to be multiplied on a higher number of processors due to lower data communications overhead.

Keywords Fast matrix multiplication · Parallel processing · Torus interconnection networks · 2D Torus · 3D Torus

1 Introduction

Matrix multiplication is one of the most prevalent operations in scientific computing and its effective parallelization is of utmost importance for being able to harness the processing power offered by massively parallel supercomputers. Matrix multiplication can be formulated according to two general classes of computational models. In the first class, the matrix product can be defined as a *row-by-column multiplication* and for $C = AB$, $C(i, j)$ is the dot product of the ith row vector of A and j^{th} column vector of B. It is also possible to define the matrix product as *sum of a series of column-row products*. Assuming that A and B are of order $n \times n$, both definitions require the same number of multiplications, namely n^3. The second class is comprised of the algorithms[1] aimed to reduce the number of multiplications such as the algorithms proposed by Strassen [6] and Winogard [7]. In this paper, we propose a new matrix multiplication algorithm for 2D/3D torus topology where matrix product

[1] A review of these methods can be found in Chap. 47, Handbook of Linear Algebra [1].

C. Baransel (✉)
Saltus Yazılım Ltd., Hacettepe University Technopolis, Ankara, Turkey
e-mail: casur@ada.net.tr

K. Imre · H. Artuner
Dept. of Computer Engineering, Hacettepe University, Ankara, Turkey

N. M. Fonseca Ferreira, J. A. Tenreiro Machado (eds.), *Mathematical Methods in Engineering,* DOI 10.1007/978-94-007-7183-3_2,

is defined as a *row-by-column multiplication.* A special routing pattern for 2D/3D tori is integrated into the proposed multiplication algorithm for efficiently exploiting the available bandwidth to provide higher scalability. Torus has proved to be the most popular topology in industry over the years and modern massively parallel supercomputers such as IBM Blue Gene®/L and CRAY XT3 employ 2D/3D torus interconnection networks to accommodate tens of thousands of processing elements (PE).

This paper is structured as follows. A short review of the previous work is provided in the next section. We will give the details of the proposed algorithm in Sect. 3. Section 4 provides the performance results. Paper ends with conclusions.

2 Previous Work

Assume two matrices A and B, both of size $n \times n$, are mapped[2] onto a 2D array of p processing elements (PEs), arranged as a $\sqrt{p} \times \sqrt{p}$ torus with $(p < n^2)$. Consequently, each PE stores a separate block of (n^2/p) entries from matrices A, B and C, where $C = AB$. Since, computing $C(i, j)$ requires the ith row vector of A and j^{th} column vector of B, a simple parallel matrix multiplication algorithm can be defined as follows;

1. perform an all-to-all broadcast within each row,
2. perform an all-to-all broadcast within each column,
3. perform required *multiply-and-add* operations.

Under all-port model, column and row broadcasts can be executed in parallel. Broadcasting a block within a row or column takes $(\log_3 \sqrt{p})$ steps to complete and there are $\sqrt{p}$ blocks to broadcast within each row or column. After the broadcasting phase is completed, each PE will have $(n^2/\sqrt{p})$ matrix entries and will perform (n^3/p) multiply-and-add operations. Assuming a multiply-and-add operation takes unit time, the parallel cost T_{par} of the algorithm is given in Eq. (1),

$$T_{par} = \sqrt{p}\,\log_3 \sqrt{p} \left(t_s + \frac{n^2}{p} t_w \right) + \frac{n^3}{p} \tag{1}$$

where t_s and t_w represent the message startup time and the time to transmit a single matrix entry, respectively. Note that it is possible to drop the (n^3/p) term from the above cost by properly interleaving and overlapping broadcast and multiply-and-add operations. However, the above algorithm is not memory efficient since it consumes $\sqrt{p}$ times more space compared to the serial implementation.

Cannon proposed a memory-efficient matrix algorithm which takes $2\sqrt{p}$ steps to complete, including the initial and final alignment steps in [2]. Since this algorithm

[2] Throughout the paper, we assume that the matrix multiplication operation is to be performed such that this initial mapping is preserved at the end of the operation.

is explained elsewhere [3], its details will not be repeated here. Cannon's algorithm has the following phases on 2D torus.

1. the initial alignment; requires at most circular ($\sqrt{p}$-*1*)-shifts which can be completed in at most $\sqrt{p}/2$ steps, since a circular *q-shift* on a p node ring takes *min {q, p-q}* steps.
2. $\sqrt{p}$ steps of two direct-neighbor shifts followed by a multiply-and-add operation,
3. the final alignment which also can be completed in at most $\sqrt{p}/2$ steps.

The parallel cost T_{par} of Cannon's algorithm is given in Eq. (2).

$$T_{par} = 2\sqrt{p}\left(t_s + \frac{n^2}{p}t_w\right) + \frac{n^3}{p} \tag{2}$$

Here, it is also possible to drop the (n^3/p) term from the cost by properly overlapping transmission and multiply-and-add operations, assuming that the multiplication of two blocks can be completed by the time the next two blocks are received by the PE. In the next section, we will show that it is possible to complete matrix multiplication in $O\,(\log_5\sqrt{p})$ time rather than $O(\sqrt{p})$, at the expense of longer messages.

DNS algorithm, proposed by Dekel, Nassimi and Sahni can be employed both in hypercube and 3D torus architectures. This algorithm can use up to n^3 processors to complete the matrix multiplication operation in *O(log n)* time. DNS algorithm has four phases on 3D torus assuming p processors are arranged into a $\sqrt[3]{p} \times \sqrt[3]{p} \times \sqrt[3]{p}$ cube.

1. Assume that the matrices to be multiplied (i.e., A and B) and the result matrix C is to be stored on the bottom face of the cube. There are $\sqrt[3]{p}$ planes of $\sqrt[3]{p} \times \sqrt[3]{p}$ processors in the cube. Copy column i of the matrix A into the column i of the ith plane and row i of the matrix B into the row i of the ith plane. Regardless of the number of processors, this phase is completed in a single step.
2. row-wise and column-wise propagation within each plane. This phase is completed in $(\log_3\sqrt[3]{p})$ steps under all-port model since row-wise and column-wise propagation can be executed in parallel.
3. multiply,
4. reduce the result onto the bottom face of the cube in $(\log_3\sqrt[3]{p})$ steps by adding relevant terms.

The parallel cost T_{par} of DNS algorithm is given in Eq. (3).

$$T_{par} = 2\log_3\sqrt[3]{p}\left(t_s + \frac{n^2}{p^{2/3}}t_w\right) + \frac{n^3}{p} \tag{3}$$

Note that, it is not possible to drop the (n^3/p) term from the cost since multiply and add operations are not to be interleaved and therefore it is not possible to overlap communication, multiply and add operations.

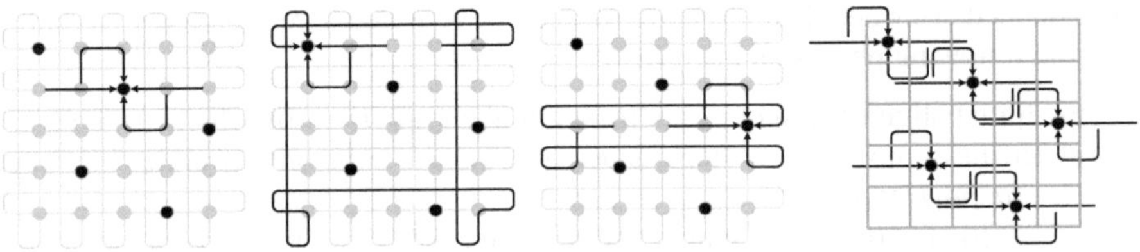

Fig. 1 Broadcast pattern

3 The Proposed Algorithms

In this section, we propose two algorithms for performing matrix multiplication on 2D torus architecture with up to n^2 processors and on 3D torus architecture with up to n^3 processors. The time complexities of both algorithms are logarithmic.

3.1 *Matrix Multiplication on 3D Torus with up to n^2 Processors*

Our 2D parallel matrix multiplication requires no alignment steps and completes in five phases. We introduce the algorithm assuming single matrix element per processor assignment; its extension to matrix blocks is trivial. For a $\sqrt{p} \times \sqrt{p}$ processor array, we define $\sqrt{p}$ *concentration processors* (CP) with no two CP are being in the same row or column. First, the *CP(i, j)* gathers the elements of the ith row of *A* and broadcasts it to the processors on the ith row in two consecutive phases. Then, *CP(i, j)* gathers the elements of the jth column of *B* and broadcasts it to the processors on the jth column, also in two consecutive phases. Since, computing *C(i, j)* requires the ith row vector of *A* and j^{th} column vector of *B*, all processors acquire the required data in four phases. The last phase is the *multiply-and-add* phase after which no data alignment is required.

The proposed algorithm differs from other proposals mainly in how row-wise and column-wise gathers and broadcasts are performed. Using the communication pattern given in Fig. 1, it is possible to complete each gather or broadcast in $(\log_5 \sqrt{p})$ steps on a wormhole-routed, all-port torus[3].

Although the proposed communication may seem somewhat complicated, the basic rule is quite simple On a 5×5 torus, *CP*(i, *j*) has two neighbors located two hops away and two immediate neighbors on row *i*. To communicate with the CP, the nodes located 2-hops away use the links within the row while immediate neighbors take a detour via row (*i-1*) and row $(i + 1)$. Since detour links are arranged to lay on the opposite direction to the within-row links, the communication pattern is contention free. Column-wise communication is arranged similarly and consequently all CPs

[3] For a good intro to routing in general and wormhole routing in particular, see [5].

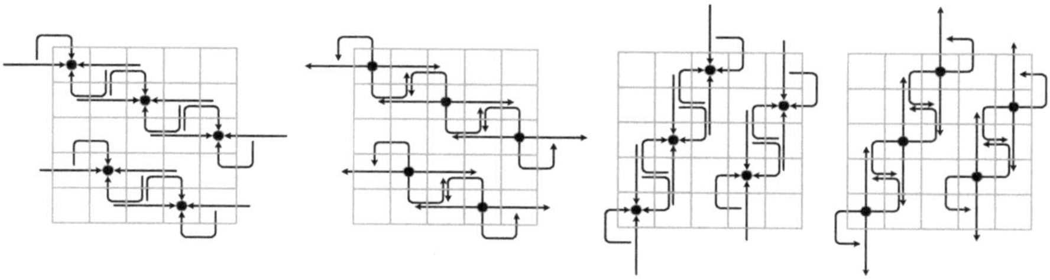

Fig. 2 Matrix multiplication on 5 × 5 torus, (**a**) Row-wise Gather (**b**) Row-wise Broadcast (**c**) Column-wise Gather (**d**) Column-wise Broadcast. Local multiply-and-add phase is not shown

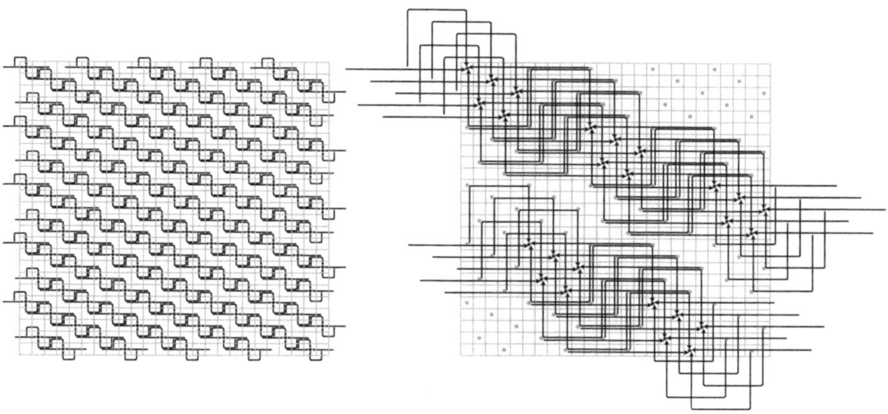

Fig. 3 Communication pattern for *25* × *25* torus

can perform row or column gathers or broadcasts in parallel. Matrix Multiplication on 5 × 5 torus, using this broadcast pattern is illustrated in Fig. 2.

The parallel cost T_{par} of the proposed algorithm is given in Eq. (4).

$$T_{par} = 4\log_5 \sqrt{p}\; t_s + \left(\frac{5^{(1+\log_5 \sqrt{p})} - 1}{2} \right) \frac{n^2}{p} t_w + \frac{n^3}{p} \tag{4}$$

Here, it is not possible to drop the (n^3/p) term from the cost since multiply and add operations are not to be interleaved and therefore it is not possible to overlap communication, multiply and add operations. Also note that message lengths are different for gather and broadcast phases.

The given seed broadcast pattern can be recursively extended to the powers of 5 by replacing each node by a *5* × *5* torus. Figure 3 shows the communication pattern for a *25* × *25* torus. Other seeds are also be defined for *2* × *2*, *3* × *3*, *4* × *4* and *5* × *5* tori (Fig. 4). These seed patterns can be used in combination to support virtually all practical matrix sizes [4].

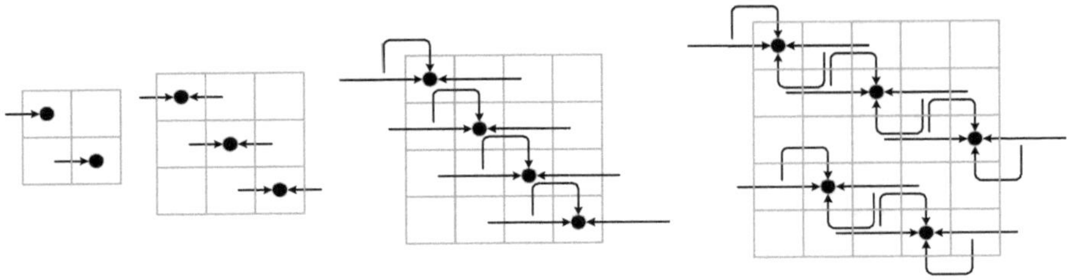

Fig. 4 Seed communication patterns for *2 × 2, 3 × 3, 4 × 4* and *5 × 5* tori

3.2 Matrix Multiplication on 3D Torus with up to n^3 Processors

The proposed 3D multiplication algorithm is basically an extension of our 2D multiplication algorithm onto the third dimension. On 3D torus, each processing element has six links compared to those four on 2D torus, and proper use of these extra two links allow the broadcast of elements of matrices A and B on the third dimension to be completed in the same phase and in $O\,(\log_5 \sqrt[3]{p})$ steps on a $\sqrt[3]{p} \times \sqrt[3]{p} \times \sqrt[3]{p}$ torus.

Figure 5 shows the communication patterns on a 5 × 5 × 5 torus for DNS algorithm and the proposed algorithm. In the figure, the elements of matrix A and matrix B are indicated with upper-case and lower-case letters, respectively. The elements of matrix A are broadcasted along the horizontal planes and the elements of matrix B along the vertical planes. The broadcast pattern of the proposed algorithm is more efficient compared to the broadcast pattern DNS algorithm since its time complexity is $O\,(\log_5 \sqrt[3]{p})$ steps, rather than $(\log_3 \sqrt[3]{p})$ of DNS on a $\sqrt[3]{p} \times \sqrt[3]{p} \times \sqrt[3]{p}$ torus. The parallel cost T_{par} of the proposed algorithm is given in Eq. (5).

$$T_{par} = 2\,\left(1 + \log_5 \sqrt[3]{p}\right)\left(t_s + \frac{n^2}{p^{2/3}} t_w\right) + \frac{n^3}{p} \quad (5)$$

4 Performance Analysis

Performance results are provided in Tables 1 and 2. In computing speedups, Strassen's algorithm with the cost of $n^{2.807}$ is taken as the best available serial implementation. The proposed 2D algorithm yields a speedup similar to Cannon's when the matrix block size is large (e.g., *753,76* vs. *726,31* when block size is *125* for *625 × 625* matrices). The proposed algorithm performs better as the block size gets smaller (e.g., 373,38 vs. *15.364,08* when block size is *1* for *625 × 625* matrices). The difference between the DNS and the proposed 3D algorithms is not as great compared to 2D case. However, the proposed algorithm is never slower than DNS and can provide up to *30 %* more speedups in some cases. The results also indicate that as the t_s/t_w ratio of the systems gets smaller, the difference between the algorithms also gets somewhat smaller both in 2D and 3D cases, but not significantly.

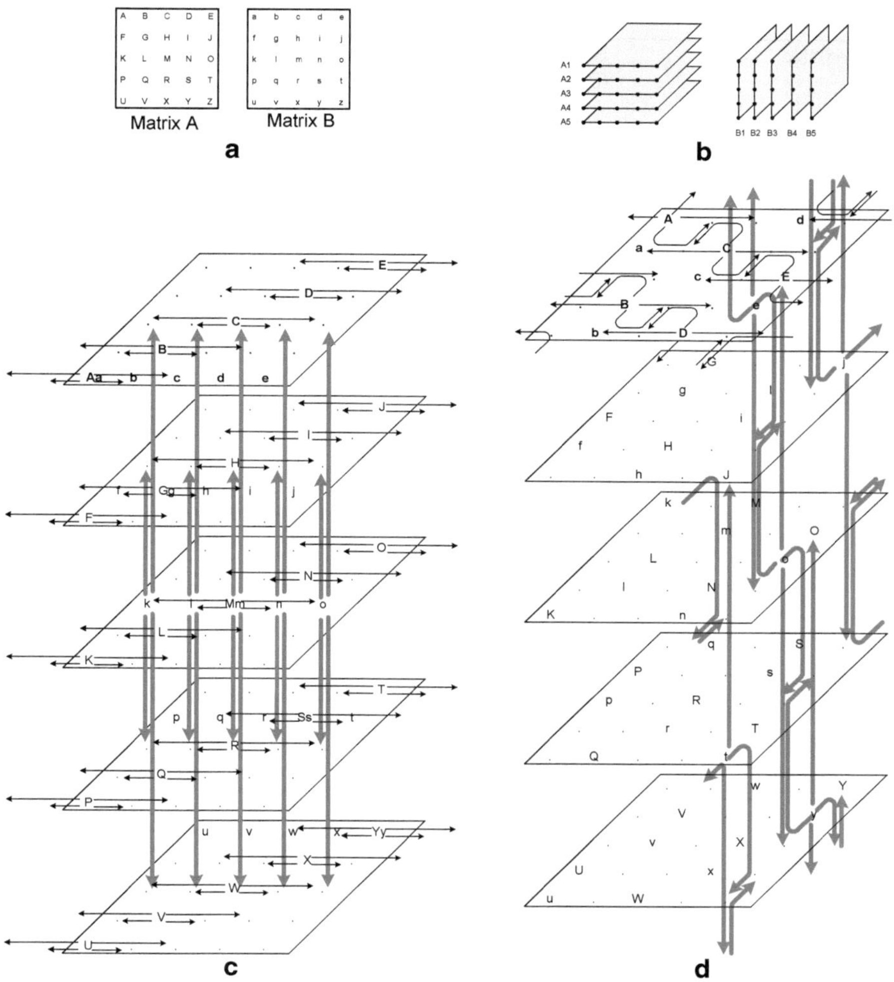

Fig. 5 Communication patterns on a 5 × 5 × 5 torus for (**a**) DNS algorithm (**b**) the proposed algorithm

5 Conclusions

In the last two decades, the number of processors in massively parallel supercomputers have been increased from tens of processors to tens of thousands of processors and, as progressively larger number of processors became available, the size of the matrix sub-blocks in matrix multiplication grew to be smaller for a given problem size. Consequently, new algorithms which can work efficiently with smaller blocks are required to exploit the processing power offered by the modern massively parallel

Table 1 Performance results for 2D algorithm

Matrix size	p	Block size (n^2/p)	$t_s/t_w = 150/1$		$t_s/t_w = 450/1$	
			Speed up for cannon	Speed up for 2D proposed	Speed up for cannon	Speed up for 2D proposed
25 × 25	25	25	4,80	5,50	1,77	3,08
	125	5	2,42	7,22	0,83	2,83
	625	1	1,11	6,52	0,37	2,28
125 × 125	25	625	9,38	8,92	9,21	8,80
	125	125	41,13	38,54	34,88	35,35
	625	25	87,91	130,92	32,39	92,95
	3125	5	44,39	272,63	15,12	132,13
	15625	1	20,38	343,84	6,82	131,78
625 × 625	25	15625	7,16	7,08	7,16	7,08
	125	3125	35,42	34,55	35,36	34,52
	625	625	171,89	163,68	168,80	162,77
	3125	125	753,76	726,31	639,12	704,53
	15625	25	1.610,86	2.793,86	593,47	2.444,93
	78125	5	813,35	8.085,38	277,08	5.456,26
	390625	1	373,38	15.364,08	125,01	7.507,73
3125 × 3125	25	390625	5,28	5,27	5,28	5,27
	125	78125	26,35	26,22	26,35	26,22
	625	15625	131,18	129,67	131,16	129,66
	3125	3125	649,05	632,92	647,96	632,73
	15625	625	3.149,85	3.003,45	3.093,27	2.998,43
	78125	125	13.812,49	13.452,24	11.711,77	13.335,56
	390625	25	29.518,61	54.001,23	10.875,28	51.917,16
	1953125	5	14.904,50	180.410,04	5.077,36	156.759,35
	9765625	1	6.842,06	463.313,24	2.290,80	323.880,05

Table 2 Performance results for 3D algorithm

Matrix size	p	Block size ($n^2/p^{2/3}$)	$t_s/t_w = 150/1$		$t_s/t_w = 450/1$	
			Speed up for DNS	Speed up for 3D proposed	Speed up for DNS	Speed up for 3D proposed
25 × 25	125	25	10,33	10,18	4,21	4,15
	15625	1	8,10	9,26	2,71	3,10
125 × 125	125	625	41,20	41,08	38,75	38,60
	15625	25	580,30	654,62	227,34	258,55
	1953125	1	519,97	636,21	174,17	213,13
625 × 625	125	15625	34,97	34,95	34,95	34,93
	15625	625	3365,34	3475,96	3064,21	3192,53
	1953125	25	38338,68	46213,13	14758,63	17955,42
	244140625	1	36673,50	46641,31	12282,96	15622,93
3125 × 3125	125	390625	26,28	26,28	26,28	26,28
	15625	15625	3132,52	3153,27	3129,40	3150,51
	1953125	625	278182,03	295862,39	246937,82	266550,94
	244140625	25	2746617,80	3443838,20	1047073,45	1324553,15
	30517578125	1	2731343,54	3561608,73	914742,69	1192905,34

processors. At the same time, torus interconnection networks gained wide-spread popularity in the industry. In this paper, we proposed a new parallel matrix multiplication algorithm for 2D and 3D torus architectures which performs better than the competitive algorithms especially as the size of the matrix sub-blocks gets smaller.

References

1. Bini DA (2007) Fast matrix multiplication. In: Hogben L (ed) Handbook of linear algebra. Chapman & Hall/CRC press, Boca Raton (Chap. 47)
2. Cannon LE (1969) A cellular computer to implement the kalman filter algorithm. Ph. D. Thesis, Montana State University
3. Grama A, Gupta A, Karypis G, Kumar V (2003) Introduction to parallel computing, 2nd edn. Addison Wesle, Eugene
4. Imre KM, Baransel C, Artuner H (2010) Efficient and scalable routing algorithms for collective communication operations on 2D All-Port Torus networks. Int J Parallel Progr 39(6):746–782
5. Ni LM, McKinley PK (1993, February) A survey of wormhole routing techniques in direct networks. Computer 26(2):62–76
6. Strassen V (1969) Gaussian elimination is not optimal. Numer Math 13:354–356
7. Winograd S (1971) On multiplication of 2×2 matrices. Linear Algebra Appl 4:381–388

References

2.5D Acoustic Wave Propagation in Shallow Water Over an Irregular Seabed Using the Boundary Element Method

A. Pereira, A. Tadeu, L. Godinho and J. A. F. Santiago

Abstract In this paper a Boundary Element formulation, in the frequency domain, is used to investigate the 2.5D acoustic wave propagation in shallow water over an irregular seabed that is assumed to have a rigid bottom and a free surface.

The problem is solved using a model which incorporates Green's functions that take into account the presence of flat surfaces. With this procedure only the irregular bottom and the vertical interface between regions of different depths are discretized. The model is implemented to obtain the 3D time domain pressure responses in a shallow water region with step or slope irregularities, originated by point pressure loads placed at different positions. Simulations are performed to identify wave propagation features that may help the assessment of the presence and shape of the bottom irregularities.

Keywords Shallow water · Irregular seabed · Boundary element method · 2.5 D wave propagation

1 Introduction

Acoustic wave propagation in the open ocean has been a topic of interest for researchers for many years. Traditional approaches to the problem include computationally efficient methods based on acoustic ray theory, normal modes or parabolic equations. The now classical book by Jensen et al. [1] presents an extensive review of

A. Pereira (✉)
CICC, Department of Civil Engineering, University of Coimbra, Coimbra, Portugal
e-mail: apereira@dec.uc.pt

A. Tadeu
CICC, Department of Civil Engineering, University of Coimbra, Coimbra, Portugal
e-mail: tadeu@dec.uc.pt

L. Godinho
CICC, Department of Civil Engineering, University of Coimbra, Coimbra, Portugal
e-mail: lgodinho@dec.uc.pt

J. A. F. Santiago
COPPE/Federal University of Rio de Janeiro, Rio de Janeiro, Brazil
e-mail: santiago@coc.ufrj.br

N. M. Fonseca Ferreira, J. A. Tenreiro Machado (eds.), *Mathematical Methods in Engineering,* DOI 10.1007/978-94-007-7183-3_3,

those techniques, describing in detail their formulation and applications. However, in the context of this work, it is important to note that each has specific limitations. Ray theory is known to be suitable for deep water wave propagation, in particular for high excitation frequencies, but it does not provide accurate results at the lower frequencies and in configurations with strong bottom interaction. Normal mode theory is efficient for the analysis of range-independent problems, but its application is not straightforward when the propagation domain is range dependent. For range dependent problems the coupled-mode model developed by Evans [2], which assumes that the waveguide is subdivided into a finite number of adjacent columns, has been widely used through the associated code COUPLE [3]. This model considers the full coupling between the modes and is able to handle the backscattering effects. However, the model has to approximate the various continuous surfaces of the problem by making use of piecewise constant sections, forming a sequence of small steps. A large number of sections are usually needed, with a corresponding increase in the computational requirements of the model.

Parabolic equation methods, initially introduced in underwater acoustics by Hardin and Tappert [4], have also been extensively applied in deepwater longrange sound propagation. The fact that they can easily account for range-dependent propagation and their very high computational efficiency have made them the preferred method for long-range ocean acoustic analysis. However, since the parabolic equation is a simplified version of the full-wave equation for the case of a pure one-way propagation, it neglects the backscattering effects which are essential in shallow waters in the presence of topographic features.

Due to the advent of high-speed computers and to the recent developments of numerical physics, sound propagation in the shallow ocean margins can be studied and quantitatively described in greater detail with the more exact wave theory. Many models have been developed based on the well-established finite difference, finite element and boundary element numerical methods.

This paper describes a Boundary Element Model developed to compute the three-dimensional wave propagation in a shallow water region with an irregular bottom. The model assumes a two-dimensional geometry, to simulate coastal regions which have little variation in the long shore direction, excited by a point pressure source. The regions of constant depth are modeled using Green's functions to avoid the discretization of the surface and bottom boundaries, which means that only the bottom irregularities and the vertical interfaces need to be discretized. The applicability of the formulation is then illustrated by analyzing the wave propagation in the vicinity of an irregular seabed. Time domain signatures computed for the case of a seabed containing a step or a slope are displayed by applying an Inverse Fast Fourier Transform and the main features of wave propagation are identified.

The 2.5D problem formulation is presented next, in Sect. 2. The Boundary Element Method and the Green's functions used are then described, followed by the verification of the model. The procedure used to obtain time domain signatures is also described. Finally the proposed model is applied to compute frequency and time domain signatures for several shallow water configurations, in order to identify wave

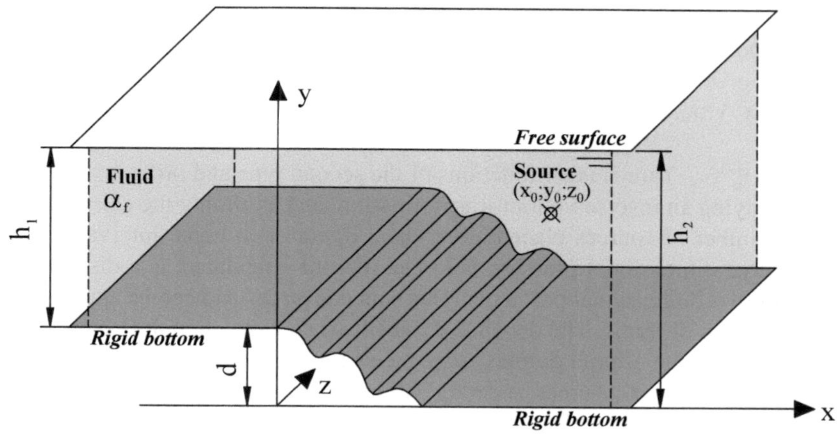

Fig. 1 Three-dimensional geometry of the problem

propagation features that may allow assessment of the presence and shape of the bottom irregularities.

2 2.5D Problem Formulation

Consider the problem of acoustic wave propagation in shallow water with an irregular seabed, displayed in Fig. 1. This figure shows a region of infinite extent along the x and z directions, limited by a rigid bottom and free surface. The normal particle velocity must be null at the rigid bottom and the pressure has null values at the free surface.

The confined acoustic medium has a mass density ρ_f, a Lamé constant λ_f and permits a constant dilatational wave velocity $\alpha_f = \sqrt{\lambda_f/\rho_f}$.

Consider the above model to be excited by a point pressure load, oscillating with an angular frequency ω, acting in the fluid medium at (x_0, y_0, z_0). In these conditions the incident pressure wave field, at (x, y, z) can be expressed by

$$\hat{\sigma}^{full}(x, y, z, x_0, y_0, z_0, \omega) = \frac{A e^{i\frac{\omega}{\alpha_f}\left(\alpha_f t - \sqrt{(x-x_0)^2+(y-y_0)^2+(z-z_0)^2}\right)}}{\sqrt{(x-x_0)^2+(y-y_0)^2+(z-z_0)^2}}, \tag{1}$$

in which A is the wave amplitude and $i = \sqrt{-1}$.

As the geometry of the model is constant along the z direction and the source is three-dimensional, it is possible to apply a Fourier transformation along z, decomposing the 3D problem into simpler 2D problems with varying wavenumbers along z. With this procedure, the responses are obtained in the frequency-wavenumber domain for varying effective wavenumbers, $k_\alpha = \sqrt{(\omega/\alpha_f)^2 - k_z^2}$, with $\mathrm{Im}\,(k_{\alpha_f}) \leq 0$ and k_z being the wavenumbers along z ($k_z = \frac{2\pi}{L_z} m$). In this k_z domain, the system is excited by spatially sinusoidal harmonic line loads acting at (x_0, y_0) whose pressure field at a point (x, y) is given by,

$$\sigma^{full}(x, y, x_0, y_0, k_z, \omega) = \frac{-\mathrm{i}A}{2}\mathrm{H}_0^{(2)}\left(k_\alpha\sqrt{(x-x_0)^2+(y-y_0)^2}\right)\mathrm{e}^{-\mathrm{i}k_z z}, \quad (2)$$

in which $\mathrm{H}_\mathrm{n}^{(2)}(\ldots)$ are Hankel functions of the second type and order n.

By applying an inverse Fourier transformation, and assuming the existence of an infinite number of sources placed along the z direction at equal intervals, L_z, the former three-dimensional pressure field can then be calculated as a discrete summation of two-dimensional problems. This sum converges and can be approximated by a finite sum of terms. The distance L_z needs to be large enough to avoid spatial contamination. The use of complex frequencies further reduces the influence of the neighbouring fictitious sources. A detailed description of the technique can be found, for example, in reference [5].

Using this technique, the scattered field caused by a point pressure load in the presence of the confined medium can likewise be obtained as a discrete summation of 2D harmonic line loads, with different values of k_z. This problem is often referred to in the literature as a 2.5D problem, because the geometry is 2D and the source is 3D.

3 Numerical Analysis

3.1 Boundary Element Method

Each two-dimensional scattered field produced by a harmonic line load aligned along the z direction, acting on the fluid medium confined by a free surface and an irregular rigid bottom, is computed in the frequency domain by using the Boundary Element Method (BEM). The formulation makes use of Green's functions that directly verify the boundary conditions at the rigid bottom and free flat surfaces. Therefore these interfaces do not need to be discretized. The two-dimensional domain is divided into two regions (Ω_1 and Ω_2 as displayed in Fig. 2) and only the vertical interface between regions and bottom irregularities are discretized (interface Γ_1 and boundary Γ_2 as depicted in Fig. 2), so that different Green's functions can be used in the two regions.

Continuity of pressure and normal velocity is ascribed along the interface Γ_1, while at the boundary Γ_2 null normal velocity must be enforced. Assuming a virtual load acting at a point $\underline{x}$ of the interface Γ_1 or boundary Γ_2, and after simplification to account for the specific boundary conditions, the Boundary Integral equations may be written as:

Region Ω_1:

$$cp(\underline{x}_0, k_z, \omega) = \int_{\Gamma_1} q(\underline{x}, \nu_n, \omega) G^{\Omega_1}(\underline{x}, \underline{x}_0, k_z, \omega) d\Gamma_1 - \int_{\Gamma_1} H^{\Omega_1}(\underline{x}, \nu_n, \underline{x}_0, k_z, \omega) p(\underline{x}, k_z, \omega) d\Gamma_1 + \upsilon p^{inc}(\underline{x}, \underline{x}_F, k_z, \omega). \quad (3)$$

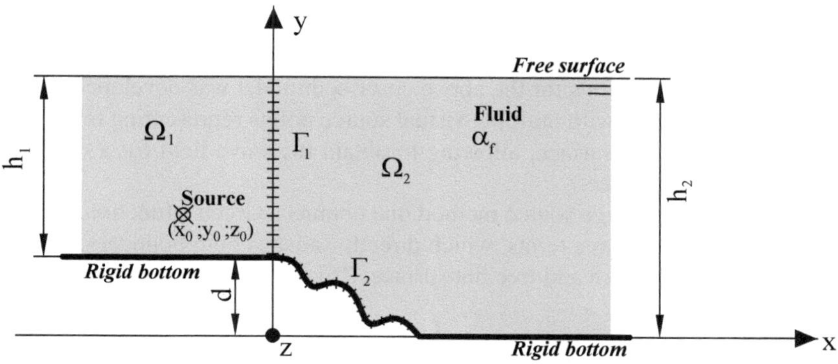

Fig. 2 Regions dividing the shallow water section and the discretization of the problem

Region Ω_2:

$$cp(\underline{x}_0, k_z, \omega) = \int_{\Gamma_1} q(\underline{x}, \nu_n, \omega) G^{\Omega_2}(\underline{x}, \underline{x}_0, k_z, \omega) d\Gamma_1 - \int_{\Gamma_1} H^{\Omega_2}(\underline{x}, \nu_n, \underline{x}_0, k_z, \omega) p(\underline{x}, k_z, \omega) d\Gamma_1 - \int_{\Gamma_2} H^{\Omega_2}(\underline{x}, \nu_n, \underline{x}_0, k_z, \omega) p(\underline{x}, k_z, \omega) d\Gamma_2 + (1 - \upsilon)\, p^{inc}(\underline{x}, \underline{x}_F, k_z, \omega) \quad (4)$$

In these equations, G^{Ω_i} and H^{Ω_i} are Green's functions for the acoustic region Ω_i (with $i = 1{,}2$) confined by a rigid bottom and a free surface, that enables the pressure (p) and normal velocity (q) to be obtained at point $\underline{x}$ of the boundary when a virtual harmonic line load acts at $\underline{x}_0$ of the boundary; c depends on the geometry of the boundary at the loaded point and equals $1/2$ if $\underline{x}_0 \in \Gamma$ and the boundary is smooth; ν_n is the unit outward normal for the boundary; $p^{inc}(\underline{x}, \underline{x}_F, k_z, \omega)$ is the incident field when the source is placed at $\underline{x}_F$, $\upsilon = 1$ if the source is placed in the region Ω_1, and $\upsilon = 0$ when the source is positioned in region Ω_2. The incident field is obtained by using the Green's functions for a confined fluid layer, described in the next section.

The Boundary Integral equations are solved after discretization of the interface Γ_1 and boundary Γ_2 into N_1+N_2 constant boundary elements. The resulting integrations are calculated using a Gaussian quadrature scheme, except for the integrations of the source terms of the Green's functions for the confined fluid layer, which are carried out analytically when the element to be integrated is the loaded element.

Solving the resulting system makes it possible to obtain the nodal solid pressure and normal velocities. The scattered wave field at any point of the domain can then be calculated by applying the Boundary Integral equation.

3.2 Green's Function

The fundamental solutions for the above described model was developed using the image source method, with multiple virtual source points representing reflections at the bottom and at the surface, allowing to obtain the wave field for a region with rigid and free flat surfaces.

By applying the image source method one obtains a Green's function, written as an infinite series of source terms, which directly satisfies both boundary conditions at the ocean rigid bottom and free flat surface. ([6]). This solution can be given as:

$$G(\omega, x, y, k_z) = -\frac{\mathrm{i}}{4}\left[H_0(k_{\alpha_f} r)\right] - \frac{\mathrm{i}}{4}\left\{\sum_{n=0}^{NS}(-1)^n\left[H_0(k_{\alpha_f} r_1) - \sum_{i=2}^{4} H_0(k_{\alpha_f} r_i)\right]\right\}, \tag{5}$$

where

$$r = \sqrt{(x - x_0)^2 + (y - y_0)^2};$$
$$r_1 = \sqrt{(x - x_0)^2 + (y + y_0 + 2hn)^2};$$
$$r_2 = \sqrt{(x - x_0)^2 + (y - 2h - y_0 - 2hn)^2};$$
$$r_3 = \sqrt{(x - x_0)^2 + (y + 2h - y_0 + 2hn)^2};$$
$$r_4 = \sqrt{(x - x_0)^2 + (y - 2h + y_0 - 2hn)^2}.$$

In these expressions, h represents the depth of the channel and NS is the number of virtual sources

It is worth noting that the above defined series exhibits a slow convergence, requiring a large number of terms to obtain the solution. However, this process can be greatly improved by using complex frequencies, with the form $\omega_c = \omega - \mathrm{i}\zeta$, with ζ defining a damping effect [7].

4 Responses in the Time Domain

The pressure field in the spatial-temporal domain is obtained by modeling a Ricker wavelet whose Fourier transform is

$$U(\omega) = A\left[2\pi^{1/2} t_o \mathrm{e}^{-\mathrm{i}\omega t_s}\right]\Omega^2 \mathrm{e}^{-\Omega^2} \tag{6}$$

in which $\Omega = \omega t_o/2$; A is the amplitude; t_s is the time when the maximum occurs and πt_o is the characteristic (dominant) period of the wavelet.

This wavelet form has been chosen because it decays rapidly in both time and frequency, thereby reducing computational effort and allowing easier interpretation of the computed time series and synthetic waveforms.

The Fourier transformations are obtained by discrete summations over wavenumbers and frequencies. Mathematically, this is achieved by adding periodic sources at spatial intervals $L_x = 2\pi/\Delta k_n$ (in the x axis) with Δk_n being the wavenumber step, and temporal intervals $T = 2\pi/\Delta\omega$ where $\Delta\omega$ is the frequency step [7]. The spatial separation, L_x, must be large enough to guarantee that the response of the fictitious sources occurs at times later than T, thereby avoiding contamination. The analysis uses complex frequencies where $\omega_c = \omega - \mathrm{i}\zeta$, with $\zeta = 0.7\Delta\omega$, which further reduce the influence of the neighboring fictitious sources and avoid the aliasing phenomena. In the time domain, this shift is later taken into account by applying an exponential window $e^{\xi t}$ to the response [8].

5 Verification of the Model

The BEM model (designated by Model 1) developed in this work was then implemented and verified using a conventional BEM model as reference. In this model (designated by Model 2), the Green's function for an infinite fluid medium was used, and therefore the rigid bottom and the free surface needed to be discretized. In order to limit the number of boundary elements used to discretize these interfaces, complex frequencies with an imaginary part are used ($\zeta = 0.7\frac{2\pi}{T}$). This considerably attenuates the contribution of the responses from the boundary elements placed at $L = 2\alpha_f T$, reducing the length of the interface to be discretized. In our calculations a value of $T = 0.25$s and a length $L_1 = 750$ m and $L_2 = 764$ m were used to define the discretization of the surface and bottom respectively.

Several verifications were performed, considering different configurations of the irregular seabed. The responses in shallow water with a seabed forming a slope 10.0 m high and 10 m length (see Fig. 3a), were chosen to illustrate the accuracy of the models. The depths of the shallower and deeper water regions are $h_2 = 10.0$ m and $h_1 = 20.0$ m, respectively. The acoustic medium with a density $\rho_f = 1000.00$ kg/m^3, allows a dilatational wave velocity of $\alpha_f = 1500.0$ m/s. The geometry was subjected to a dilatational harmonic line load applied at point (−25.0 m; 3.0 m) of the fluid medium with $k_z = 0.2$rad/m. The responses were calculated at a receiver R1 with coordinates (−4.0 m; 1.0m), placed in region Ω_2 and at a receiver R2 placed within region Ω_1((14.0 m; 11.0 m)). which contains the source. Computations are performed in the frequency range between 4.0 Hz to 256.0 Hz, with a frequency step of 4.0 Hz. The BEM model using Green's functions for a full space assumes surfaces discretized with 1514 boundary elements. Using the BEM model (Model 1) described in this paper, the slope and the vertical interface are discretized, using 34 boundary elements. In order to illustrate the responses obtained in the verification, Fig. 3b, 3c display the pressure recorded at receivers R1 and R2 obtained with the three models. Analysis of the results confirms a good agreement between the solutions.

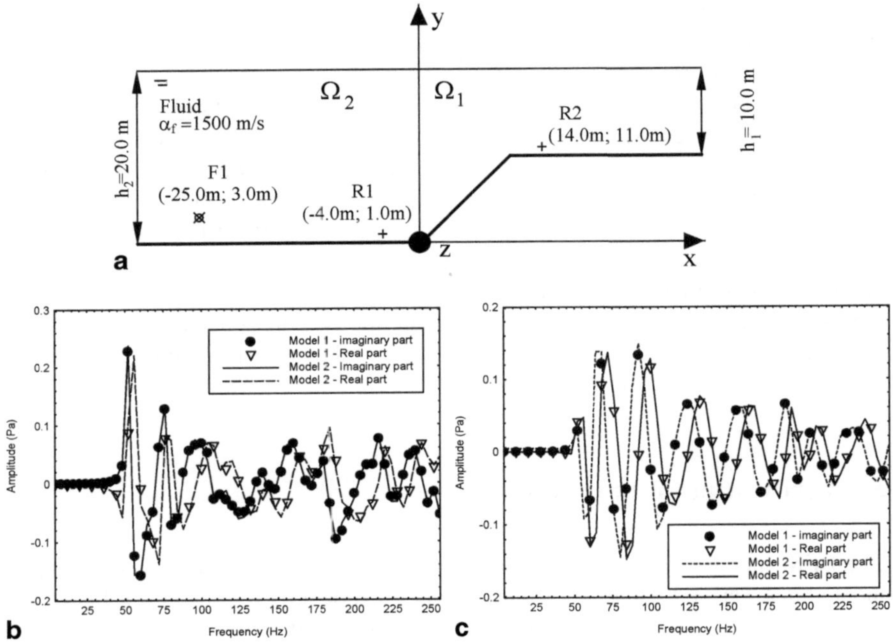

Fig. 3 Verification of the BEM Model: **a** Geometry; **b** Response provided by and harmonic line load with $k_z = 0.2$rad/m at receiver *R1*; **c** Response provided by an harmonic line load with $k_z = 0.2$rad/m at receiver *R2*

6 Applications

The BEM model presented has been implemented to compute the 3D pressure wave field generated in a shallow-water configuration with an irregular seabed (see Fig. 4). Two different geometries are analyzed in this section: an irregular bottom consisting of two flat seabed regions separated by a 10.0 m high step (see Fig. 4a); the same two regions separated by a smoother transition, consisting of a 45° slope (see Fig. 4b). The depths of the two flat regions are $h_1 = 20.0$ m and $h_2 = 10.0$ m, respectively. In all cases the seabed is assumed perfectly rigid, while null pressures are assumed along the water surface.

The responses were computed for a point pressure source placed at the bottom of the deeper region (F1) at $(-25.0; 0.0; 0.0)$. The acoustic medium is assumed to be water, with a density $\rho_f = 1000.0\,\text{kg/m}^3$ and allowing a dilatational wave velocity of $\alpha_f = 1500.0$ m/s.

The irregular seabed and the vertical interface were modelled using a number of boundary elements that was defined according to the excitation frequency of the harmonic source. A ratio of 15 was adopted between the wavelength of the incident waves and the length of the boundary element. The minimum number of boundary elements used was 30.

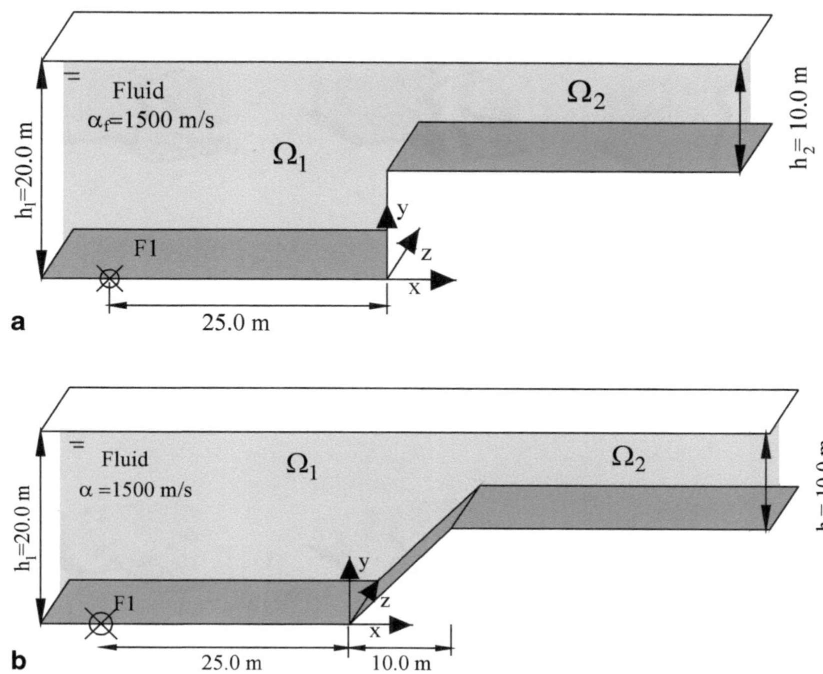

Fig. 4 Geometry of the numerical applications: **a** shallow water seabed with a step; **b** shallow water seabed with a slope

The calculations were performed over a frequency range between 4.0 Hz and 512.0 Hz, assuming a frequency step of 4.0 Hz, which gives a total time of $T = 250.0$ ms. Time domain signals are computed by means of an inverse Fourier transform, using the methodology described before. For this case, the source is modeled as a Ricker wavelet with a characteristic frequency of 200.0 Hz.

The pressure field was computed over three grids of receivers, equally spaced of $\Delta x = 1.0$ m, $\Delta y = 1.0$ m and $\Delta z = 1.0$ m, placed between: $x = -25.5$ m and $x = 25.5$ m; $y = 1.0$ m and $y = 19.0$ m, $z = 0.0$ m and $z = 39.0$ m. A sequence of snapshots displaying the pressure wave field over the grids of receivers at different instants is presented to illustrate the results. The responses provided by a flat seabed with the same depth as the source region are also displayed and used as a reference.

Figure 5 illustrates the propagation of the Ricker pulse, generated by a source located at position F1, for the cases of a flat seabed (first column), a seabed with a step (second column) and forming a slope (third column). In all cases, at time t = 0.00 ms, the point load creates a spherical pressure wave that propagates away from it. In the first set of snapshots (Fig. 5a1–5a3) this incident pulse is visible (identified as P1), although it is already combined with a first reflection from the rigid bottom of the waveguide. A second reflection generated at the free surface can also be identified (P2), with inverted polarity. In Fig. 5a2 and 5a3 it is also possible to observe a first

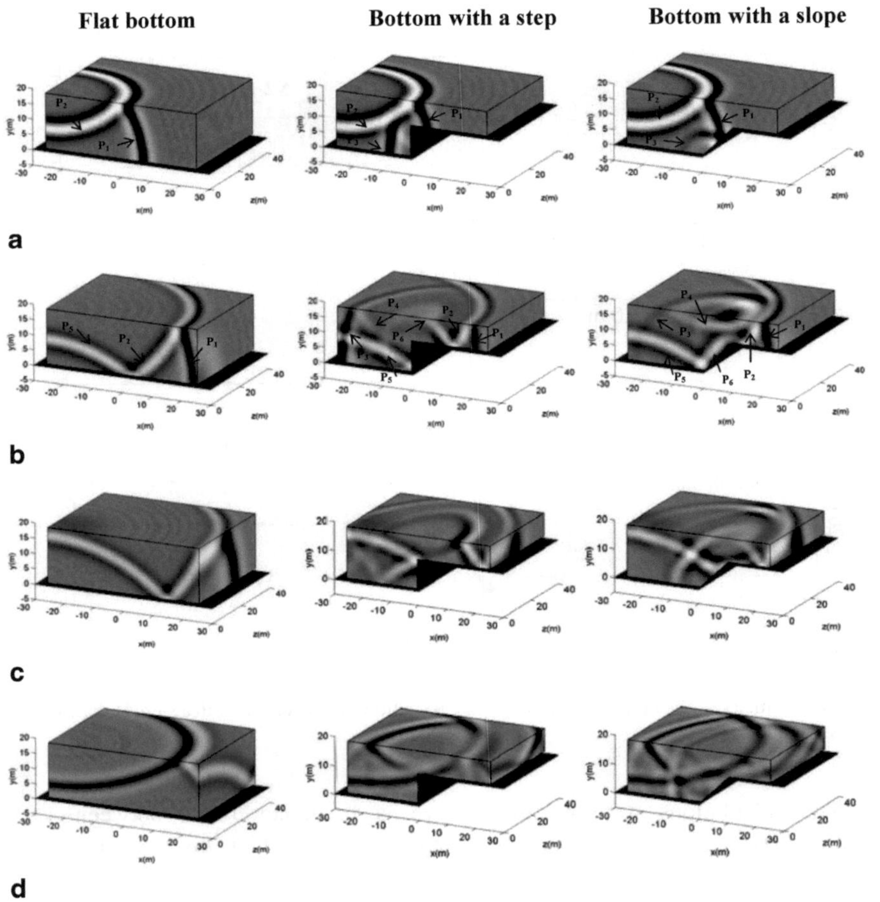

Fig. 5 3D time domain responses in a shallow water region when a point source is placed at position $F1$: **a** $t = 21.97$ ms; **b** $t = 32.96$ ms; **c** $t = 36.62$ ms; **d** $t = 50.04$ ms

reflection from the bottom discontinuity (P3), with significantly higher amplitude when the obstacle is a step. In fact, for this set of snapshots, the wavefront reflected from the step exhibits a very strong amplitude, evidencing the existence of a strong geometrical discontinuity in the propagation path. By contrast, when the obstacle is a slope, the pressure wave energy is spread over a larger area to produce a wavefront of lower amplitudes. At a later time (see Fig. 5b1–5b3, computed for t = 31.74 ms), higher order reflections start to occur (identified as P4, P5 and P6), originated at the horizontal free surface and at the irregular bottom. At this time, the pulses generated at the bottom discontinuity also become visible for receivers located near the surface placed further away along the z axis. It also becomes clear, in these figures, that the simple wave pattern exhibited by Fig. 5b1, when the bottom is horizontal, is greatly disturbed by the presence of the obstacles. This disturbance is further enhanced for later times, as can be easily observed in Figs. 5c1–5c3 and 5d1–5d3. Interestingly, comparison between the snapshots of Figs. 5d1–5d3 reveals that the presence of the slope gives rise to the most complex pattern, due to the intricate sequence of reflections that are generated, and to the energy-spreading effect of the inclined bottom.

7 Conclusions

In this paper, a model based on the Boundary Element Method has been proposed and successfully applied to predict acoustic wave propagation in a shallow water region with an irregular seabed, excited by a point pressure load. In this model, Green's functions that take into account the flat free surface and the flat bottom are used, while seabed irregularities are discretized with boundary elements. Its applicability has been demonstrated by computing the time domain signatures provided by a seabed with a step or a slope, when a point pressure load is placed either in the shallower region or in the deeper region. When the source is in the deeper region, wave propagation features were identified both in the time and in the frequency domain signatures that allowed a clear identification of the presence and type of an irregularity on the seabed.

Acknowledgments The financial support by CAPES (*Coordenação de Aperfeiçoamento de Pessoal de Nivel Superior*), and FCT (*Fundação para a Ciê ncia e Tecnologia*, within the scope of the FCT-CAPES convenium) is greatly acknowledged.

References

1. Jensen FB, Kuperman WA, Porter MB, Schmidt H (2000) Computational ocean acoustics, American Institute of Physics, Woodbury
2. Evans RB (1983) A coupled mode solution for the acoustic propagation in a waveguide with stepwise depth variations of a penetrable bottom. J Acoust Soc Am 74:188–195
3. Evans RB (1986) COUPLE: A user's manual, NORDA TN–332
4. Hardin RH, Tappert FD (1973) Applications of the split step fourier method to the numerical solution of nonlinear and variable coefficient wave equa-tions. SIAM 15:423
5. Branco F, Godinho L, Tadeu A (2002) Propagation of pressure waves inside a confined fluid channel with an irregular floor. J Comput Acoust 10:183–194
6. Tadeu A, Pereira A, Godinho L (2000) Three dimensional wave scattering by rigid circular pipelines submerged in an acoustic waveguide. J Comput Model Eng Sci 2(1):49–61
7. Kausel E, Roesset JM (1992) Frequency domain analysis of undamped systems. J Eng Mech ASCE 118: 721–734
8. Bouchon M, Aki K (1977) Discrete wave-number representation of seismic source wavefields. Bulletin Seismol Soc Am 67:259–277

Towards Replacing Lyapunov's "Direct" Method in Adaptive Control of Nonlinear Systems

József K. Tar

Abstract In adaptive nonlinear control Lyapunov's 2nd or "Direct" method became a fundamental tool in control design due to the typical practical difficulties viz. a) most of the control problems do not have closed analytical solutions; b) from numerical calculations "well behaving within a finite period" the stability cannot be taken for granted. According to Lyapunov, guaranteeing negative time-derivative of the Lyapunov function by relatively simple estimations the stability of the solution can theoretically be guaranteed. However, finding an appropriate Lyapunov function to a given problem is rather an "art" that cannot algorithmically be automated. Adaptivity normally requires slow tuning of numerous model parameters. This process is sensitive to unknown external disturbances, and the tuning rule is determined by numerous other, more or less arbitrary "adaptive control parameters". Furthermore, making the necessary estimations is a laborious, tedious work that normally results in "very strange conditions" to be met for guaranteeing stability of the solution. In the present paper the application of "Robust Fixed Point Transformations" is proposed instead of the Lyapunov technique. It can find the proper solution without any parameter tuning and depends on the setting only of three "adaptive control parameters". As application example direct control of a "Single Input—Single Output (SISO)" system, and a novel version of the "Model Reference Adaptive Control (MRAC)" of a "Multiple Input—Multiple Output (MIMO)" system is presented. Since this method cannot automatically guarantee global stability, as a novelty, a possible adaptive tuning of one of the adaptive control parameters is proposed for SISO systems to keep the control within the local basin of attraction of the proper convergence. Its operation is presented via simulations at first time in this paper.

Keywords Robust fixed point transformation-based adaptive control · Model reference adaptive control · Control parameter tuning · Local stability · Lyapunov's direct method

---

J. K. Tar (✉)
Institute of Applied Mathematics, Óbuda University, Budapest, Hungary
e-mail: tar.jozsef@nik.uni-obuda.hu

N. M. Fonseca Ferreira, J. A. Tenreiro Machado (eds.), *Mathematical Methods in Engineering*, DOI 10.1007/978-94-007-7183-3_4,

1 Introduction

Lyapunov's 2nd Method is a widely used technique in the analysis of the stability of the motion of the *non-autonomous dynamic systems* of equation of motion as $\dot{x} = f(x,t)$. The typical stability proofs provided by Lyapunov's original method published in 1892 [5] (and later on in e.g. [6]) have the great advantage that they do not require to analytically solve the equations of motion. Instead of that the *uniformly continuous nature* and non-positive time-derivative of a positive definite Lyapunov-function V constructed of *quadratic terms* of the tracking and modeling errors of the system's parameters are assumed in the $t \in [0,\infty)$ domain. From that the convergence $\dot{V} \to 0$ can be concluded according to Barbalat's lemma [4] utilizing the uniform continuity of $\dot{V}$. It used to be guaranteed by showing that $\ddot{V}$ is bounded. Due to the positive definite nature of V from that it normally follows that the tracking errors have to remain bounded, or in certain special cases, has to converge to 0. To illustrate the difficulties related to the "orthodox use of Lyapunov's direct method", on the basis of [4, 11, 14], and [13] a brief summary will be given in the next subsection.

1.1 Example for Orthodox Use of Lyapunov Functions

The most "historical" adaptive controllers used in *robotics* are the methods of "Adaptive Inverse Dynamics" and the "Adaptive Slotine–Li" controllers [4]. Since similar observations can be done for both of them, in the present considerations we recapitulate only the latter one. It utilizes subtle details of the Euler–Lagrange equations of motion, viz. that the terms quadratic in the generalized velocity components can specially be symmetrized. In this approach the exerted generalized torque/force components are constructed by the use of the *actual model marked by the symbol* ^ and causes $\ddot{q}$ according to the exact model values:

$$
\begin{aligned}
Q = \hat{H}(q)\dot{v} + \hat{C}(q,\dot{q})v + \hat{g} + K_D r &= H(q)\ddot{q} + C(q,\dot{q})\dot{q} + g \\
e := q^N - q, v := \dot{q}^N + \Lambda e, \quad r &:= \dot{e} + \Lambda e, \tilde{p} := \hat{p} - p \\
C_{ij} = \tfrac{1}{2}\sum_z \dot{q}_z \left(-\tfrac{\partial \hat{H}_{zj}}{\partial q_i} + \tfrac{\partial \hat{H}_{ij}}{\partial q_z} + \tfrac{\partial \hat{H}_{iz}}{\partial q_j} \right), \quad Q &= Y(q,\dot{q},v,\dot{v})\hat{p} + K_D r
\end{aligned}
\tag{1}
$$

in which q^N and q denote the generalized co–ordinates of the *nominal* and the *actual* motion, K_D and Λ are symmetric positive definite matrices, matrices H, C, and g denote the system's inertia matrix, the Coriolis, and the gravitational terms. The possession of the exact form of the dynamical model makes it possible to linearly separate the system's dynamic parameters p in the expression of the physically interpreted *generalized forces* Q by the use of matrix Y that exclusively consists of known kinematical data. The Lyapunov function of this method is $V = r^T H r + \tilde{p}^T \Gamma \tilde{p}$, with positive definite symmetric matrix Γ. For guaranteeing negative derivative of the Lyapunov function the *skew symmetry* of the C_{ij} matrix and the parameter tuning rule $\dot{\hat{p}} = \Gamma^{-1}Y^T r$ are utilized. The above results well exemplify the difficulties with the application of the Lyapunov function: (a) no unknown external perturbations can be present; (b) for a complex Classical Mechanical System $\hat{p}$ may have many (say m) independent components; besides the elements of the positive definite matrices Λ, K_D we have further $m+(m^2-m)/2$ independent elements in the positive definite matrix Γ (the main diagonals plus the parameters of the arbitrary *orthogonal matrix* O that can transform a *positive definite diagonal matrix* D into a more general non-diagonal form $\Gamma = O^T D O$); (c) the tuning process is too slow, since it happens according to the matrix Γ in spite of the fact that more explicit information can be obtained for the parameter errors if we subtract $\hat{H}\ddot{q}$, $\hat{C}\dot{q}$, and $\hat{g}$ from both sides of (1) [13]: $\hat{H}\dot{r} + \hat{C}r + K_D r = (H - \hat{H})\ddot{q} + (C - \hat{C})\dot{q} + g - \hat{g} = \Upsilon(q, \dot{q}, \ddot{q})(p - \hat{p})$. Since both the LHS of this equation and Υ are known, an SVD-based generalized inverse of Υ can provide direct information for optimal parameter tuning. Regarding the variation of the "error metrics" from both sides of the 1st line of (1) $H\dot{v}$, $C\dot{v}$, $K_D r$, and g can be subtracted so again some information can be obtained on the modeling errors: $Y\tilde{p} = -K_D r - H\dot{r} - Cr$. The fragment of the Lyapunov function $r^T H r$ itself can serve as a metrics for r. It has the time-derivative $d(r^T H r)/dt = 2r^T H\dot{r} + r^T \dot{H} r = r^T(\dot{H} - 2C)r - 2r^T K_D r - 2r^T Y\tilde{p} = -2r^T K_D r - 2r^T Y\tilde{p}$. That is this metrics is kept at bay during the new tuning process by the negative quadratic term and it is perturbed only by a linear one with a coefficient $\tilde{p}$ that converges to zero as the tuning proceeds. That is asymptotic stability can be also maintained without using the original Lyapunov function.

1.2 *Adaptive Control Based on Robust Fixed Point Transformations*

Certain control tasks can be formulated by using the concept of the appropriate *"excitation"* U of the controlled system to which it is expected to respond by some *"desired response"* r^d. The appropriate excitation can be computed by the use of the *inverse dynamic model of the system* as $U = \varphi(r^d)$. Since normally this inverse model is neither complete nor exact, the actual response determined by the system's dynamics, ψ, results in a *realized response* r^r that differs from the desired one: $r^r \equiv \psi(\varphi(r^d)) \equiv f(r^d) \neq r^d$. The controller normally can manipulate or "deform" the input value from r^d to $r^d_\star$ so that $r^d \equiv \psi(r^d_\star)$. Such a situation can be maintained by the use of some local deformation that can properly "drag" the system's state in time while it meanders along some trajectory. To realize this idea a fixed point transformation was introduced in [12] that is quite "robust" as far as the dependence of the resulting function on the behavior of $f(\bullet)$ is concerned. This robustness can approximately be investigated by the use of an affine approximation of $f(x)$ in the vicinity of $r^d_\star$ and it is the consequence of the strong nonlinear saturation of the sigmoid function tanh (x):

$$
\begin{aligned}
G(r|r^d) &:= (r+K)\left[1+B\tanh\left(A[f(r)-r^d]\right)\right]-K \\
G(r^d_\star|x^d) &= r^d_\star,\ \text{if } f(r^d_\star)=r^d \text{ then } G(-K|r^d)=-K, \\
G(r|r^d)' &= \frac{(r+K)ABf'(r)}{\cosh\left(A[f(r)-r^d]\right)^2}+1+B\tanh\left(A[f(r)-r^d]\right), \\
G(r^d_\star|r^d)' &= (r^d_\star+K)ABf'(r^d_\star)+1.
\end{aligned}
\tag{2}
$$

It is evident that the transformation defined in (2) has a proper ($r^d_\star$) and a false ($-K$) fixed point, but by properly manipulating the control parameters A, B, and K the good fixed point can be located within the basin of attraction of the iteration obtained by the repetitive use of function $r_{n+1} := G(r_n|r^d)$ if the requirement of $|G'(r|r^d)| < 1$ can be guaranteed in the vicinity of $r^d_\star$: if $|G'| \leq H$ $[0 \leq H < 1]$ can be maintained then a Cauchy sequence is obtained via the iteration that is convergent in the real numbers and it converges to the solution of the *Fixed Point Problem* $r_n \to r^d_\star = G(r^d_\star)$ [12]. Instead of the function tanh any sigmoid function, i.e. any bounded, monotone increasing, smooth function $\sigma(x)$ with the property of $\sigma(0)=0$ can naturally be used (e.g. $\sigma(x) := x/(1+|x|)$), too. A possibility for applying the same idea outlined in (2) of adaptivity is the application of a sigmoid function projected to the direction of the response-error defined in the nth control cycle as $\vec{h} := \vec{f}(\vec{r}_n)-\vec{r}^d$, $\vec{e} := \vec{h}/||\vec{h}||$, $\tilde{B} = B\sigma(A||\vec{h}||)$, so that $\vec{r}_{n+1} = (1+\tilde{B})\vec{r}_n + \tilde{B}K\vec{e}$. If $||\vec{h}||$ is very small, instead of normalizing with it the approximation $\vec{r}_{n+1} = \vec{r}_n$ can be applied since then the system already is in the very close vicinity of the fixed point.

This idea can be used in the following manner for SISO systems: on the basis of the available rough system model a simple PID controller can be simulated that reveals the order of magnitude of the occurring responses. Parameter K can be so chosen for which the $r+K$ values are considerable negative numbers. Depending on $sign(f')$ let $B\pm 1$ and let $A>0$ be a small number for which $|\partial G(r|r^d)/\partial r| \approx 1-\varepsilon$ for a small $\varepsilon > 0$. For r^d varying in time the following estimation can be done in the vicinity of the fixed point when $|r_n - r_{n-1}|$ is small: $r_{n+1}-r_n = G(r_n|r^d_n) - G(r_{n-1}|r^d_{n-1}) \approx \frac{\partial G(r_{n-1}|r^d_{n-1})}{\partial r}(r_n - r_{n-1}) + \frac{\partial G(r_{n-1}|r^d_{n-1})}{\partial r^d}(r^d_n - r^d_{n-1})$. Since from the analytical form of $\sigma(x)$ $\frac{\partial G(r_{n-1}|r^d_{n-1})}{\partial r^d}$ is known, and the past "desired" inputs as well as the arguments of function G are also known, this equation can be used for realtime estimation of $\frac{\partial G(r_{n-1}|r^d_{n-1})}{\partial r}$. ε can be tried to be fixed around -0.25 by a slow tuning of parameter A as $\dot{A} = \alpha(\varepsilon_{est}+0.25)A$ $(\alpha > 0)$ to keep the system in the local basin of attraction. The simulations revealed that increasing A resulted in smooth control, decreasing A caused small fluctuations. To avoid the occurrence of such fluctuations instead of a single α different values were chosen for "slow increase" (α_{incr}) and "very fast decrease" was prescribed by $\alpha_{decr} = 20\alpha_{incr}$. In the sequel a simple possible application is outlined for a strongly nonlinear system, the Electrostatic Microactuator (EμA). In connection with this problem in [10] the possibility of tuning A was not considered.

2 A Potential Example for Tuning the Adaptive Control Parameter

Paper [10] was inspired by the work by Vagia, Nikolakopoulos and Tzes who suggested the application of a robust switching PID controller coupled to a feed-forward compensator for controlling an electrostatic micro-actuator (EμA) in [15]. In their approach the precise non-linear model of a given EμA was linearized in certain set-points as typical operating points and the LMI technique was used in the design phase to stabilize separate PID controllers that were determined in the vicinity of these set points. Such kinds of controllers have to switch at the boundaries within which static PID parameters are set. The design typically was made by minimization of quadratic cost functions. The EμA corresponds to a micro-capacitor whose one plate is attached to the ground while its other moving plate is floating in the air. In the present paper the model considered was taken from [15]. Accordingly, the equation of motion of the system is given as follows

$$\ddot{q} = \frac{-b\dot{q} - kq + \varepsilon A_{Pl} U^2/(2(\eta_{max} - q)^2) + Q_d}{m} \tag{3}$$

in which $b = 1.4 \times 10^{-5}$ kg $\cdots$ is the viscous damping of the motion of the EμA in air, $k = 0.816$ N/m is a spring constant, $A_{Pl} = (400 \times 10^{-6}\ \mathrm{m})^2$ denotes the area of the plate, $m = 7.096 \times 10^{-10}$ kg is its mass, $\eta_{max} = 4 \times 10^{-6}$ m is the distance between the plates when the spring is relaxed, q is the displacement of the plates from the relaxed position, $\varepsilon = 9 \times 10^{-12}\ \mathrm{C}^2/(\mathrm{N} \cdot \mathrm{m}^2)$ is the dielectric constant, Q_d denotes the external disturbance forces, and U denotes the control voltage e.g. the physical agent by the help of which the plate's displacement can be controlled. It can be seen that (3) is singular near $q = \eta_{max}$, therefore for controllability allowable displacements of the micro-capacitor's plate in the vertical axis were $q \in [0.1, 1.3] \times 10^{-6}$ m that was deemed necessary in order to guarantee the stability of the linearized open-loop system in [15]. In that paper only responses to step-like inputs were considered.

In the present simulations continuous variation of the nominal motion was prescribed by 3rd order spline functions in which the 2nd derivatives linearly vary with the time within neighboring intervals. A the boundaries of these intervals the accelerations are continuous functions. To study the effect of the modeling errors in the simulations the controller assumed the *approximate model parameters* as follows: $\hat{A}_{Pl} = 0.8\mathrm{A_{Pl}}$, $\hat{m} = 1.2$ m, $\hat{b} = 1.2\mathrm{b}$, $\hat{k} = 1.2\mathrm{k}$, $\hat{\eta}_{max} = 0.8\eta_{\mathrm{max}}$, and $\hat{\varepsilon} = 0.8\varepsilon$. Their effects can well be traced in the first row of Fig. 1 that reveals erroneous trajectory and acceleration (response) tracking in the case of a common PID controller defining the prescribed relaxation as $\ddot{q}^{Des} = \ddot{q}^N + 3\Lambda^2(q^N - q) + 3\Lambda(\dot{q}^N - \dot{q}) + \Lambda^3 \int_{t_0}^{t} (q^N(\xi) - q(\xi))d\xi$ with $\Lambda = 8500$/s in which $q^N(t)$ denotes the nominal trajectory. The *adaptive controller* used the following parameters: $K = -500\mathrm{m/s}^2$, $B = 1$, and as an initial estimation $A_{ini} = 3 \times 10^{-3}\ \mathrm{s}^2/\mathrm{m}$, and $\alpha_{incr} = 10^3$/s. As it can be seen from the 2nd and 3rd rows of Fig. 1, the tracking errors quickly relaxed, and in the non-transient phase of the motion fine tracking and acceleration tracking were realized. It is also clear that the initial A_{ini} parameter was "overestimated", it

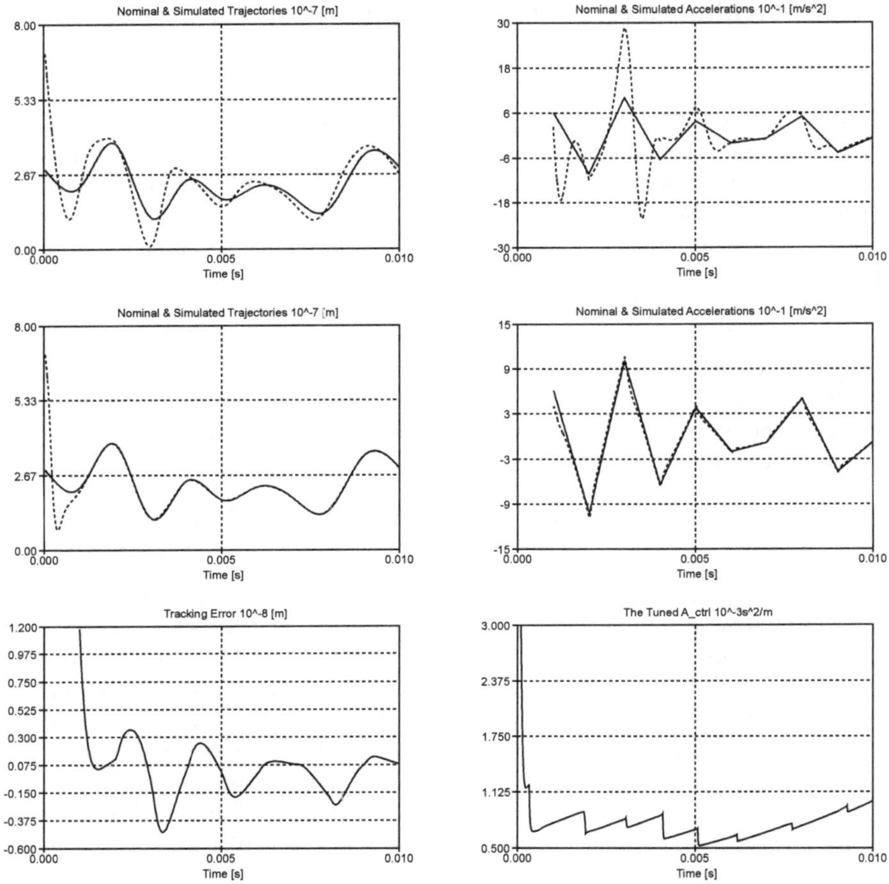

Fig. 1 The operation of the *non–adaptive* controller (1st row), and that of the *adaptive one*: the trajectory and acceleration tracking (2nd row), the tracking error and the tuned adaptive control parameter A (3rd row)

was quickly decreased in the initial phase of the motion and later was finely tuned to keep the control near the center of the local basin of attraction by decreasing and increasing sessions. *The simulations well exemplify the expected behavior of the simple adaptive controller.*

3 The Traditional and the Novel MRAC Approaches

The *MRAC* technique is a popular and efficient approach in the adaptive control of nonlinear systems e.g. in robotics. A great manifold of appropriate papers can be found for the application of MRAC from the early nineties (e.g. [4]) to our days (e.g. [2]). One of its early applications was a breakthrough in adaptive control. In [7] C. Nguyen presented the implementation of a joint-space adaptive control scheme that was used for the control of a non-compliant motion of a Stewart platform-based manipulator that was used in the *Hardware Real-Time Emulator* developed at Goddard Space Flight Center to emulate space operations. The mainstream of the adaptive control literature at that time used some parametric models and applied Lyapunov's "direct method" for parameter tuning (e.g. [4, 3]). The essence of the idea of the *MRAC* is the transformation of the actual system under control into a well behaving reference system (reference model) for which simple controllers can be designed. In the practice the *reference model* used to be stable linear system of constant coefficients. To achieve this simple behavior normally special adaptive loops have to be developed.

In our particular case the reference model can be *the nonlinear analytical model of the system built up of its nominal parameters*. Assume that on purely kinematical basis we prescribe a trajectory tracking policy that needs a desired acceleration of the mechanical system as $\ddot{q}^D$. From the behavior of the reference model for that acceleration we can calculate the physical agent that could result in the response $\ddot{q}^D$ for the reference model (in our case the generalized force components U^D). The direct application of this U^D for the actual system could result in different response since its physical behavior differs from that of the reference model. Therefore it can be "deformed" into a "*required*" U^{Req} value that directly can be applied to the *actual system*. Via substituting the realized response of the actual system $\ddot{q}$ into the reference model the "realized control action" U^R can be obtained instead of the "desired one" U^D. Our aim is to find the proper deformation by the application of which U^R well approaches U^D, that is at which the controlled system seems to behave as the reference system. The proper deformation may be found by the application of an iteration as follows. Consider the iteration generated by some function G as $U^{Req}_{n+1} = G(U^{Req}_n, U^R_n, U^D_{n+1})$ in which n is the index of the control cycle. For slowly varying desired value U^D can be considered to be constant. In this case the iteration is reduced to $U^{Req}_{n+1} = G(U^{Req}_n, U^R_n|U^D)$ that must be made convergent to $U^{Req}_\star$. It is evident that the same function G and the same considerations can be applied in this case as that detailed in Sect. 1.2. In the sequel an possible application is outlined via simulation.

3.1 Novel MRAC Control of a 3 DOF System

The sketch of the system considered is given in Fig. 2. In the dynamical model it was assumed that the hamper is assembled to the end of the beam at its mass center point. The Euler-Lagrange equations of motion also given in Fig. 2 are valid in this case.

The dynamic parameters of the *actual system* were assumed to be $M = 30$ kg (the mass of the cart moving in the horizontal direction), $m = 10$ kg (the mass of the hamper), $L = 2$ m (the length of the beam), $\Theta = 20\ \mathrm{kg} \cdot \mathrm{m}^2$ (the inertia momentum

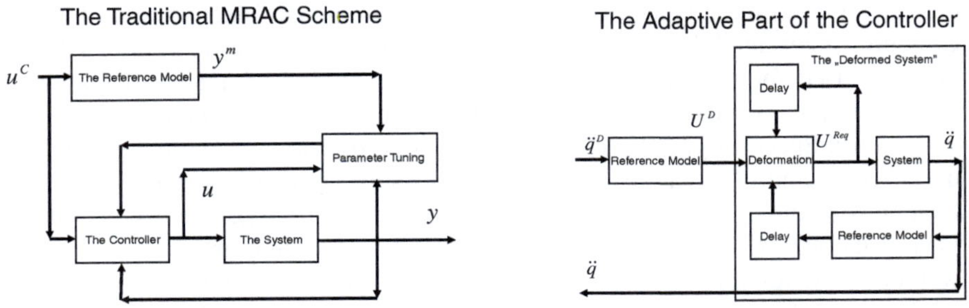

Fig. 2 The "traditional" MRAC scheme (LHS) operated by some Lyapunov function based parameter tuning, and the novel one (RHS) based on "Robust Fixed Point Transformations"

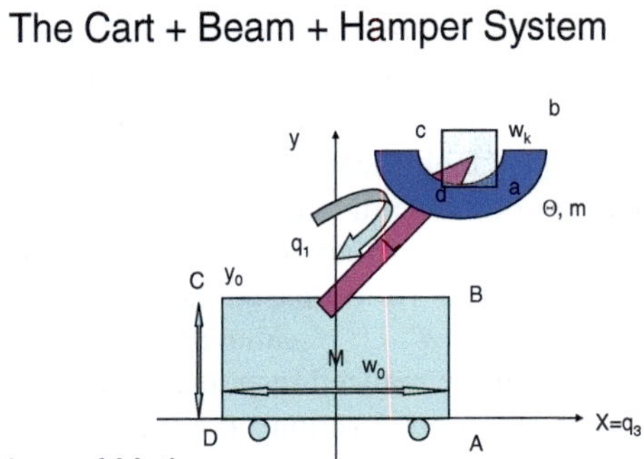

Equations of Motion:

$$\begin{bmatrix} (mL^2+\Theta) & \Theta & mL\cos q_1 \\ \Theta & \Theta & 0 \\ mL\cos q_1 & 0 & (m+M) \end{bmatrix} \begin{bmatrix} \ddot{q}_1 \\ \ddot{q}_2 \\ \ddot{q}_3 \end{bmatrix} + \begin{bmatrix} -mgL\sin q_1 \\ 0 \\ -mL\sin q_1 \dot{q}_1^2 \end{bmatrix} = \begin{bmatrix} Q_1 \\ Q_2 \\ Q_3 \end{bmatrix}$$

Fig. 3 The paradigm considered: the cart + beam + hamper system

of the hamper), and $g = 10\,\mathrm{m/s^2}$ the gravitational acceleration. For the simplicity the mass and the inertial momentum of the beam was neglected. The appropriate "nominal parameters of the reference model" quite considerably differed from the actual one as follows: $\hat{M} = 60$ kg, $\hat{m} = 20$ kg, $\hat{L} = 2.5$ m (only within the *dynamic model*, the *kinematic model* that is responsible for trajectory tracking used the exact value), $\hat{\Theta} = 50\,\mathrm{kg \cdot m^2}$, and $\hat{g} = 8\,\mathrm{m/s^2}$. The appropriate results for the *non-adaptive* and *adaptive* approaches for the PID type prescribed tracking error relaxation as $\ddot{q}^D = \ddot{q}^N + 3\Lambda^2(q^N - q) + 3\Lambda(\dot{q}^N - \dot{q}) + \Lambda^3 \int_{t_0}^{t} (q^N(\xi) - q(\xi))d\xi$ with a small $\Lambda = 1/s$ resulting in lose tracking [$q^N(t)$ denotes the nominal trajectory that was generated by 3rd order spline functions to produce linearly varying nominal acceleration within well defined intervals with continuous connection at their boundaries]. The fixed *adaptive parameters* were: $K = 7000$, $B = -1$, and $A = 10^{-5}$. Representative results are given in Fig. 3. The improvement due to the adaptivity is quite illustrative:

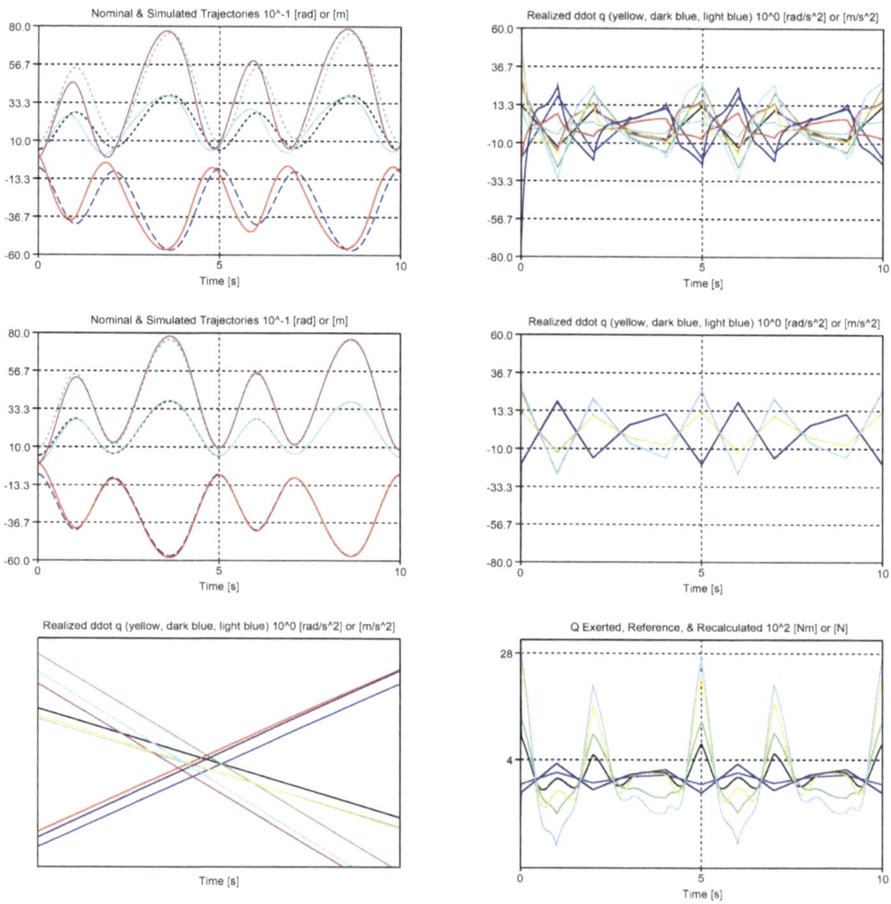

Fig. 4 The operation of the *non–adaptive* controller (1st row), and that of the *adaptive one* (2nd and 3rd rows) (color coding for the trajectories: $q_1^N = black$, $q_2^N = blue$, $q_3^N = green$, $q_1 = bright\quad blue$, $q_2 = red$, $q_3 = magenta$); for the accelerations: $\ddot{q}_1^N = black$, $\ddot{q}_2^N = blue$, $\ddot{q}_3^N = green$ for the nominal values, $\ddot{q}_1^D = bright\quad blue$, $\ddot{q}_2^D = red$, $\ddot{q}_3^D = magenta$ for the kinematically corrected *"desired"* values, and $\ddot{q}_1 = yellow$, $\ddot{q}_2 = dark\quad blue$, $\ddot{q}_3 = light\quad blue$ for the realized values)]; generalized forces: exerted: $Q_1 = black$, $Q_2 = blue$, $Q_3 = green$; calculated from the reference model: $Q_1 = bright\quad blue$, $Q_2 = red$, $Q_3 = magenta$, recalculated from the realized acceleration and the parameters of the reference model: $Q_1 = yellow$, $Q_2 = dark\quad blue$, $Q_3 = light\quad blue$)

the *nominal trajectories* are well approximated by the *"realized"* (simulated) ones while the Q^D forces calculated from the *reference model* are well approximated by the recalculated Q^R values and both considerably differ from the really exerted forces Q^{Req} that is needed for properly manipulating the actual physical system under control. *This altogether proves that the MRAC controller works, and the "deformed part" in the RHS of Fig. 4 really behaves like the reference model.*

4 Concluding Remarks

In this paper possible substitution of Lyapunov's "Direct Method" by the application of "Robust Fixed Point Transformations (RFPT)" in the adaptive control of nonlinear dynamic systems was suggested. For this purpose two typical frameworks, the "Model Based Computed Force Control" using approximate model and the "Model Reference Adaptive Controllers" were considered for a SISO and a MIMO system to be controlled. It was shown that this latter method is far less complicated and works with far less "arbitrary" parameters than the Lyapunov function based tuning approaches. Illustrative examples obtained by simulation have shown that in spite of the fact that this latter method cannot guarantee global asymptotic stability, it can work for a wide set of physical systems to be controlled. For compensating this deficiency for the case of SISO systems additional tuning of one of the altogether three control parameters proposed to keep the control near the center of the local basin of attraction of the RFPT transformation. Its operation was illustrated via simulations. In the next phase of the research this tuning is expected to be extended for the adaptive control of MIMO systems.

Acknowledgements The author gratefully acknowledges the support by the *National Office for Research and Technology (NKTH)* using the resources of the *Research and Technology Innovation Fund* within the project *OTKA No. CNK-78168*.

References

1. Golub GH, Kahan W (1965) Calculating the singular values and pseudoinverse of a matrix. SIAM J Numer Anal 2:205–224
2. Hosseini-Suny K, Momeni H, Janabi-Sharifi F (2010) Model reference adaptive control design for a teleoperation system with output prediction. J Intell Robot Syst. doi:10.1007/s10846-010-9400-4 (pp 1–21)
3. Isermann R, Lachmann KH, Matko D (1992) Adaptive control systems. Prentice-Hall, New York
4. Jean-Jacques S, Weiping L (1991) Applied nonlinear control. Prentice Hall International, Inc., Englewood Cliffs
5. Lyapunow AM (1892) A general task about the stability of motion . PhD Thesis (in Russian)
6. Lyapunov AM (1966) Stability of motion. Academic Press, New York
7. Nguyen CC, Antrazi SS, Zhou Z-L, Campbell CE (1993) Adaptive control of a stewart platform-based manipulator. J Robot Syst 10(5):657–687
8. Somló J, Lantos B, Cát PT (2002) Advanced robot control. Akadémiai Kiadó, Budapest
9. Stewart GW 1992 On the early history of singular value decomposition. Technical Report TR-92-31, Institute for Advanced Computer Studies, University of Mariland
10. Tar JK (2010) Robust fixed point transformations based adaptive control of an electrostatic microactuator. Acta Electrotechn et Inform 10(1):18–23
11. Tar JK, Bitó JF, Rudas IJ, Preitl S, Precup RE (2009) An SVD based modification of the adaptive inverse dynamics controller. In: Proceedings of 5th international symposium on applied computational intelligence and informatics, Timişoara 193–198 May 2009
12. Tar JK, Bitó JF, Rudas IJ, Kozłowski KR, Tenreiro Machado JA (2008) Possible adaptive control by tangent hyperbolic fixed point transformations used for controlling the Φ^6-Type Van der

Pol oscillator. In: Proceedings of the 6th IEEE International Conference on Computational Cybernetics (ICCC 2008), Stará Lesná, 15–20 Nov 2008
13. Tar JK, Rudas IJ, Gáti J (2009) Improvements of the adaptive Slotine & Li controller – comparative snalysis with solutions using local robust fixed point transformations. In: Proceedings of the 14th WSEAS International Conference on Applied Mathematics (MATH'09), Puerto De La Cruz 14–16 Dec 2009 (pp 305–311)
14. Tar JK, Rudas IJ, Hermann G, Bitó JF, Tenreiro Machado JA (2009) On the robustness of the Slotine-Li and the FPT/SVD-based adaptive controllers. WSEAS Trans Syst Control 3(9): 686–700
15. Vagia M, Nikolakopoulos G, Tzes A (2008) Design of a robust PID-control switching scheme for an electrostatic microactuator. Control Eng Pract 16:1321–1328

Fractional Particle Swarm Optimization

E. J. Solteiro Pires, J. A. Tenreiro Machado and P. B. de Moura Oliveira

Abstract The paper addresses new perspective of the PSO including a fractional block. The local gain is replaced by one of fractional order considering several previous positions of the PSO particles. The algorithm is evaluated for several well known test functions and the relationship between the fractional order and the convergence of the algorithm is observed. The fractional order influences directly the algorithm convergence rate demonstrating a large potential for developments.

Keywords Fractional calculus · Particle swarm optimization

1 Introduction

In the last decade particle swarming optimization (PSO) has been applied in a plethora of fields such as social modeling, computer graphics, simulation and animation of natural flocks or swarms, pattern recognition, color image quantization and computational biology [1]. PSO has motivated considerable interest from the natural computing research, where important work has been enforced in the study of its convergence.

Fractional Calculus (FC) is a natural extension of the classical mathematics. Since the beginning of theory of differential and integral calculus, several mathematicians investigated the calculation of noninteger order derivatives and integrals. Nevertheless, the application of FC has been scarce until recently, but the recent scientific advances motivated a renewed interest in this field.

E. J. Solteiro Pires (✉)
Centro de Investigação e de Tecnologias Agro-Ambientais e Biológicas,
Universidade de Trás-os-Montes e Alto Douro, Vila Real, Portugal
e-mail: epires@utad.pt

J. A. Tenreiro Machado
Instituto Superior de Engenharia, Instituto Politécnico do Porto, Porto, Portugal
e-mail: jtm@isep.ipp.pt

P. B. de Moura Oliveira
Universidade de Trás-os-Montes e Alto Douro, Vila Real, Portugal
e-mail: oliveira@utad.pt

N. M. Fonseca Ferreira, J. A. Tenreiro Machado (eds.), *Mathematical Methods in Engineering*, DOI 10.1007/978-94-007-7183-3_5,

Bearing these ideas in mind, this work uses a fractional derivative to control the convergence rate of the PSO. The article is organized as follows. Section 2 introduces the FC. Section 3 presents the PSO and its working principles. Based on this formulation, Sect. 4 generalizes the PSO to a fractional order. Section 5 presents the results for the PSO with fractional velocity. Finally, Sect. 6 outlines the main conclusions.

2 Introduction to Fractional Calculus

FC goes back to the beginning of the theory of differential calculus. Nevertheless, the application of FC just emerged in the last two decades, due to the progresses in the area of nonlinear and complex systems that revealed subtle relationships with the FC concepts. In the field of dynamics systems theory some work has been carried out, but the proposed models and algorithms are still in a preliminary stage of establishment.

The fundamentals aspects of FC theory are addressed in [2–5]. Concerning FC applications research efforts can be mentioned in the area of viscoelasticity, chaos, fractals, biology, electronics, signal processing, diffusion, wave propagation, modeling, control and irreversibility [6–10].

FC is a branch of mathematical analysis that extends to real, or even complex, numbers the order of the differential and integral operators. Since its foundation, the generalization of the concept of derivative and integral to a non-integer order α has been the subject of distinct approaches. A formulation based on the concept of fractional differential, is the Grünwald–Letnikov definition given by the equation:

$$D^{\alpha}\left[x(t)\right] = \lim_{h \to 0} \left[\frac{1}{h^{\alpha}} \sum_{k=0}^{+\infty} \frac{(-1)^k \Gamma(\alpha+1) x(t-kh)}{\Gamma(k+1)\Gamma(\alpha-k+1)} \right] \tag{1}$$

where $\Gamma()$ is the Euler function.

An important property revealed by expression (1) is that while an integer-order derivative just implies a finite series, the fractional-order derivative requires an infinite number of terms. Therefore, integer derivatives are 'local' operators in opposition with fractional derivatives which have, implicitly, a 'memory' of all past events.

Often, in discrete time implementations expression (1) is approximated by:

$$D^{\alpha}\left[x(t)\right] = \frac{1}{T^{\alpha}} \sum_{k=0}^{r} \frac{(-1)^k \Gamma(\alpha+1) x(t-kT)}{\Gamma(k+1)\Gamma(\alpha-k+1)} \tag{2}$$

where T is the sampling period and r is the truncation order.

The $\mathscr{Z}$−transform formulation of a derivative of fractional order $\alpha \in \mathbb{C}$ of the signal $x(t)$, $D^{\alpha}[x(t)]$, is a 'direct' generalization of the classical integer-order scheme yielding, for zero initial conditions:

$$\mathscr{Z}\{D^{\alpha}[x(t)]\} = \left(\frac{1-z^{-1}}{T}\right)^{\alpha} X(z) \tag{3}$$

where z is the $\mathscr{Z}-$transform variable.

The characteristics revealed by fractional-order models make this mathematical tool well suited to describe phenomena such as irreversibility and chaos because of its inherent memory property. In this line of thought, the propagation of perturbations and the appearance of long-term dynamic phenomena in a population of individuals subjected to an evolutionary process configure a case where FC tools fit adequately [11].

3 Particle Swarm Optimization Algorithm

Evolutionary algorithms have been successfully adopted to solve many complex optimization engineering applications. Together with genetic algorithms, the PSO algorithm, proposed by Kennedy and Eberhart [12], has achieved considerable success in solving optimization problems.

The PSO algorithm was proposed originally in [12]. This optimization technique is inspired in the way swarms behave and its elements move in a synchronized way, both as a defensive tactic and for searching food. An analogy is established between a particle and a swarm element. The particle movement is characterized by two vectors, representing its current position x and velocity v. Since 1995, many techniques were proposed to refine and/or complement the original canonical PSO algorithm, namely by analyzing the tuning parameters [13] and by considering hybridization with other evolutionary techniques [14].

In literature, some work embedding FC and PSO algorithms can be found. Pires et al. [15] studies the fractional dynamics during the evolution of a PSO. Reis et al. [16] propose a PSO, for logic and circuit design, where is implemented a proportional-derivative fitness function to guide the optimization. Pires et al. [17] study the convergence of a PSO with a fractional order velocity.

Algorithm 1 illustrates a standard PSO algorithm. The basic algorithm begins by initializing the swarm randomly in the search space. As it can be seen in the pseudo-code, were t and $t+1$ represent two consecutive iterations, the position x of each particle is updated during the iterations by adding a new velocity v term. This velocity is evaluated by summing an increment to the previous velocity value. The increment is a function of two components representing the cognitive and the social knowledge.

The cognitive knowledge of each particle is included by evaluating the difference between its best position found so far b and the current position x. On the other hand, the social knowledge, of each particle, is incorporated through the difference between the best swarm global position achieved so far g and its current position x. The cognitive and the social knowledge factors are multiplied by random uniformly generated terms ϕ_1 and ϕ_2, respectively.

Algorithm 1: Particle swarm optimization

```
Initialize Swarm;
repeat
    forall particles do
        calculate fitness f
    end
    forall particles do
        v_{t+1} = v_t + φ1.(b − x) + φ2.(g − x);
        x_{t+1} = x_t + v_{t+1};
    end
    t = t + 1
until stopping criteria ;
```

PSO is a optimization algorithm that proves to be efficient, robust and simple. However, if no care is taken the velocities may attain large values, particularly when particles are far away from local and global bests. Some approaches were carried out in order to eliminate this drawback. Eberhat et al. [18] proposed a clamping function (4) to limit the velocity, through the expression:

$$v_{ij}(t+1) = \begin{cases} v'_{ij}(t+1) & \text{if} \quad v'_{ij}(t+1) < V_{\max j} \\ V_{\max j} & \text{if} \quad v'_{ij}(t+1) \geq V_{\max j} \end{cases} \tag{4}$$

where $v'_{ij}(t+1)$ is given by $v'_{ij}(t+1) = v_{ij}(t) + \phi_1.(b-x) + \phi_2.(g-x)$ for the parameter j of particle i at iteration $t+1$.

Later, a constant, the inertia weight, was introduced [13] to control the velocity from exploding (5). The inertia weight ω is very important to ensure convergence behavior over evolution by adopting the equation:

$$v_{t+1} = \omega.v_t + \phi_1.(b-x) + \phi_2.(g-x) \tag{5}$$

Some empirical and theoretical studies were made to determine the best inertia value [19] in order to obtain better PSO behavior.

Oliveira et al. [20] represent the PSO as a feedback control loop and establishing an analogy between the particle dynamics with a feedback control loop. They present a proportional and integral controller based on particle swarm algorithm, which does not require any parameter to regulate the swarm convergence over time.

4 Fractional PSO

In this section the PSO is modeled through Z block diagram. Additionally, the local unit feedback of x_t signal is replaced by a fractional one, represented by the pink block in Fig. 1.

The position term $b - x$ is substituted by a fractional version given in expression (5). In fact, considering the first $r = 4$ terms of the fractional derivative series, the $b - x$ term is replaced by:

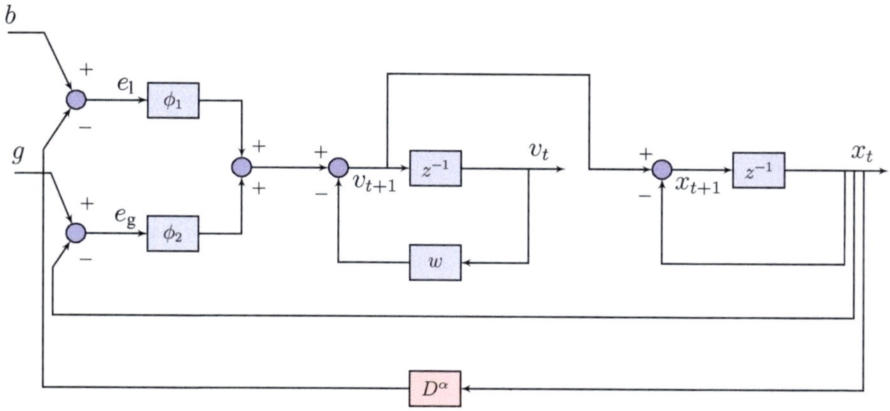

Fig. 1 PSO Diagram Block, e_l—local error, e_g—global error

$$b - x_{t+1} = b - \Big[\alpha x_t + \frac{\alpha}{2}(1-\alpha)x_{t-1} + \frac{\alpha}{6}(1-\alpha)(2-\alpha)x_{t-2} + \frac{\alpha}{24}(1-\alpha)(2-\alpha)(3-\alpha)x_{t-3}\Big] \quad (6)$$

Therefore, the expression (5) can be rewritten as:

$$v_{t+1} = v_t + \phi_1.\Big[b - \alpha x_t - \frac{\alpha}{2}(1-\alpha)x_{t-1} - \frac{\alpha}{6}(1-\alpha)(2-\alpha)x_{t-2} - \frac{\alpha}{24}(1-\alpha)(2-\alpha)(3-\alpha)x_{t-3}\Big] + \phi_2.(g - x_t) \quad (7)$$

In the next section, several distinct values of r are tested.

5 Test Functions

This section introduces the optimization functions that are adopted during the tests of PSO with fractional velocity update (7). The objective consists in minimizing several well known functions [19]. These functions have n parameters, $i = \{1, \ldots, n\}$ and their global optimum value is f^*. The algorithm adopts a real encoding scheme. In this line of thought are considered: (1) Rosenbrock's valley (also known as Banana function), (2) Drop wave, (3) Easom and (4) Michalewicz's, represented in the following expressions:

(1) *Rosenbrock's valley function:*

$$f_1(x) = \sum_{j=1}^{n-1} 100\left(x_{i+1} - x_i^2\right)^2 \quad (8)$$

with $x_i \in [-2.048, 2.048]$, $i = \{1, \ldots, 4\}$ and $f^*(x) = 0.0$.

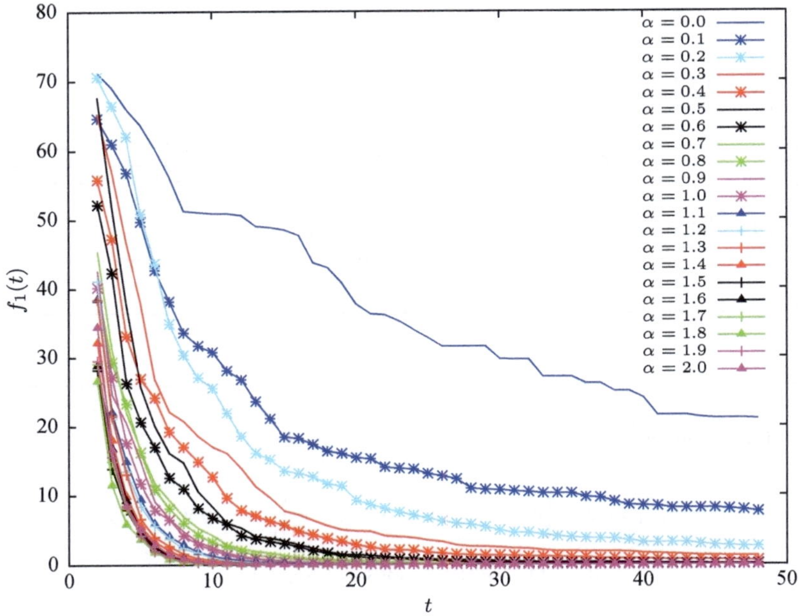

Fig. 2 Rosenbrock's function, evolution of the median of best PSO solution *versus* iteration for $\alpha = \{0, 0.1, \ldots, 2.0\}$

(2) *Drop wave function:*

$$f_2(x) = -\frac{1 + \cos\left(12\sqrt{x_1^2 + x_2^2}\right)}{0.5\left(x_1^2 + x_2^2\right) + 2} \tag{9}$$

with $x_i \in [-10, 10]$, $i = \{1, 2\}$ and $f^*(x) = -1.0$.
(3) *Easom function:*

$$f_3(x) = -\cos(x_1)\cos(x_2)\mathrm{e}^{-(x_1-\pi)^2-(x_2-\pi)^2} \tag{10}$$

with $x_1, x_2 \in [-100, 100]$ and $f^*(x) = -1.0$.
(4) *Michalewicz's function:*

$$f_4(x) = \sum_{j=1}^{n} -\sin(x_j)\left[\sin\frac{(j+1)x_j^2}{\pi}\right]^{2\,m} \tag{11}$$

with $n = 2$, $m = 1$, $x_i \in [0, \pi]$, $i = \{1, 2\}$ and $f^*(x) = -1.84$.

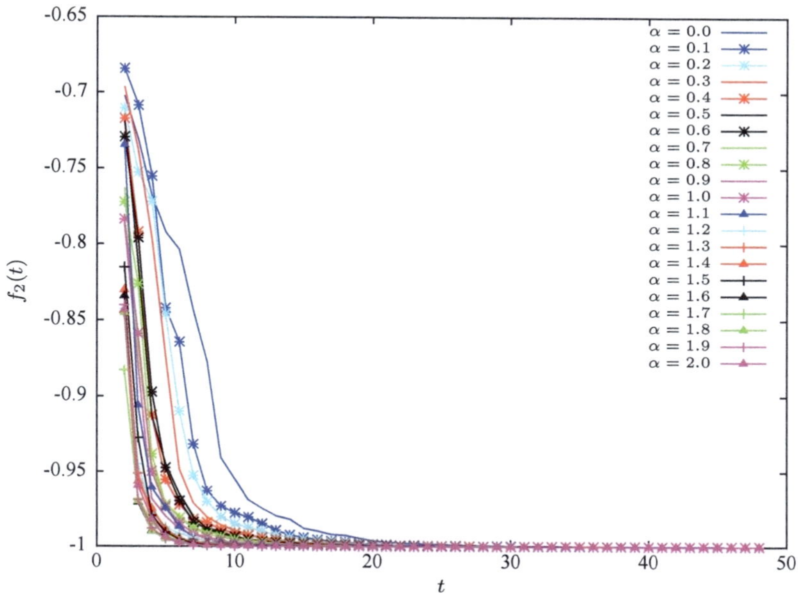

Fig. 3 Drop wave, evolution of the median of best PSO solution *versus* iteration for $\alpha = \{0, 0.1, \ldots, 2.0\}$

6 Simulation Results

To study the influence of the fractional feedback effect in the algorithm, several tests are now developed. A 10–population size PSO is executed during a period of 200 iterations with $\{\phi_1, \phi_2\} \sim U[0,1]$, where U represents the function that generates numbers with a uniform distribution in the specified range. The fitness evolution of the best global particle is taken as the system output.

Since PSO is a stochastic algorithm, every time it is executed it leads to a different trajectory convergence. Therefore, a test group of 201 simulation was considered, and the median is taken as the final output, for each value in the set of fractional order $\alpha = \{0, 0.1, \ldots, 2.0\}$. In Figs. 2, 3, 4, and 5 are depicted results for the optimization functions f_j, $j = \{1, \ldots, 4\}$.

It can be verified that the convergence of the algorithm depends directly upon the fractional order α. Normally, values of $\alpha = 1.8$ reaches faster convergence results. One the other hand, for low values of α the algorithm reveals convergence problems. This is due to the local error, $e_1 = \phi_1[b - \alpha x_t - 0.5\alpha(1-\alpha)x_{t-1} - \ldots] \simeq \phi_1 b$, that does not weights adequately the error between the particle actual position and the best position found so far by the particle. Therefore, the algorithm becomes inefficient and the algorithms takes more time to find the optimum.

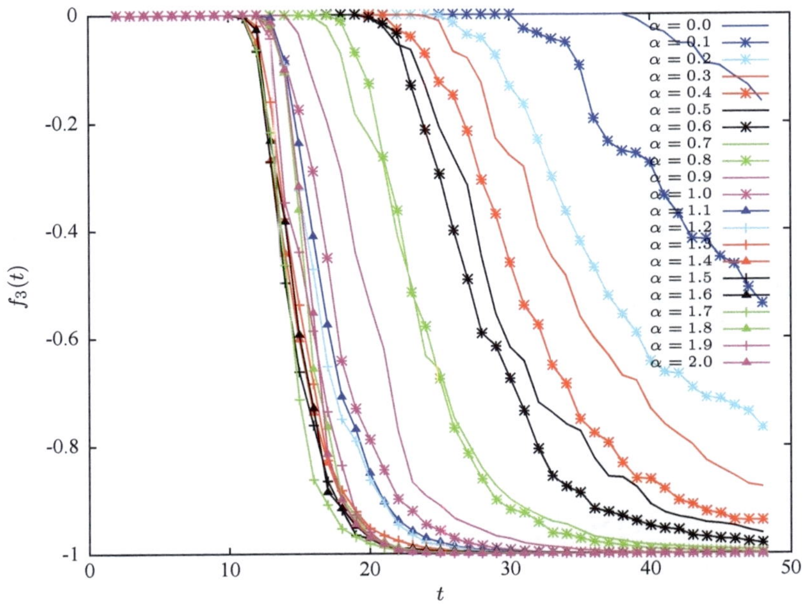

Fig. 4 Easom function, evolution of the median of best PSO solution *versus* iteration for $\alpha = \{0, \ldots, 1\}$

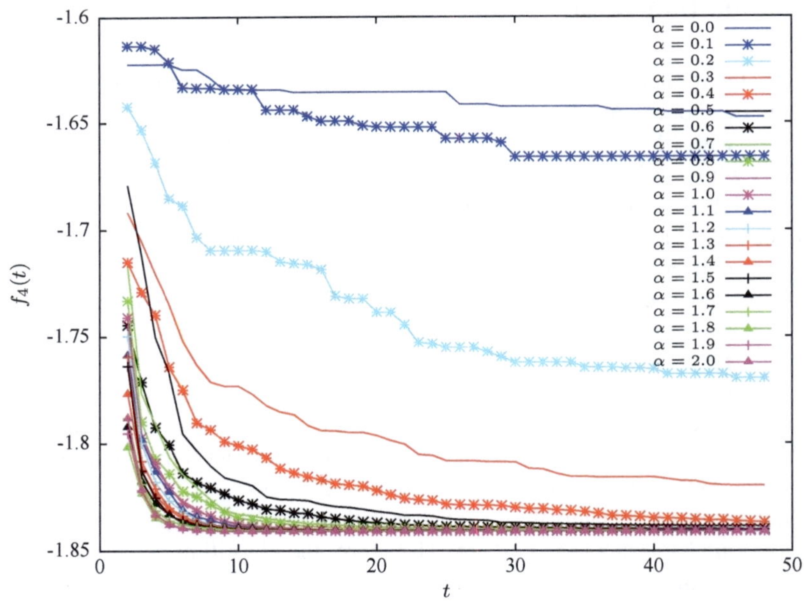

Fig. 5 Michalewicz's function, evolution of the median of best PSO solution *versus* iteration for $\alpha = \{0, \ldots, 1\}$

7 Conclusions

Block diagram and $\mathscr{Z}$−transform are engineering tools that lead the designer to a better understanding of the PSO in a control perspective. On the other hand, FC is a mathematical tool that enables an efficient generalization of the PSO algorithm. Bearing these facts in mind, the fractional order position error was analyzed showing that it influences directly the algorithm convergence. Moreover, the results are consistent representing an important step towards understanding the relationship between the system position and the convergence behavior. In conclusion, the FC concepts open new perspectives towards the development of more efficient evolutionary algorithms.

References

1. Banks A, Vincent J, Anyakoha C (2008) A review of particle swarm optimization. Part II: Hybridisation, combinatorial, multicriteria and constrained optimization, and indicative applications. Nat Comput 7(1):109–124
2. Gement A (1938) On fractional differentials. Proc Philos Mag 25:540–549
3. Oustaloup A (1991) La Commande CRONE: Commande Robuste d'Ordre Non Intier. Hermes, Paris
4. Méhauté Alain Le (1991) Fractal geometries: theory and applications. Penton Press, London, UK
5. Podlubny I (1999) Fractional differential equations. Academic, San Diego
6. Tenreiro Machado JA (1997) Analysis and design of fractional-order digital control systems. J Syst Anal-Modelling-Simul 27:107–122
7. Tenreiro Machado JA (2001) System modeling and control through fractional-order algorithms. FCAA – J Fract Calculus & Appl Anal 4:47–66
8. Vinagre BM, Petras I, Podlubny I, Chen YQ (July 2002) Using fractional order adjustment rules and fractional order reference models in model-reference adaptive control. Nonlinear Dyn 29(1–4):269–279
9. Torvik PJ, Bagley RL (June 1984) On the appearance of the fractional derivative in the behaviour of real materials. ASME J Appl Mech 51:294–298
10. Westerlund S (2002) Dead matter has memory! Causal Consulting, Kalmar
11. Solteiro Pires EJ, de Moura Oliveira PB, Tenreiro Machado JA, Jesus IS (Oct. 2007) Fractional order dynamics in a particle swarm optimization algorithm. In seventh international conference on intelligent systems design and applications, ISDA 2007. IEEE Computer Society, Washington, DC, pp 703–710
12. Kennedy J, Eberhart RC (1995) Particle swarm optimization. Proceedings of the 1995 IEEE international conference on neural networks, vol 4. IEEE Service Center, Perth, pp 1942–1948
13. Shi Y, Eberhart R (1998) A modified particle swarm optimizer. In evolutionary computation proceedings, 1998. IEEE world congress on computational intelligence, The 1998 IEEE international conference on computational intelligence, pp 69–73
14. Løvbjerg M, Rasmussen TK, Krink T 2001 Hybrid particle swarm optimiser with breeding and subpopulations. In: Spector L , Goodman ED, Wu A, Langdon WB, Voigt H-M , Gen M, Sen S, Dorigo M, Pezeshk S, Garzon MH, Burke E (eds) Proceedings of the Genetic and Evolutionary Computation Conference (GECCO-2001). Morgan Kaufmann, San Francisco, pp 469–476, 7–11 July 2001
15. Solteiro Pires EJ, Tenreiro Machado JA, de Moura Oliveira PB, Reis C 2007 Fractional dynamics in particle swarm optimization. In ISIC IEEE international conference on systems, man and cybernetics, Montréal, pp 1958–1962, 7–10 October 2007

16. Reis C, Tenreiro Machado JA, Galhano AMS, Cunha JB (Aug. 2006) Circuit synthesis using particle swarm optimization. In ICCC—IEEE international conference on computational cybernetics, pp 1–6
17. Solteiro Pires EJ, Tenreiro Machado JA, de Moura Oliveira PB, Boaventura Cunha J, Mendes L (2010) Particle swarm optimization with fractional-order velocity. Nonlinear Dyn 61:295–301
18. Eberhart R, Simpson P, Dobbins R (1996) Computational intelligence PC tools. Academic, San Diego
19. Van den Bergh F, Engelbrecht AP (2006) A study of particle swarm optimization particle trajectories. Inf Sci 176(8):937–971
20. de Moura Oliveira PB, Boaventura Cunha J, Solteiro Pires EJ 2008 Controlling the particle swarm optimization algorithm. Proceedings of the CONTROLO 2008 conference, 8th Portuguese conference on automatic control, pp 23–28, Vila Real, Portugal, 21–23 July 2008

Toward the Concept of Robot Society: A Multi-Robot SLAM Case Study

Micael S. Couceiro, Andria R. Lopes, N. M. Fonseca Ferreira, Anabela G. Ferreira and Rui Rocha

Abstract Over time, biological societies such as humans, ants or bees have shown us the advantages inherent to the collective work. It is based on such results that many researchers have been trying to successfully develop new approaches in Multi-Robot Systems. Nevertheless, several assumptions need to be assured for collective work to emerge. In this paper, it is presented the significance and the advantages of cooperation in the different societies bridging the gap to the concept of robot society. In order to compare the advantages of cooperative robots, it is considered essential the development of computational simulation based on the robotic cooperation in unstructured environments. Hence, a Multi-Robot Simultaneous Localization and Mapping (SLAM) using Rao-Blackwellized particle filter is implemented in a simulation environment developed in the Player/ Stage platform for robot and sensor applications.

Keywords Robot · Society · Cooperation · Multi-robot slam

M. S. Couceiro (✉) · R. Rocha
Institute of Systems and Robotics, University of Coimbra,
Pólo II, 3030-290 Coimbra, Portugal
e-mail: micaelcouceiro@isr.uc.pt; micael@isec.pt

R. Rocha
e-mail: rprocha@isr.uc.pt

M. S. Couceiro · N. M. Fonseca Ferreira
RoboCorp, Department of Electrotechnics Engineering, Engineering Institute of Coimbra, Rua Pedro Nunes, 3030-199 Coimbra, Portugal

N. M. Fonseca Ferreira
e-mail: nunomig@isec.pt

A. R. Lopes
Faculty of Economics of Coimbra, University of Coimbra,
Av. Dias da Silva 165, 3004-512 Coimbra, Portugal
e-mail: andilope@gmail.com

A. G. Ferreira
Coimbra School of Education, Pólo I, Praça Heróis do Ultramar,
Solum, 3030-329 Coimbra, Portugal
e-mail: anabelagoncalves@esec.pt

N. M. Fonseca Ferreira, J. A. Tenreiro Machado (eds.), *Mathematical Methods in Engineering,* DOI 10.1007/978-94-007-7183-3_6,

1 Introduction

The concept of robot society soon appeared showing the inherent advantages when compared to single solutions [1]. Since societies are formed as collaborative structures to execute tasks which are not possible or are difficult for individuals alone, having societies formed by robots would bring at least two advantages: fault tolerance and parallelism. At first glance, having multiple robots performing a task or a set of common tasks may seem more problematical (and challenging) than useful. Why not use a single and complex robot capable of performing all these tasks? The answer is all around us in nature. Much of the work developed in the area of cooperative robots mention biological systems as a source of inspiration. The collective behavior of ants, bees, birds and other similar societies provide strong evidence that systems composed of simple agents can perform complex tasks in the real world. The robustness and adaptability of biological systems represent a powerful motivation to replicate these mechanisms in an attempt to generate software and hardware with features comparable to those of biological systems. These and many other reasons will be addressed in this study showing the benefits of cooperative robots over a single robot.

Cooperative robots, or Multi-Robot Systems (*MRS*), describe the situation in which a group of robots get an overall benefit. A first key issue in cooperation is whether robots should be identical (homogeneous groups) or different (heterogeneous grouping) and if the efficiency should come into consideration for the performance of the whole group or only to each individual robot. This kind of cooperation in robotics can vary from having only two robots to perform a simple task together (*e.g.*, two industrial arms manipulating a large object) [2] to a group of heterogeneous robotic agents that can connect and form a more complex structure [3].

More recently, some studies have been focused interest in *MRS* incorporating algorithms of localization and mapping [4] thus enjoying all the advantages of the cooperation between robots in unstructured environments. Many applications in robotics, such as search and rescue, surveillance, exploration, among others, require the exact location in unknown environments. When robots are operating in unstructured environments, in order to obtain their exact location, we need to create and analyze the map of the environment. The concept of robot society will show us the improvements of systems that require robots to operate in unstructured environment.

Section 2 highlights the importance of cooperation in societies focusing on cooperation and sociological systems. Section 3 gives a brief survey of Simultaneous Localization and Mapping (*SLAM*) applied to single robots and multiple robots and in order to demonstrate the advantages of cooperative robots over a single robot, a Multi-Robot *SLAM* algorithm inspired in the work of Andrew Howard [5] is implemented in Sect. 4 using the *Player/ Stage* platform for robot and sensor applications. Finally, in Sect. 5 outlines the main conclusions.

2 Cooperative Systems

"*Man is a natural animal and, inevitably, selfish*" has been the beginning of all the discussions about capitalism since the early stages reinforced by the powerful and seemingly scientific notion *survival of the fittest*. Charles Darwin defended that in what turned out to be one of the most important works in the history of science: *The Origin of Species* [6]. The theory of natural selection defended by Darwin concluded that not all organisms, at birth, offered the same survival conditions and that only those who better adapt to the environment survive. Put in a less complicated way, Darwin believed that the evolution of the species was like the "*law of the jungle*", where only the brightest would survive and evolve, while all the others disappear or hardly survive. Darwin's theory has been applied mainly to the biological level, but over the years, turned out to be applied also in the economic and social competition.

Nevertheless, the survival of a particular member of a society may depend on the cooperation with other members of this society or even other societies. The Britain's Kevin Foster [7], which has given the continuity to the work of William Hamilton, proved that there are situations of cooperation between individuals who do not fit the basic principle of Darwin arguing that altruism is a way that nature has to assert itself. In June 2008, Kevin Foster said, at the Institute of Molecular Pathology and Immunology in the University of Porto, Portugal, that cooperation is everywhere: "*The genes have joined in the genomes, the cells work together in multicellular organisms and animals cooperate in societies*". Also, thousands of years before, the *King Solomon*, who was a student of the nature, observed the humble ant, and wrote: "*Go to the ant, you sluggard; consider its ways and be wise! It has no commander, no overseer or ruler, yet it stores its provisions in summer and gathers its food at harvest.*" [8]. In fact the ants are a perfect example of cooperation, diligence and order. In addition to work together and help each others, the ants seem to be able to find their paths (the nest to a food source and back or just getting around an obstacle) with relative ease, despite being virtually blind. Several studies have found that in many cases this capacity is the result of the interaction of chemical communication between ants (for a substance called pheromone) and emergent phenomena caused by the presence of many ants. This is the concept of *stigmergy* [9]. This mechanism is so efficient that there are algorithms that use this principle as is the case of the heuristic principle *Ant System* that simulates the behavior of a group of ants that work together to solve an optimization problem using a simple communications [10] and the case of *Brood Sorting* (group selection) used in swarms of robots [11].

Another very similar principle can be seen in other optimization algorithms such as genetic algorithms, evolutionary strategies and the well-known *Particle Swarm Optimization* (*PSO*) initially proposed by Kennedy and Eberhart [12], based on the behavior of social organisms such as birds or fishes. On cooperation and competition among the potential solutions, the optimal complex problems can be achieved more quickly. In *PSO* algorithms each individual of the population is called a particle and the position of these individuals is modified over time. Thus, the particles wander through the multidimensional search space. Along the way, each particle adjusts its position according to their experience and the experience of the other members of the population, taking advantage of the best position of each particle and the best position of the whole group.

Suppose the following scenario: a group of birds are randomly looking for food in an area where there is only one type of food. Although birds don't know where the food is, they know how close to the food they are at each iteration. So what is the best strategy to find the food? The most efficient one is to follow the bird that is closer to the food.

The *PSO* has been successfully used in many applications such as robotics [13–15] and electrical systems [16].

Another interesting engineering example based on biological cooperation is reflected in the flight of pelicans. Researchers discover that the pelicans that fly in formation earn extra boost when compared to the ones flying forward, resulting in a 15 % reduction in the heart rate. In order to validate this concept, a group of engineers prepared a flight test with electronic equipment that enabled the pilot to keep the plane at a distance of 90 m (with a small tolerance of 30 cm) over the plane that was ahead. What was the outcome? The plane suffered an air resistance 20 % lower and it consumed 18 % less fuel. These results can be used on military or civilian planes, but also in the concept of robotics to improve the dynamics of flying robots to monitor forest fires [17] or biologically inspired robots for spying [18].

However, when we speak about cooperation we should say *Cooperative Systems*. The cooperation is just one of the indispensable tools for the Cooperative Systems since without the collaboration between different members of a particular group or society Cooperative Systems cannot survive. On the other hand, to cooperate, the communication is essential between group members and this communication must be familiar to all of them. The coordination also plays an important tool in cooperative systems, since it organizes the group to prevent that communication and cooperation efforts are lost and that tasks are performed in the correct order, at the correct time and meeting the constraints and objectives. The Cooperative Systems has been studied in several areas including computer science [19] and [20] and robotics [4, 21, 22].

Inspired by the results of the existing cooperation in various societies (*e. g.*, ants, bees, plants, humans), researchers have placed a great emphasis on developing robots that can cooperate with each other and perform multiple tasks. The *Cooperative Multi-Robot Systems* (*CMRS*) are based on the interception of the contribution of each member (*i.e.*, robot): if we have a group of robots cooperating to perform a given task, they need to communicate with each other in order to coordinate their actions and obtain the desired result. This concept offers a countless number of advantages similar to the benefits of Cooperative Systems in other societies that may be described in the following key factor: time. One way to circumvent the limitations inherent to the concept of time is to perform simultaneous procedures: if we have multiple robots instead of one they can act on multiple places at the same time (spatial distribution) and they can perform multiple tasks simultaneously (temporal distribution).

3 Multi-Robot SLAM

The search for a solution to the *SLAM* problem has been one of the notable successes of the robotics community over the past decade. The *SLAM* has been formulated and solved as a theoretical problem in a number of different forms being implemented in a number of different domains from indoor robots to outdoor, underwater, and airborne systems. Basically, *SLAM* is a process by which a mobile robot can build a map of an environment and at the same time use this map to deduce its location. So, in a probabilistic form, the *SLAM* problem requires that the probability distribution (1) be computed for all times k.

$$P(x_k, m | Z_{0:k}, U_{0:k}, x_0) \tag{1}$$

This probability distribution describes the joint posterior density of the landmark locations and vehicle state (at time k) given the recorded observations and control inputs up to and including time k together with the initial state of the vehicle.

The *SLAM* approach for a single robot began to receive attention in 1990 [23]. The majority of the solutions to the *SLAM* problem are based on the implementation of the extended Kalman filter (*EKF*) that correlates the pose estimation relative to different landmarks [24, 25]. Although the *EKF* is one of the most effective approaches for map estimation, [26] proved that the *FastSLAM* performance was substantially higher than those obtained by the *EKF*. The *FastSLAM* algorithm was used for the construction of indoor maps in [27, 28]. They used an algorithm based on occupancy grids in order to build a metric map of the environment.

A variant of the *FastSLAM* was proposed [29] combining the Rao-Blackwellized particle filter (*RBPF*) for samples of the trajectory of the robot and an *EKF* to represent the map. This algorithm contains many elements of the standard Monte-Carlo localization algorithm [30]. The challenge lies in maximizing the per-particle update speed while minimizing the corresponding storage requirements, so that the filter may run in real time and in bounded memory with a relatively large number of particles. As always, the speed and storage demands tend to conflict, and our implementation favors the former over the latter.

Based on the previous single robot *SLAM* algorithm Andrew Howard developed a similar algorithm applied to multiple robots [5].

This algorithm has two important assumptions: (i) robots are able to detect, identify and measure the relative pose of other robots at some time during the exploration task (when those robots are both nearby and within line-of-sight, for example). Such encounters allow robots to fuse their subsequent observations into a common map, using the measured relative pose to initialize the filter (note, however, that only the first such encounter is used; subsequent encounters between robots are ignored); and (ii) the particle-filter based *SLAM* algorithm supports time-reversed updates; this generalization allows robots to incorporate observations that occurred prior to the first encounter, by treating those observations as if they came from additional "virtual" robots travelling backwards in time.

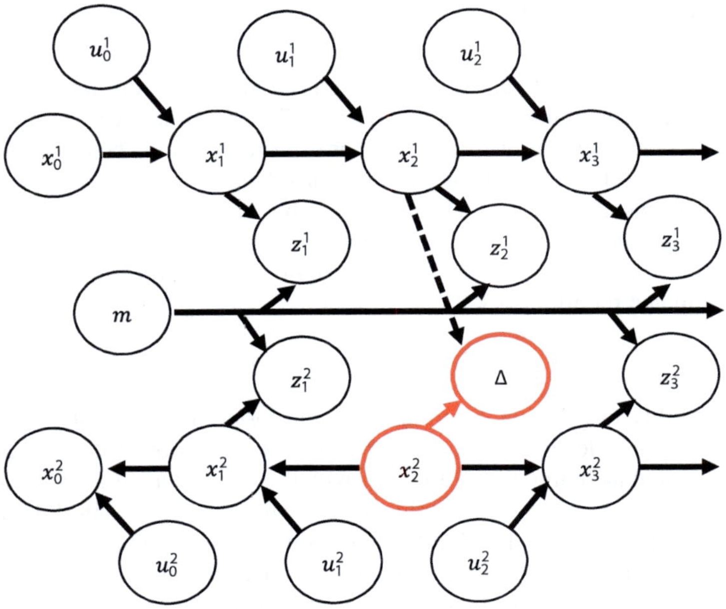

Fig. 1 Bayes net for multi-robot SLAM with unknown initial poses [5]. The robots first encounter one another at time s, recording the relative pose Δ_s^2

As an illustration, consider the following example: two robots are exploring an environment from distant and unknown initial locations. When robots encounter one another they measure their relative pose constructing a filter in which robot 1 has an initial pose of zero, and robot 2 has the measured relative pose. Subsequent measurements from the two robots are fed to the filter, and thereby fused into a common map. At the same time, two virtual robots are added to the filter with poses initialized as above where the previously recorded measurements are fed to the filter in reverse time-order, such that these virtual robots appear to be driving backwards through the environment. Thus, the filter incrementally fuses data from both robots, recorded both before and after the encounter, into a single map.

Let Δ_s^2 denote the relative pose of robot 2 as measured by robot 1 at time s. We wish to estimate the posterior over maps and trajectories given by:

$$\begin{aligned} & p\left(x_{1:t}^1, x_{s+1:t}^2, m | z_{1:t}^1, u_{0:t-1}^1, x_0^1, z_{s+1:t}^2, u_{s:t-1}^2, \Delta_s^2\right) \\ & = p\left(m | x_{1:t}^1, z_{1:t}^1, x_{s+1:t}^2, z_{s+1:t}^2\right) \\ & \quad p\left(x_{1:t}^1 \, | z_{1:t}^1, u_{0:t-1}^1, x_0^1\right) p\left(x_{s+1:t}^2 | z_{s+1:t}^2, u_{s:t-1}^2, x_s^1, \Delta_s^2\right) \end{aligned} \tag{2}$$

where $x_{1:t}^1$*and* $x_{s+1:t}^2$*denotes a sequence of robot 1 and 2 poses at times 1; 2;. . . ; t,* and $s+1; s+2; \ldots; t$, respectively. $z_{1:t}^1$and $z_{s+1:t}^2$denotes the corresponding sequence of observations, and $u_{0:t-1}^1$and $u_{s:t-1}^2$denotes the sequence of actions executed by the robots (Fig. 1).

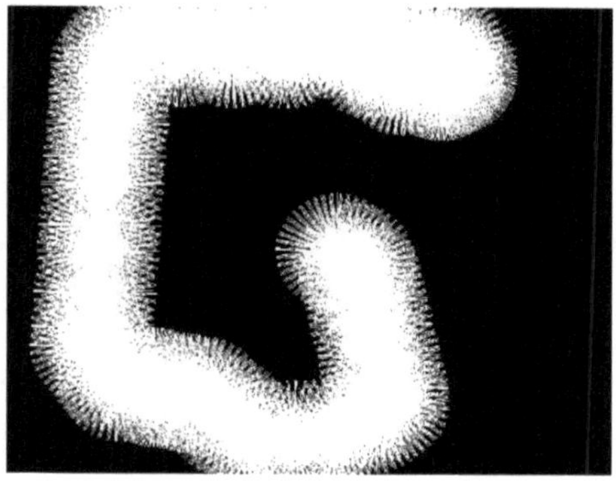

Fig. 2 Map generated using the single-robot algorithm; the map is 16 m by 16 m with a resolution of 0.50 m

This algorithm has number of attractive features. First, it is able to fuse all data from all robots into a single map, without knowing the initial robot poses. Second, it inherits the bounded-time, bounded-memory properties of the single robot *SLAM* algorithm (CPU and memory requirements do not increase with path length). Third and finally, the algorithm is fast: our implementation can fuse data from two robots in real time. Collectively, these features make the algorithm highly suitable for on-line, in-the-loop applications, such as multi-robot exploration and search tasks.

4 Experimental Results

In order to demonstrate the advantages of cooperative robots over a single robot, we implemented a single and Multi-Robot *RBPF-SLAM* algorithm in the *Player/ Stage* platform based on the work of Andrew Howard [5].

The filter update step requires two ray-tracing operations on the occupancy grid for each and every particle: one to evaluate the sensor model and another to update the map. Since these operations are expensive, we approximate the ray-tracing step by considering only the ray end-points, and decimate the laser scans by using only one scan for every 0.50 m of distance traveled. These approximations improve processing speed by an order of magnitude or more, thereby allowing real-time operation. For each particle, we maintain a complete occupancy grid map, generally with a resolution of 0.50 m and covering an area of between 2,000 and 8,000 m^2. The robots used are the *Pioneer II* with odometry and 2D laser (horizontal plane) with 1° of resolution and retro-reflective markers (for mutual recognition).

Figure 2 shows a typical map generated by the single-robot algorithm, with all three loops closed correctly. Processing time for this map is 126 s on a 1.6 GHz *Intel Centrino* using 150 particles. Figure 3 shows the results produced by the multi-robot algorithm for an autonomous exploration task. Two robots were deployed into this environment at distant locations, from which they executed a cooperative, but largely reactive, exploration strategy.

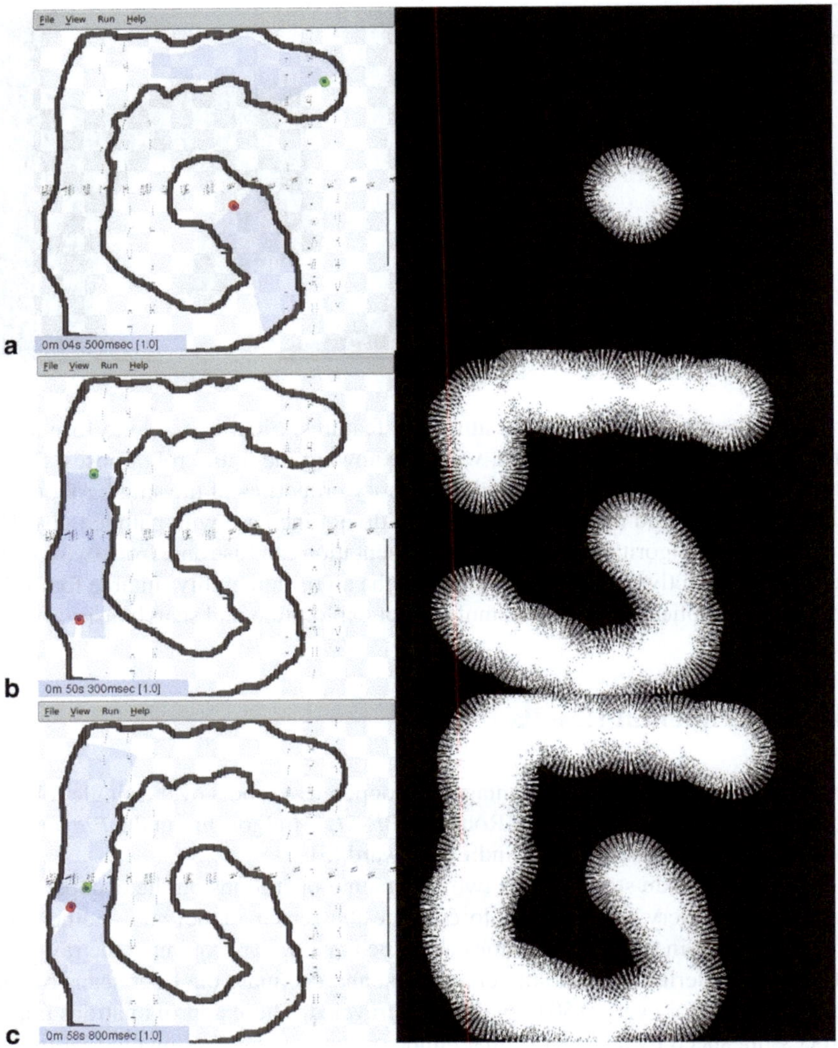

Fig. 3 Sequence of events events: **a** robots starts at distant locations and the global map begins being generated by the robot 1 (*red*) considering its initial position zero; **b** robot 1 (*red*) encounters robot 2 (*green*) (due to the retro-reflective markers) at time $t = 54$ s and uses the combined information adding it to the global map; **c** at time $t = 62$ s the entire map is obtained with a resolution of 0.50 m

In the final map all the major topological features have been properly extracted and the map quality is uniformly high. The processing time for this map is 62 s on a computer 1.6 GHz *Intel Centrino* using 150 particles.

5 Conclusions and Discussion

The concept of robot society shows potential in applications where the space and time distribution of single robots are restricted and also as an alternative to more complex robots. To demonstrate possible advantages of the cooperation in robotics, a single and a multi-robot *SLAM* algorithm based on a *RBPF* was implemented on the *Player/ Stage* platform. One of the attractive features of the multi-robot SLAM algorithm is that it's easy to implement after the implementation of the single robot algorithm. The basic elements of the algorithm (*i.e.*, the sensor and action models, occupancy grids and ray-tracing) are easily adapted from the Monte-Carlo location algorithm. Despite possible improvements to this algorithm, the results show that a cooperative exploration strategy becomes far superior to the individual one. The processing time of the map for the single-robot solution is greater than twice the processing time of the two-robots solution. Therefore, the use of cooperative strategies in robotics offers several attractive features since robots are constantly interacting and communicating with each other with the dynamic environment and with other members of different societies (*e.g.*, man). The collective intelligence emerging from cooperative strategies in robotics gives a reason to call these systems, at their highest level, as robot society.

Acknowledgement This work was supported by a PhD scholarship (SFRH/BD/73382/2010) granted by the Portuguese Foundation for Science and Technology (FCT), the Institute of Systems and Robotics (ISR) and RoboCorp.

References

1. Halme A, Jakubik P, Schönberg T, Vainio M (1993) The concept of robot society and its utilization. In: Proceedings of IEEE international workshop on advanced robotics, Espoo, Finland
2. Ferreira NMF (2006) Sistemas Dinâmicos e Controlo de Robôs Cooperantes. The Phd thesis (in 5 of September) University of Trás-os-Montes e Alto Douro
3. Fukuda T, Nakagawa S, Kawauchi Y, Buss M (1989) Structure decision method for self organizing robots based on cell structures—CEBOT. In: Proceedings of IEEE international conference on robotics and automation, pp 695–700, Scottsdale, AZ
4. Rocha R (2006) Building volumetric maps with cooperative mobile robots and useful information sharing: a distributed control approach based on entropy. PhD thesis, Faculty of Engineering of University of Porto, Portugal, May 2006
5. Howard, A (2006) Multi-robot SL, mapping using particle filters. Int J Robot Res 25(12): 1243–1256
6. Darwin, C (1872) The origin of species by means of natural selection, or the preservation of favoured races in the struggle for life. John Murray, London
7. Foster KR, Xavier JB (2007) Cooperation: bridging ecology and sociobiology. Curr Biol 17:R319–R321
8. Dean, M (1913) Book of Proverbs. Catholic encyclopedia. Adapted from Holman Bible Handbook on Proverbs
9. Aras R, Dutech A, Charpillet F (2004) Stigmergy in multi agent reinforcement learning. Loria, Inst. Nat de Recherche en Inf et Autom, Nancy
10. Dorigo, M, Stützle T (2004) Ant colony optimization. MIT Press, Cambridge

11. Wilson M, Melhuish C, Sendova-Franks A, Scholes S (2004) Algorithms for building annular structures with minimalist robots inspired by brood sorting in ant colonies. Auton Robot 17 (2–3):115–136
12. Kennedy J, Eberhart R (1995) A new optimizer using particle swarm theory. In: Proceedings of the IEEE sixth international symposium on micro machine and human science, pp 39–43, Nagoya
13. Tang J, Zhu J, Sun Z (2005) A novel path planning approach based on AppART and particle swarm optimization. Advances in Neural Networks–ISNN 2005. Springer Berlin Heidelberg, pp 253–258
14. Pires EJS, Oliveira PBM, Machado JAT, Cunha JB (2006) Particle swarm optimization versus genetic algorithm in manipulator trajectory planning. In: 7th Portuguese conference on automatic contol, Instituto Superior Técnico, Lisbon, Portugal, 11–13 Sept 2006
15. Couceiro MS, Mendes R, Ferreira NMF, Machado JAT (2009) Control optimization of a robotic bird. EWOMS '09, Lisbon, Portugal, 4–6 June, 2009
16. Alrashidi MR, El-Hawary MEA (2006) Survey of particle swarm optimization applications in power system operations. Electric Power Compon Syst 34(12):1349–1357
17. Martinez JR, Merino L, Caballero F, Ollero A, Viegas DX (2006) Experimental results of automatic fire detection and monitoring with UAVs. For Ecol Manage 234:232
18. Couceiro MS, Figueiredo CM, Ferreira NMF, Machado JAT (2009) Biological inspired flying robot. In: Proceedings of IDETC/CIE 2009 ASME 2009 international design engineering technical conferences & computers and information in engineering conference, San Diego, 30 Aug–2 Sept 2009
19. Borghoff UM, Schlighter JH (2000) Computer-supported cooperative work: introduction to distributed applications. Springer, USA
20. Fuks H, Raposo AB, Gerosa MA, Lucena CJPO (2003) Modelo de Colaboração 3C e a Engenharia de Groupware. Pontifícia Universidade Católica, Rio de Janeiro (PUC-Rio)
21. Cao Y, Fukunaga A, Kahng A (1997) Cooperative mobile robotics: antecedents and directions. Auton Robot 4:1–23
22. Jung, D (1998) An architecture for cooperation among autonomous agents. PhD thesis, Department of Computer Science, University of Wollongong, Australia
23. Smith R, Self M, Cheeseman P (1990) Estimating uncertain spatial relationships in robotics. In: Ingemar JC, Gordon TW (eds) Autonomous robot vehicles. Springer, New York, pp 167–193
24. Dissanayake M, Newman P, Clark S, Durrant-Whyte H, Csorba M (2001) A solution to the simultaneous localization and map building (SLAM) prob-lem. IEEE Trans Robot Autom, 17(3):229–241
25. Thrun S, Dirk H, David F, Montemerlo D, Rudolph T, Wolfram B, Christopher B, Zachary O, Scott T, William W (2003) A system for volumetric robotic mapping of abandoned mines. Robotics and Automation, 2003. Proceedings. ICRA'03. IEEE International Conference on ,vol. 3, pp. 4270–4275. IEEE
26. Thrun S, Dirk H, David F, Montemerlo D, Rudolph T, Wolfram B, Christopher B, Zachary O, Scott T, William W (2003) A system for volumetric robotic mapping of abandoned mines. Robotics and Automation, 2003. Proceedings. ICRA'03. IEEE International Conference on ,vol. 3, pp. 4270–4275. IEEE
27. Stachniss C, Hahnel D, Burgard W (2004) Exploration with active loop-closing for FastSLAM. In: Proceedings of IEEE/RSJ international conference on intelligent robots and systems, Department of Computer Science, Freiburg University, Germany
28. Stachniss C, Grisetti G, Burgard W (2005) Recovering particle diversity in a Rao-Blackwellized particle filter for SLAM after actively closing loops. In: Proceedings of IEEE international conference on robotics and automation, Freiburg, Germany
29. Hahnel D, Burgard W, Fox D (2003) An efficient FastSLAM algorithm for generating maps of large-scale cyclic environments from raw laser range measurements. In: IEEE/RSJ international conference on intelligent robots and systems, Las Vegas, Nevada, USA, Oct 2003
30. Thrun S, Fox D, Burgard W (2001) Robust Monte Carlo localization for mobile robots. Artif Intell J 128(1–2):99–141

Generalized State-Space Modeling for m Level Diode-Clamped Multilevel Converters

Miguel Chaves, Elmano Margato, J. Fernando Silva and Sónia F. Pinto

Abstract Multilevel power converter structures have been introduced as the solutions for high power high voltage applications and also for grid interface connection of renewable energy sources systems, where they have several advantages, namely low distortion voltages and currents, low switching losses resulting in higher efficiency. As a consequence of the increasing interest on multilevel converter applications, accurate models of these power converters are essential for computer simulation studies. This paper presents a systematic modeling approach suitable to obtain generalized state-space models for m level diode-clamped multilevel converters supplying AC loads. In particular, for m = 5, the proposed model is compared to the corresponding model using general purpose Simulink blocks and SimPowerSystems toolbox.

Keywords Power electronics · Power drives · Multilevel converter modeling

1 Introduction

Recently, multilevel converters are being used in applications such as high-voltage high-power DC-AC drives and renewable energy sources grid connection.

The first multilevel converter structures, also known as neutral point clamped (NPC) converters, had three levels [1]. These converter structures consist of two DC

M. Chaves (✉) · E. Margato
Centro de Electrotecnia e Electrónica Industrial, ISEL, R. Conselheiro Emídio Navarro 1, 1950-062 Lisboa, Portugal
e-mail: mchaves@deea.isel.ipl.pt

E. Margato
e-mail: efmargato@isel.ipl.pt

J. F. Silva · S. F. Pinto
Instituto Superior Técnico, DEEC, TULisbon, Av. Rovisco Pais, 1049-001 Lisboa, Portugal
e-mail: fernandos@alfa.ist.utl.pt

S. F. Pinto
e-mail: soniafp@ist.utl.pt

M. Chaves · E. Margato · J. F. Silva · S. F. Pinto
Cie3, Av. Rovisco Pais, 1049-001 Lisboa, Portugal

N. M. Fonseca Ferreira, J. A. Tenreiro Machado (eds.), *Mathematical Methods in Engineering*, DOI 10.1007/978-94-007-7183-3_7,

bus capacitor voltages in series being the central tap the neutral level. Each converter leg has two pairs of switching devices being the center of each device pair tied to the neutral level by clamping diodes. The diode clamped method can be applied and generalized for *m* level converters.

Multilevel converters have several attractive features when compared to conventional voltage source inverters such as: staircase output voltage waveforms, meaning lower distortion and smaller *dv/dt*, along with lower voltage rated switches (often IGBTs); common-mode voltage is generally lower and can be eliminated using advanced modulations strategies benefiting the converter load [2, 3]. The converter output voltages waveforms approach the ideal sinusoidal waveforms, as converter *m* level number increases, allowing load currents with low distortion. Multilevel converters can operate with relatively low switching frequency which means lower switching losses and higher efficiency.

The basic multilevel concept uses appropriate switch combinations to obtain an output voltage which can be one of the *(m-1)* DC voltages provided by the capacitive voltage divider connected to the DC bus voltage U_{dc} [4]. The converter staircase like output waveforms can reach high voltage values, while the power semiconductors must withstand only a reduced voltage normally $U_{dc}/(m\text{-}1)$.

Still, there are some problems to solve in multilevel converters. One is a generalized strategy to balance all *(m*1) DC capacitor voltages, which requires additional modulation techniques to face the capacitive voltage divider drift, otherwise some capacitors become overcharged and others discharged.

Also, accurate models of power converters are essential for computer simulation. These studies are an important tool, allowing time and cost savings, when it is aimed the design, behavior analysis, capacitor voltage equalization and converter controller synthesis, since multilevel power converters are non-linear structures that can establish, in a discrete way imposed by the switching strategy, several appropriate electrical connections between the DC source and AC load.

Converter structure complexity increases for high *m* level number converters. It can become cumbersome when associations of multilevel power converters are used and operated in a cooperative way. This is the case of grid interface connection of renewable energy sources using back-to-back multilevel converters. This type of power structures needs several control loops and linear [5–7] and non-linear [8, 9] controllers. Thus, for converter computer modeling implementation and controller synthesis, it is useful to have a systematic modeling approach, valid for m level converters, in order to ease the modeling task and to design suitable controllers.

This paper presents a systematic modeling methodology suitable to develop generalized switched state-space models for *m* level diode-clamped multilevel power converters whose general structure is shown in Fig. 1.

The converter interconnects capacitor divided DC bus voltages to the AC load. The capacitor number is equal to *(m1)*. All the three converter legs, one per output phase, have *(m-1)* complementary state switch pairs with anti-parallel and series connected diode. Each leg has also *(m-2)* clamping diodes pairs in order to guarantee the circuit topological constraints. The converter legs are parallel connected to the DC bus voltage and each leg midpoint is connected to the respective output load phase.

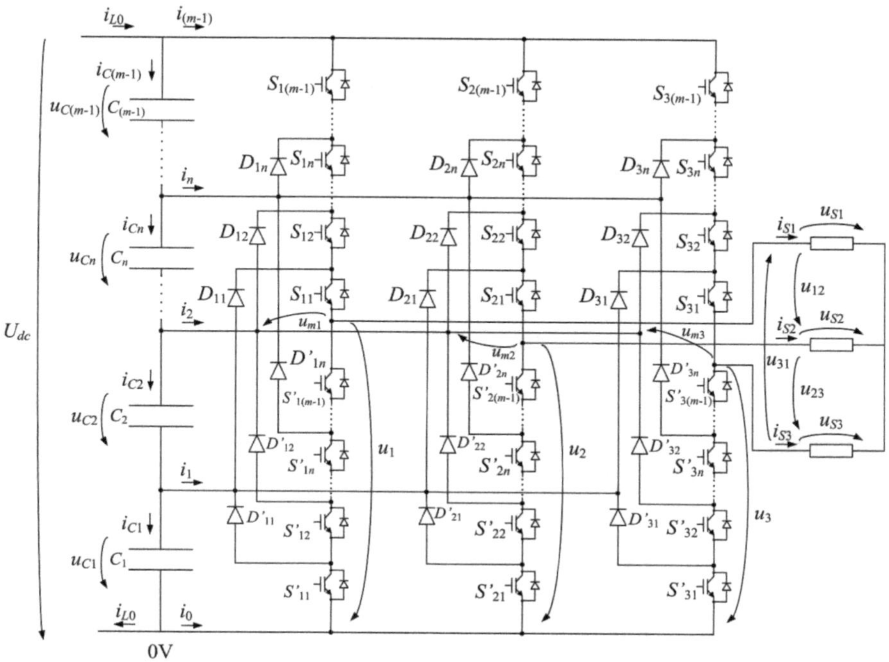

Fig. 1 Three phase *m* level diode-clamped converter supplying an AC load

In both modeling approaches, electrical and semiconductor devices are considered ideal components (zero ON voltages, zero OFF currents, zero switching times).

The proposed modeling is presented in Sect. 2 and compared to the existing one [8, 9] in Sect. 3. It is shown that the existing modeling has an easier code implementation if used for multilevel converters with few levels ($m = 3$), while the generalized modeling is useful for converters having higher number of levels.

The generalized state space model developed for a *m* level neutral point clamped multilevel converter is applied to $m = 5$ and implemented using general purpose Simulink blocks. Simulations results are compared with the model of the same converter obtained using SimPowerSystems toolbox, Sect. 4.

2 Generalized State Space Converter Model in Phase Coordinates

To obtain a general model for a *m* level converter, it is advantageous, for each *k* leg ($k \in \{1,2,3\}$), to start numbering the upper IGBT switches S_{k1}, S_{k2}, ... S_{kn}, ... $S_{k(\mathrm{m1})}$ from the leg midpoint, and S'_{k1}, S'_{k2}, ... S'_{kn}, ... $S'_{k(\mathrm{m1})}$ up from the zero voltage point.

Using this numbering methodology, the switching strategy for a m level multilevel converter assures that upper leg switches $[S_{k1}\ S_{k2}\ \ldots\ S_{kn}\ \ldots\ S_{k(m\text{-}1)}]$ and the corresponding ones on the lower side $[S'_{k1}\ S'_{k2}\ \ldots\ S'_{kn}\ \ldots\ S'_{k(m\text{-}1)}]$ are always in complementary states. Consequently: if $S_{kj} = 1$, meaning that the specified S_{kj} switch is ON, then S'_{kj} must be equal to 0 (S'_{kj} switch is OFF).

For each leg the output voltage u_k, defined from the k leg midpoint to the zero voltage, can be written in terms of leg switches logical state and DC bus capacitor voltages by (1):

$$u_k = \sum_{j=1}^{m-1} S_{kj} u_{cj} \tag{1}$$

Where u_{cj} is the voltage of a considered n level capacitor ($j = n$) that is connected in series with the previous capacitor voltage $u_{c(n\text{-}1)}$ by the action of turning ON switch S_{kn}.

Equation (1) can also be written in matrix form as (2).

$$\begin{bmatrix} u_1 \\ u_2 \\ u_3 \end{bmatrix} = \begin{bmatrix} S_{11} & S_{12} & \ldots & S_{1n} & \ldots & S_{1(m-1)} \\ S_{21} & S_{22} & \ldots & S_{2n} & \ldots & S_{2(m-1)} \\ S_{31} & S_{32} & \ldots & S_{3n} & \ldots & S_{3(m-1)} \end{bmatrix} \begin{bmatrix} u_{C1} \\ u_{C2} \\ \vdots \\ u_{Cn} \\ \vdots \\ u_{C(m-1)} \end{bmatrix} \tag{2}$$

The DC bus n level current i_n, $n \in \{1,2, \ldots, m\text{-}1\}$, can be related to load phase currents i_{Sk} by (3).

$$i_n = \sum_{k=1}^{3} \Gamma_{nk} i_{Sk} \tag{3}$$

Where Γ_{nk} is a time dependent switching variable, written in terms of the k leg switches logical state as (4).

$$\Gamma_{nk} = S_{k1} S_{k2} \ldots S_{kn} (1 - S_{k(n+1)})(1 - S_{k(n+2)}) \ldots (1 - S_{k(m-1)}) \tag{4}$$

Equation (3) can also be written in matrix form:

$$\begin{bmatrix} i_1 \\ i_2 \\ \vdots \\ i_n \\ \vdots \\ i_{(m-1)} \end{bmatrix} = \begin{bmatrix} \Gamma_{11} & \Gamma_{12} & \Gamma_{13} \\ \Gamma_{21} & \Gamma_{22} & \Gamma_{23} \\ \ldots & \ldots & \ldots \\ \Gamma_{n1} & \Gamma_{n2} & \Gamma_{n3} \\ \ldots & \ldots & \ldots \\ \Gamma_{(m-1)1} & \Gamma_{(m-1)2} & \Gamma_{(m-1)3} \end{bmatrix} \begin{bmatrix} i_{S1} \\ i_{S2} \\ i_{S3} \end{bmatrix} \tag{5}$$

In Eq. (5) each column of the matrix can carry a maximum of one nonzero element, since each load phase current i_{Sk} (k leg) is connected to a n DC bus level when $\Gamma_{nk} = 1$, or to the zero voltage bus when $\sum_{j=1}^{(m-1)} \Gamma_{jk} = 0$.

Each DC bus n level current capacitor i_{Cn} can be related to the corresponding voltage u_{Cn} by (6).

$$i_{Cn} = C_n \frac{du_{Cn}}{dt} \tag{6}$$

The above current i_{cn} can be expressed in terms of the upper capacitor current $i_{c\,(n+1)}$ and the corresponding DC bus n level current i_n. In the case of capacitor $C_{(m\text{-}1)}$, a generic source i_{L0} current (7) (8) is considered to contribute to the upper capacitor current.

$$i_{Cn} = i_{C(n+1)} - i_n = C_n \frac{du_{Cn}}{dt} \tag{7}$$

$$i_{C(m-1)} = i_{L0} - i_{(m-1)} = C_{(m-1)} \frac{du_{C(m-1)}}{dt} \tag{8}$$

Rewriting Eqs. (7) and (8) for the voltage capacitor state variables, the matrix equation (9) is obtained.

$$\frac{d}{dt}\begin{bmatrix} u_{C1} \\ u_{C2} \\ \vdots \\ u_{Cn} \\ \vdots \\ u_{C(m-1)} \end{bmatrix} = \begin{bmatrix} -\frac{1}{C_1} & -\frac{1}{C_1} & \cdots & -\frac{1}{C_1} & \cdots & -\frac{1}{C_1} & \frac{1}{C_1} \\ 0 & -\frac{1}{C_2} & \cdots & -\frac{1}{C_2} & \cdots & -\frac{1}{C_2} & \frac{1}{C_2} \\ \cdots & \cdots & \cdots & \cdots & \cdots & \cdots & \cdots \\ 0 & 0 & \cdots & -\frac{1}{C_n} & \cdots & -\frac{1}{C_n} & \frac{1}{C_n} \\ \cdots & \cdots & \cdots & \cdots & \cdots & \cdots & \cdots \\ 0 & 0 & \cdots & 0 & \cdots & -\frac{1}{C_{(m-1)}} & \frac{1}{C_{(m-1)}} \end{bmatrix} \begin{bmatrix} i_1 \\ i_2 \\ \vdots \\ i_n \\ \vdots \\ i_{(m-1)} \\ i_{L0} \end{bmatrix} \tag{9}$$

Using (5) it is possible to express equation (9) in terms of load phase currents i_{sk} (10).

$$\frac{d}{dt}\begin{bmatrix} u_{C1} \\ u_{C2} \\ \vdots \\ u_{Cn} \\ \vdots \\ u_{C(m-1)} \end{bmatrix} = \begin{bmatrix} -\frac{\Gamma_{C11}}{C_1} & -\frac{\Gamma_{C12}}{C_1} & -\frac{\Gamma_{C13}}{C_1} & \frac{1}{C_1} \\ -\frac{\Gamma_{C21}}{C_2} & -\frac{\Gamma_{C22}}{C_2} & -\frac{\Gamma_{C23}}{C_2} & \frac{1}{C_2} \\ \cdots & \cdots & \cdots & \cdots \\ -\frac{\Gamma_{Cn1}}{C_n} & -\frac{\Gamma_{Cn2}}{C_n} & -\frac{\Gamma_{Cn3}}{C_n} & \frac{1}{C_n} \\ \cdots & \cdots & \cdots & \cdots \\ -\frac{\Gamma_{C(m-1)1}}{C_{(m-1)}} & -\frac{\Gamma_{C(m-1)2}}{C_{(m-1)}} & -\frac{\Gamma_{C(m-1)3}}{C_{(m-1)}} & \frac{1}{C_{(m-1)}} \end{bmatrix} \begin{bmatrix} i_{S1} \\ i_{S2} \\ i_{S3} \\ i_{L0} \end{bmatrix} \tag{10}$$

Where the k column matrix element Γ_{Cnk} (k leg) is determined adding from n to (m-1) the values of time dependent switching variables Γ_{nk}, as in (11):

$$\Gamma_{Cnk} = \sum_{i=n}^{m-1} \Gamma_{ik} \tag{11}$$

As in existing three levels voltage source inverters modeling, the k load phase voltage u_{sk} can be related to all leg output voltages u_k and, using (1), expressed as a function of DC bus capacitor voltages as (12).

$$u_{Sk} = \frac{1}{3}\sum_{j=1}^{m-1}\left(2S_{kj} - \sum_{\substack{i=1\\ i\neq k}}^{3} S_{ij}\right)u_{Cj} \tag{12}$$

Considering a standard R-L balanced load with electromotive force e_{sk} and isolated neutral, the converter load equation can be written as (13).

$$u_{Sk} = Ri_{Sk} + L\frac{di_{Sk}}{dt} + e_{Sk} \tag{13}$$

From Eqs. (12) and (13) it is possible to obtain the phase currents dynamic model in terms of DC bus capacitor voltages using matrix form (14).

$$\frac{d}{dt}\begin{bmatrix} i_{S1} \\ i_{S2} \\ i_{S3} \end{bmatrix} = \begin{bmatrix} -\frac{R}{L} & 0 & 0 & \frac{S_{C11}}{L} & \frac{S_{C12}}{L} & \cdots & \frac{S_{C1n}}{L} & \cdots & \frac{S_{C1(m-1)}}{L} \\ 0 & -\frac{R}{L} & 0 & \frac{S_{C21}}{L} & \frac{S_{C22}}{L} & \cdots & \frac{S_{C2n}}{L} & \cdots & \frac{S_{C2(m-1)}}{L} \\ 0 & 0 & -\frac{R}{L} & \frac{S_{C31}}{L} & \frac{S_{C32}}{L} & \cdots & \frac{S_{C3n}}{L} & \cdots & \frac{S_{C3(m-1)}}{L} \end{bmatrix} \begin{bmatrix} i_{S1} \\ i_{S2} \\ i_{S3} \\ u_{C1} \\ u_{C2} \\ \vdots \\ u_{Cn} \\ \vdots \\ u_{C(m-1)} \end{bmatrix} + \\ + \begin{bmatrix} -\frac{1}{L} & 0 & 0 \\ 0 & -\frac{1}{L} & 0 \\ 0 & 0 & -\frac{1}{L} \end{bmatrix} \begin{bmatrix} e_{S1} \\ e_{S2} \\ e_{S3} \end{bmatrix} \tag{14}$$

Matrix elements S_{Ckj} are determined by (15).

$$s_{Ckj} = \frac{1}{3}\left(2S_{kj} - \sum_{\substack{i=1\\ i\neq k}}^{3} S_{ij}\right) \tag{15}$$

The complete m level multilevel converter system model, in phase coordinates, can be written in matrix form by (16).

This generalized switched state-space model, for m level diode clamped multilevel converters, is a detailed nonlinear model suited for computer implementation.

$$
\frac{d}{dt}\begin{bmatrix} i_{S1} \\ i_{S2} \\ i_{S3} \\ u_{C1} \\ u_{C2} \\ \vdots \\ u_{Cn} \\ \vdots \\ u_{C(m-1)} \end{bmatrix}
=
\begin{bmatrix}
-\frac{R}{L} & 0 & 0 & \frac{S_{C11}}{L} & \frac{S_{C12}}{L} & \dots & \frac{S_{C1n}}{L} & \dots & \frac{S_{C1(m-1)}}{L} \\
0 & -\frac{R}{L} & 0 & \frac{S_{C21}}{L} & \frac{S_{C22}}{L} & \dots & \frac{S_{C2n}}{L} & \dots & \frac{S_{C2(m-1)}}{L} \\
0 & 0 & -\frac{R}{L} & \frac{S_{C31}}{L} & \frac{S_{C32}}{L} & \dots & \frac{S_{C3n}}{L} & \dots & \frac{S_{C3(m-1)}}{L} \\
-\frac{\Gamma_{C11}}{C_1} & -\frac{\Gamma_{C12}}{C_1} & -\frac{\Gamma_{C13}}{C_1} & 0 & 0 & \dots & 0 & \dots & 0 \\
-\frac{\Gamma_{C21}}{C_2} & -\frac{\Gamma_{C22}}{C_2} & -\frac{\Gamma_{C23}}{C_2} & 0 & 0 & \dots & 0 & \dots & 0 \\
\dots & \dots & \dots & \dots & \dots & \dots & \dots & \dots & \dots \\
-\frac{\Gamma_{Cn1}}{C_n} & -\frac{\Gamma_{Cn2}}{C_n} & -\frac{\Gamma_{Cn3}}{C_n} & 0 & 0 & \dots & 0 & \dots & 0 \\
\dots & \dots & \dots & \dots & \dots & \dots & \dots & \dots & \dots \\
-\frac{\Gamma_{C(m-1)1}}{C_{(m-1)}} & -\frac{\Gamma_{C(m-1)2}}{C_{(m-1)}} & -\frac{\Gamma_{C(m-1)3}}{C_{(m-1)}} & 0 & 0 & \dots & 0 & \dots & 0
\end{bmatrix}
\begin{bmatrix} i_{S1} \\ i_{S2} \\ i_{S3} \\ u_{C1} \\ u_{C2} \\ \vdots \\ u_{Cn} \\ \vdots \\ u_{C(m-1)} \end{bmatrix}
+
$$

$$
+\begin{bmatrix}
-\frac{1}{L} & 0 & 0 & 0 \\
0 & -\frac{1}{L} & 0 & 0 \\
0 & 0 & -\frac{1}{L} & 0 \\
0 & 0 & 0 & \frac{1}{C_1} \\
0 & 0 & 0 & \frac{1}{C_2} \\
\dots & \dots & \dots & \dots \\
0 & 0 & 0 & \frac{1}{C_n} \\
\dots & \dots & \dots & \dots \\
0 & 0 & 0 & \frac{1}{C_{(m-1)}}
\end{bmatrix}
\begin{bmatrix} e_{S1} \\ e_{S2} \\ e_{S3} \\ i_{L0} \end{bmatrix}
\tag{16}
$$

3 State Space 5 level Converter Model using Existing approach

The existing approach, [8, 9], will be here applied to the five level ($m = 5$) multilevel converter.

To write this model a time dependent switching variable γ_k is defined and written in terms of the switches state logical value (17). This variable is used to define the leg output voltage u_k.

$$\gamma_k = \frac{\sum_{j=1}^{4} S_{kj}}{4} = \begin{cases} 1 \Leftarrow \begin{bmatrix} S_{k1} & S_{k2} & S_{k3} & S_{k4} \end{bmatrix} = \begin{bmatrix} 1 & 1 & 1 & 1 \end{bmatrix} \\ \frac{3}{4} \Leftarrow \begin{bmatrix} S_{k1} & S_{k2} & S_{k3} & S_{k4} \end{bmatrix} = \begin{bmatrix} 1 & 1 & 1 & 0 \end{bmatrix} \\ \frac{1}{2} \Leftarrow \begin{bmatrix} S_{k1} & S_{k2} & S_{k3} & S_{k4} \end{bmatrix} = \begin{bmatrix} 1 & 1 & 0 & 0 \end{bmatrix} \\ \frac{1}{4} \Leftarrow \begin{bmatrix} S_{k1} & S_{k2} & S_{k3} & S_{k4} \end{bmatrix} = \begin{bmatrix} 1 & 0 & 0 & 0 \end{bmatrix} \\ 0 \Leftarrow \begin{bmatrix} S_{k1} & S_{k2} & S_{k3} & S_{k4} \end{bmatrix} = \begin{bmatrix} 0 & 0 & 0 & 0 \end{bmatrix} \end{cases} \tag{17}$$

The leg output voltage u_k is expressed as function of γ_k by (18) and written as (19):

$$u_k = \begin{cases} u_{C1} + u_{C2} + u_{C3} + u_{C4} & \Leftarrow \gamma_k = 1 \\ u_{C1} + u_{C2} + u_{C3} & \Leftarrow \gamma_k = \frac{3}{4} \\ u_{C1} + u_{C2} & \Leftarrow \gamma_k = \frac{1}{2} \\ u_{C1} & \Leftarrow \gamma_k = \frac{1}{4} \\ 0 & \Leftarrow \gamma_k = 0 \end{cases} \tag{18}$$

$$\begin{aligned} u_k = {} & (\Gamma_{1k} + \Gamma_{2k} + \Gamma_{3k} + \Gamma_{4k})u_{C1} + (\Gamma_{2k} + \Gamma_{3k} + \Gamma_{4k})u_{C2} \\ & + (\Gamma_{3k} + \Gamma_{4k})u_{C3} + \Gamma_{4k}u_{C4} \end{aligned} \tag{19}$$

In this case, Γ_{nk} is a time dependent switching variable written as a γ_k dependent function (20) that, as stated before, it will be used to define the connection state between each k phase and the n DC bus level.

Beyond stated in references [8] and [9], it is now presented a systematic way to write the Γ_{nk} switching functions. Functions Γ_{nk} ($n \in \{1; 2; 3; \ldots (m\text{-}1)\}$) can be written as a product of (m-1) factors. Each factor, itself written in terms of γ_k,, must assure two conditions: the first is that each factor must be equal to zero for one specified value of γ_k, excluding $\gamma_k = n/4$ ($\gamma_k \in \{0;\ 1/(m\text{-}1);\ 2/(m\text{-}1);\ \ldots\ (m\text{-}1)/(m\text{-}1)\}$); the second condition states that all the factors must be equal to one when $\gamma_k = n/4$. The systematic form of the switching variable Γ_{nk} can be expressed as (20b).

$$\Gamma_{nk} = \begin{cases} \Gamma_{1k} = 4\gamma_k\left(-4\gamma_k+2\right)\left(-2\gamma_k+\dfrac{3}{2}\right)\left(-\dfrac{4}{3}\gamma_k+\dfrac{4}{3}\right) = \begin{cases} 1 \Leftarrow \gamma_k = 1/4 \\ 0 \Leftarrow \gamma_k \neq 1/4 \end{cases} \\ \Gamma_{2k} = 2\gamma_k\left(4\gamma_k-1\right)\left(-4\gamma_k+3\right)\left(-2\gamma_k+2\right) = \begin{cases} 1 \Leftarrow \gamma_k = 1/2 \\ 0 \Leftarrow \gamma_k \neq 1/2 \end{cases} \\ \Gamma_{3k} = \dfrac{4}{3}\gamma_k\left(2\gamma_k-\dfrac{1}{2}\right)\left(4\gamma_k-2\right)\left(-4\gamma_k+4\right) = \begin{cases} 1 \Leftarrow \gamma_k = 3/4 \\ 0 \Leftarrow \gamma_k \neq 3/4 \end{cases} \\ \Gamma_{4k} = \gamma_k\left(\dfrac{4}{3}\gamma_k-\dfrac{1}{3}\right)\left(2\gamma_k-1\right)\left(4\gamma_k-3\right) = \begin{cases} 1 \Leftarrow \gamma_k = 1 \\ 0 \Leftarrow \gamma_k \neq 1 \end{cases} \end{cases} \tag{20a}$$

$$\Gamma_{nk} = \prod_{\begin{cases} j=0 \\ \wedge \\ j\neq n \end{cases}}^{(m-1)} \left(\frac{\gamma_k - \frac{j}{(m-1)}}{\frac{n}{(m-1)} - \frac{j}{(m-1)}} \right) \tag{20b}$$

The load phase voltage u_{Sk} can be written in terms of converter legs output voltage u_k as in (21).

$$u_{Sk} = \begin{cases} u_{S1} = \frac{1}{3}(2u_1 - u_2 - u_3) \\ u_{S2} = \frac{1}{3}(-u_1 + 2u_2 - u_3) \\ u_{S3} = \frac{1}{3}(-u_1 - u_2 + 2u_3) \end{cases} \tag{21}$$

Using (19) and (21) the output phase voltages u_{Sk} can be rewritten in terms of the capacitors voltages as (22).

$$\begin{bmatrix} u_{S1} \\ u_{S2} \\ u_{S3} \end{bmatrix} = \begin{bmatrix} S_{C11} & S_{C12} & S_{C13} & S_{C14} \\ S_{C21} & S_{C22} & S_{C23} & S_{C24} \\ S_{C31} & S_{C32} & S_{C33} & S_{C34} \end{bmatrix} \begin{bmatrix} u_{C1} \\ u_{C2} \\ u_{C3} \\ u_{C4} \end{bmatrix} \tag{22}$$

Where matrix elements S_{Ckj} are functions of Γ_{nk}, as shown in the following expressions (24), which could also be obtained using (15).

$$
\begin{aligned}
S_{C11} &= \frac{1}{3}(2(\Gamma_{11}+\Gamma_{21}+\Gamma_{31}+\Gamma_{41})-(\Gamma_{12}+\Gamma_{22}+\Gamma_{32}+\Gamma_{42})-(\Gamma_{13}+\Gamma_{23}+\Gamma_{33}+\Gamma_{43}))\\
S_{C21} &= \frac{1}{3}(-(\Gamma_{11}+\Gamma_{21}+\Gamma_{31}+\Gamma_{41})+2(\Gamma_{12}+\Gamma_{22}+\Gamma_{32}+\Gamma_{42})-(\Gamma_{13}+\Gamma_{23}+\Gamma_{33}+\Gamma_{43}))\\
S_{C31} &= \frac{1}{3}(-(\Gamma_{11}+\Gamma_{21}+\Gamma_{31}+\Gamma_{41})-(\Gamma_{12}+\Gamma_{22}+\Gamma_{32}+\Gamma_{42})+2(\Gamma_{13}+\Gamma_{23}+\Gamma_{33}+\Gamma_{43}))\\
S_{C12} &= \frac{1}{3}(2(\Gamma_{21}+\Gamma_{31}+\Gamma_{41})-(\Gamma_{22}+\Gamma_{32}+\Gamma_{42})-(\Gamma_{23}+\Gamma_{33}+\Gamma_{43}))\\
S_{C22} &= \frac{1}{3}(-(\Gamma_{21}+\Gamma_{31}+\Gamma_{41})+2(\Gamma_{22}+\Gamma_{32}+\Gamma_{42})-(\Gamma_{23}+\Gamma_{33}+\Gamma_{43}))\\
S_{C32} &= \frac{1}{3}(-(\Gamma_{21}+\Gamma_{31}+\Gamma_{41})-(\Gamma_{22}+\Gamma_{32}+\Gamma_{42})+2(\Gamma_{23}+\Gamma_{33}+\Gamma_{43}))\\
S_{C13} &= \frac{1}{3}(2(\Gamma_{31}+\Gamma_{41})-(\Gamma_{32}+\Gamma_{42})-(\Gamma_{33}+\Gamma_{43}))\\
S_{C23} &= \frac{1}{3}(-(\Gamma_{31}+\Gamma_{41})+2(\Gamma_{32}+\Gamma_{42})-(\Gamma_{33}+\Gamma_{43}))\\
S_{C33} &= \frac{1}{3}(-(\Gamma_{31}+\Gamma_{41})-(\Gamma_{32}+\Gamma_{42})+2(\Gamma_{33}+\Gamma_{43}))\\
S_{C14} &= \frac{1}{3}(2\Gamma_{41}-\Gamma_{42}-\Gamma_{43})\\
S_{C24} &= \frac{1}{3}(-\Gamma_{41}+2\Gamma_{42}-\Gamma_{43})\\
S_{C34} &= \frac{1}{3}(-\Gamma_{41}-\Gamma_{42}+2\Gamma_{43})
\end{aligned}
\tag{23}
$$

Considering now a standard *R-L* balanced load with electromotive force e_{sk} and isolated neutral, and using the converter load voltages (13), the i_{sk} currents dynamic model, can be written in the matrix form (24).

$$
\frac{d}{dt}\begin{bmatrix} i_{S1}\\ i_{S2}\\ i_{S3}\end{bmatrix} = \begin{bmatrix} -\frac{R}{L} & 0 & 0 & \frac{S_{C11}}{L} & \frac{S_{C12}}{L} & \frac{S_{C13}}{L} & \frac{S_{C14}}{L}\\ 0 & -\frac{R}{L} & 0 & \frac{S_{C21}}{L} & \frac{S_{C22}}{L} & \frac{S_{C23}}{L} & \frac{S_{C24}}{L}\\ 0 & 0 & -\frac{R}{L} & \frac{S_{C31}}{L} & \frac{S_{C32}}{L} & \frac{S_{C33}}{L} & \frac{S_{C34}}{L}\end{bmatrix}\begin{bmatrix} i_{S1}\\ i_{S2}\\ i_{S3}\\ u_{C1}\\ u_{C2}\\ u_{C3}\\ u_{C4}\end{bmatrix} + \begin{bmatrix} -\frac{1}{L} & 0 & 0\\ 0 & -\frac{1}{L} & 0\\ 0 & 0 & -\frac{1}{L}\end{bmatrix}\begin{bmatrix} e_{S1}\\ e_{S2}\\ e_{S3}\end{bmatrix} \tag{24}
$$

The DC bus n level current i_n can also be expressed in terms of load phase currents i_{Sk} by (25).

$$
\begin{bmatrix} i_1\\ i_2\\ i_3\\ i_4\end{bmatrix} = \begin{bmatrix} \Gamma_{11} & \Gamma_{12} & \Gamma_{13}\\ \Gamma_{21} & \Gamma_{22} & \Gamma_{23}\\ \Gamma_{31} & \Gamma_{32} & \Gamma_{33}\\ \Gamma_{41} & \Gamma_{42} & \Gamma_{43}\end{bmatrix}\begin{bmatrix} i_{S1}\\ i_{S2}\\ i_{S3}\end{bmatrix} \tag{25}
$$

Capacitors currents i_{Cn} can be written in terms of DC bus n^{th} level current i_n as (26).

$$i_{Cn} = \begin{cases} i_{C1} = i_{L0} - i_1 - i_2 - i_3 - i_4 \\ i_{C2} = i_{L0} - i_2 - i_3 - i_4 \\ i_{C3} = i_{L0} - i_3 - i_4 \\ i_{C4} = i_{L0} - i_4 \end{cases} \tag{26}$$

Using (25) and the generic capacitor current Eqs. (6), (26) can be rewritten and the matrix equation (27) is obtained.

$$\frac{d}{dt}\begin{bmatrix} u_{C1} \\ u_{C2} \\ u_{C3} \\ u_{C4} \end{bmatrix} = \begin{bmatrix} -\frac{\Gamma_{11}+\Gamma_{21}+\Gamma_{31}+\Gamma_{41}}{C_1} & -\frac{\Gamma_{12}+\Gamma_{22}+\Gamma_{32}+\Gamma_{42}}{C_1} & -\frac{\Gamma_{13}+\Gamma_{23}+\Gamma_{33}+\Gamma_{43}}{C_1} & \frac{1}{C_1} \\ -\frac{\Gamma_{21}+\Gamma_{31}+\Gamma_{41}}{C_2} & -\frac{\Gamma_{22}+\Gamma_{32}+\Gamma_{42}}{C_2} & -\frac{\Gamma_{23}+\Gamma_{33}+\Gamma_{43}}{C_2} & \frac{1}{C_2} \\ -\frac{\Gamma_{31}+\Gamma_{41}}{C_3} & -\frac{\Gamma_{32}+\Gamma_{42}}{C_3} & -\frac{\Gamma_{33}+\Gamma_{43}}{C_3} & \frac{1}{C_3} \\ -\frac{\Gamma_{41}}{C_4} & -\frac{\Gamma_{42}}{C_4} & -\frac{\Gamma_{43}}{C_4} & \frac{1}{C_4} \end{bmatrix} \begin{bmatrix} i_{S1} \\ i_{S2} \\ i_{S3} \\ i_{L0} \end{bmatrix} \tag{27}$$

Equation (27) compares to Eq. (10) where the matrix elements can be related as it shows (28).

$$\begin{aligned} \Gamma_{C11} &= \Gamma_{11} + \Gamma_{21} + \Gamma_{31} + \Gamma_{41} \\ \Gamma_{C12} &= \Gamma_{12} + \Gamma_{22} + \Gamma_{32} + \Gamma_{42} \\ \Gamma_{C13} &= \Gamma_{13} + \Gamma_{23} + \Gamma_{33} + \Gamma_{43} \\ \Gamma_{C21} &= \Gamma_{21} + \Gamma_{31} + \Gamma_{41} \\ \Gamma_{C22} &= \Gamma_{22} + \Gamma_{32} + \Gamma_{42} \\ \Gamma_{C23} &= \Gamma_{23} + \Gamma_{33} + \Gamma_{43} \\ \Gamma_{C31} &= \Gamma_{31} + \Gamma_{41} \\ \Gamma_{C32} &= \Gamma_{32} + \Gamma_{42} \\ \Gamma_{C33} &= \Gamma_{33} + \Gamma_{43} \\ \Gamma_{C41} &= \Gamma_{41} \\ \Gamma_{C42} &= \Gamma_{42} \\ \Gamma_{C43} &= \Gamma_{43} \end{aligned} \tag{28}$$

Using (28), Eq. (27) can be rewritten in terms of Γ_{Cjk} as (29).

$$\frac{d}{dt}\begin{bmatrix} u_{C1} \\ u_{C2} \\ u_{C3} \\ u_{C4} \end{bmatrix} = \begin{bmatrix} -\frac{\Gamma_{C11}}{C_1} & -\frac{\Gamma_{C12}}{C_1} & -\frac{\Gamma_{C13}}{C_1} & \frac{1}{C_1} \\ -\frac{\Gamma_{C21}}{C_2} & -\frac{\Gamma_{C22}}{C_2} & -\frac{\Gamma_{C23}}{C_2} & \frac{1}{C_2} \\ -\frac{\Gamma_{C31}}{C_3} & -\frac{\Gamma_{C32}}{C_3} & -\frac{\Gamma_{C33}}{C_3} & \frac{1}{C_3} \\ -\frac{\Gamma_{C41}}{C_4} & -\frac{\Gamma_{C42}}{C_4} & -\frac{\Gamma_{C43}}{C_4} & \frac{1}{C_4} \end{bmatrix} \begin{bmatrix} i_{S1} \\ i_{S2} \\ i_{S3} \\ i_{LO} \end{bmatrix} \tag{29}$$

The global system model in phase coordinates can be written in matrix form (30) using (24) and (29).

$$\frac{d}{dt}\begin{bmatrix} i_{S1} \\ i_{S2} \\ i_{S3} \\ u_{C1} \\ u_{C2} \\ u_{C3} \\ u_{C4} \end{bmatrix} = \begin{bmatrix} -\frac{R}{L} & 0 & 0 & \frac{S_{C11}}{L} & \frac{S_{C12}}{L} & \frac{S_{C13}}{L} & \frac{S_{C14}}{L} \\ 0 & -\frac{R}{L} & 0 & \frac{S_{C21}}{L} & \frac{S_{C22}}{L} & \frac{S_{C23}}{L} & \frac{S_{C24}}{L} \\ 0 & 0 & -\frac{R}{L} & \frac{S_{C31}}{L} & \frac{S_{C32}}{L} & \frac{S_{C33}}{L} & \frac{S_{C34}}{L} \\ -\frac{\Gamma_{C11}}{C_1} & -\frac{\Gamma_{C12}}{C_1} & -\frac{\Gamma_{C13}}{C_1} & 0 & 0 & 0 & 0 \\ -\frac{\Gamma_{C21}}{C_2} & -\frac{\Gamma_{C22}}{C_2} & -\frac{\Gamma_{C23}}{C_2} & 0 & 0 & 0 & 0 \\ -\frac{\Gamma_{C31}}{C_3} & -\frac{\Gamma_{C32}}{C_3} & -\frac{\Gamma_{C33}}{C_3} & 0 & 0 & 0 & 0 \\ -\frac{\Gamma_{C41}}{C_4} & -\frac{\Gamma_{C42}}{C_4} & -\frac{\Gamma_{C43}}{C_4} & 0 & 0 & 0 & 0 \end{bmatrix} \begin{bmatrix} i_{S1} \\ i_{S2} \\ i_{S3} \\ u_{C1} \\ u_{C2} \\ u_{C3} \\ u_{C4} \end{bmatrix} + $$

$$+ \begin{bmatrix} -\frac{1}{L} & 0 & 0 & 0 \\ 0 & -\frac{1}{L} & 0 & 0 \\ 0 & 0 & -\frac{1}{L} & 0 \\ 0 & 0 & 0 & \frac{1}{C_1} \\ 0 & 0 & 0 & \frac{1}{C_2} \\ 0 & 0 & 0 & \frac{1}{C_3} \\ 0 & 0 & 0 & \frac{1}{C_4} \end{bmatrix} \begin{bmatrix} e_{S1} \\ e_{S2} \\ e_{S3} \\ i_{LO} \end{bmatrix} \tag{30}$$

If a zero voltage reference in the DC bus midpoint has been selected (u_{mk} voltages reference instead of u_k), slightly simpler equations for (23) and (27) would have been obtained, as is shown in (31) and (32) respectively.

$$
\begin{aligned}
S_{C11} &= \frac{1}{3}(-2\Gamma_{01} + \Gamma_{02} + \Gamma_{03}) \\
S_{C21} &= \frac{1}{3}(\Gamma_{01} - 2\Gamma_{02} + \Gamma_{03}) \\
S_{C31} &= \frac{1}{3}(\Gamma_{01} + \Gamma_{02} - 2\Gamma_{03}) \\
S_{C12} &= \frac{1}{3}(-2(\Gamma_{01} + \Gamma_{11}) + (\Gamma_{02} + \Gamma_{12}) + (\Gamma_{03} + \Gamma_{13})) \\
S_{C22} &= \frac{1}{3}((\Gamma_{01} + \Gamma_{11}) - 2(\Gamma_{02} + \Gamma_{12}) + (\Gamma_{03} + \Gamma_{13})) \\
S_{C32} &= \frac{1}{3}((\Gamma_{01} + \Gamma_{11}) + (\Gamma_{02} + \Gamma_{12}) - 2(\Gamma_{03} + \Gamma_{13})) \\
S_{C13} &= \frac{1}{3}(2(\Gamma_{31} + \Gamma_{41}) - (\Gamma_{32} + \Gamma_{42}) - (\Gamma_{33} + \Gamma_{43})) \\
S_{C23} &= \frac{1}{3}(-(\Gamma_{31} + \Gamma_{41}) + 2(\Gamma_{32} + \Gamma_{42}) - (\Gamma_{33} + \Gamma_{43})) \\
S_{C33} &= \frac{1}{3}(-(\Gamma_{31} + \Gamma_{41}) - (\Gamma_{32} + \Gamma_{42}) + 2(\Gamma_{33} + \Gamma_{43})) \\
S_{C14} &= \frac{1}{3}(2\Gamma_{41} - \Gamma_{42} - \Gamma_{43}) \\
S_{C24} &= \frac{1}{3}(-\Gamma_{41} + 2\Gamma_{42} - \Gamma_{43}) \\
S_{C34} &= \frac{1}{3}(-\Gamma_{41} - \Gamma_{42} + 2\Gamma_{43})
\end{aligned} \tag{31}
$$

$$
\frac{d}{dt}\begin{bmatrix} u_{C1} \\ u_{C2} \\ u_{C3} \\ u_{C4} \end{bmatrix} = \begin{bmatrix} \frac{\Gamma_{01}}{C_1} & \frac{\Gamma_{02}}{C_1} & \frac{\Gamma_{03}}{C_1} & \frac{1}{C_1} \\ \frac{\Gamma_{01} + \Gamma_{11}}{C_2} & \frac{\Gamma_{02} + \Gamma_{12}}{C_2} & \frac{\Gamma_{03} + \Gamma_{13}}{C_2} & \frac{1}{C_2} \\ -\frac{\Gamma_{31} + \Gamma_{41}}{C_3} & -\frac{\Gamma_{32} + \Gamma_{42}}{C_3} & -\frac{\Gamma_{33} + \Gamma_{43}}{C_3} & \frac{1}{C_3} \\ -\frac{\Gamma_{41}}{C_4} & -\frac{\Gamma_{42}}{C_4} & -\frac{\Gamma_{43}}{C_4} & \frac{1}{C_4} \end{bmatrix} \begin{bmatrix} i_{S1} \\ i_{S2} \\ i_{S3} \\ i_{L0} \end{bmatrix} \tag{32}
$$

This modeling approach also conducts to a detailed non-linear state space model which has an easier code implementation, in case of converters with low level number (specially $m = 3$). Otherwise for $m > 3$, establishing analytical functions for time dependent switching variable (Γ_{nk}) and for matrix elements S_{Ckj} could become a hard and fastidious task.

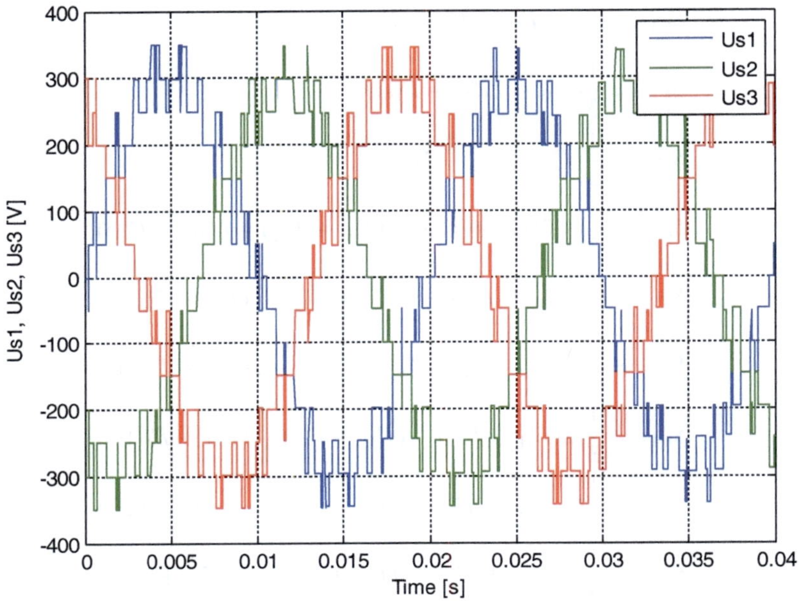

Fig. 2 Simulation of three phase output voltages using the u_{Sk} voltage equations from Sect. 2 and 3 implemented in simulink

4 Simulations Results

To evaluate the performance of presented models both were used in the modeling of a five level converter supplying an AC load. Simulation results provided by models were compared with the corresponding ones obtained using SimPowerSystems toolbox.

Multilevel converter was driven using a sinusoidal pulse width modulation (SPWM) technique. The simulation conditions were: constant capacitor voltage $u_{Cn} = 150$ V; *RL* balanced load with $R = 10\ \Omega$ and $L = 20$ mH; carrier frequency $f_c = 650$ Hz; modulation frequency $f_m = 50$ Hz.

Figure 2 shows the obtained simulation results of the three phase output voltages u_{sk} using models presented in Sects. 2 and 3, implemented in Matlab/simulink. The results of both models are overlapped and cannot be separated.

The same output voltages obtained using the IGBT/diode models of SimPowerSystems toolbox are shown in Fig. 3. It is possible to verify that results match the corresponding ones in Fig. 2.

Figure 4 and 5 present the simulated start-up transient of output load currents obtained using models in Sect. 2, 3 and SimPowerSystems toolbox, respectively. These dynamic results also match.

Figure 6 and 7 present output voltage u_{s1} and load current i_{s1} frequency spectrum, using the proposed model implemented in simulink

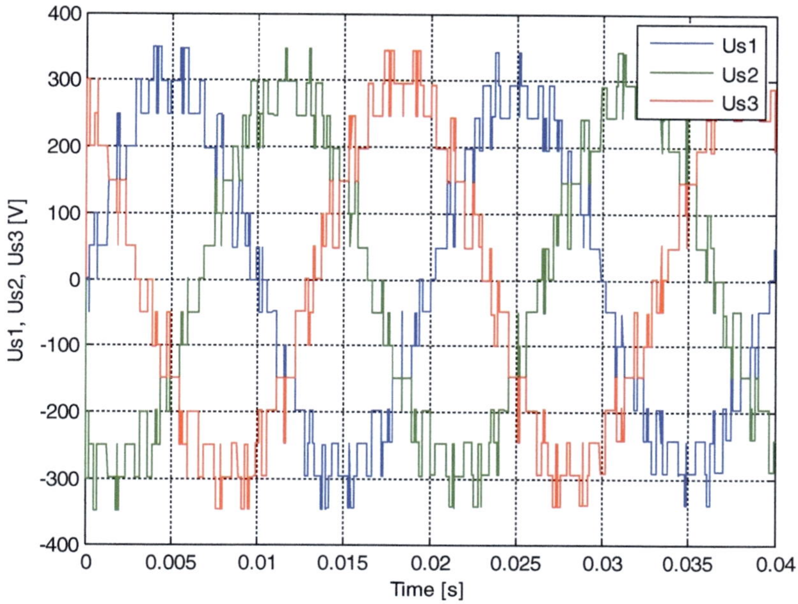

Fig. 3 Simulation of three phase output voltages using SimPowerSystems toolbox

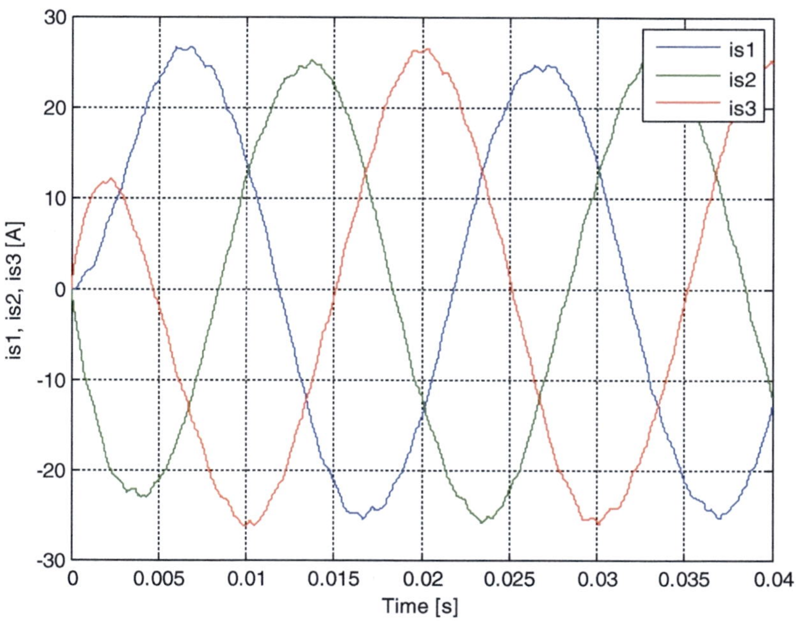

Fig. 4 Simulation of three phase load currents using the i_{Sk} current equations from Sect. 2 and 3 implemented in simulink

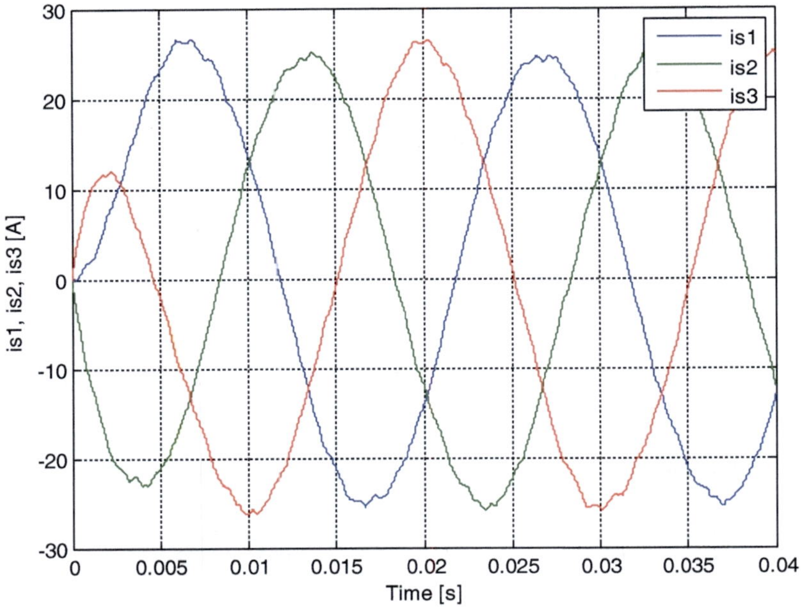

Fig. 5 Simulation of three phase load currents using SimPowerSystems toolbox

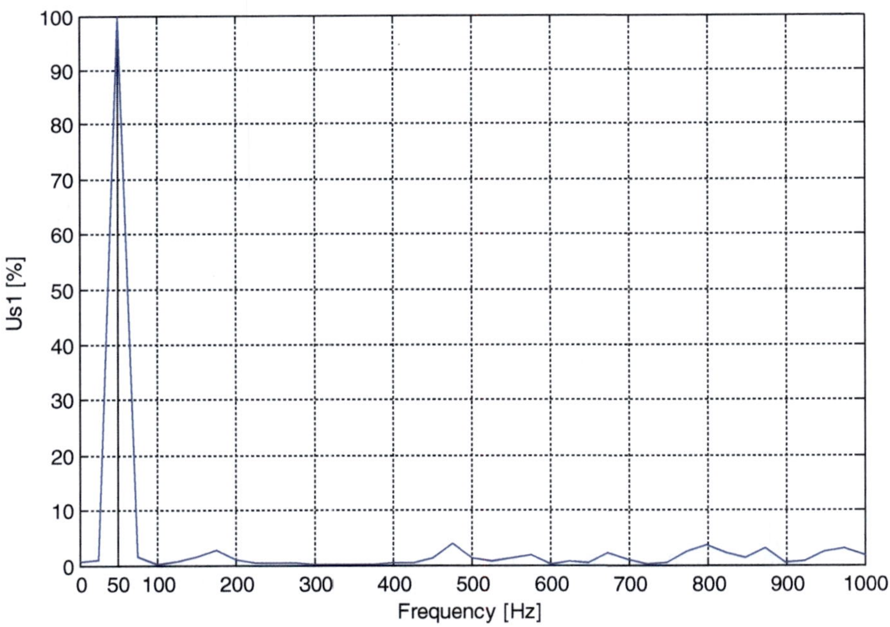

Fig. 6 Phase output voltage u_{s1} spectrum using the proposed model implemented in simulink

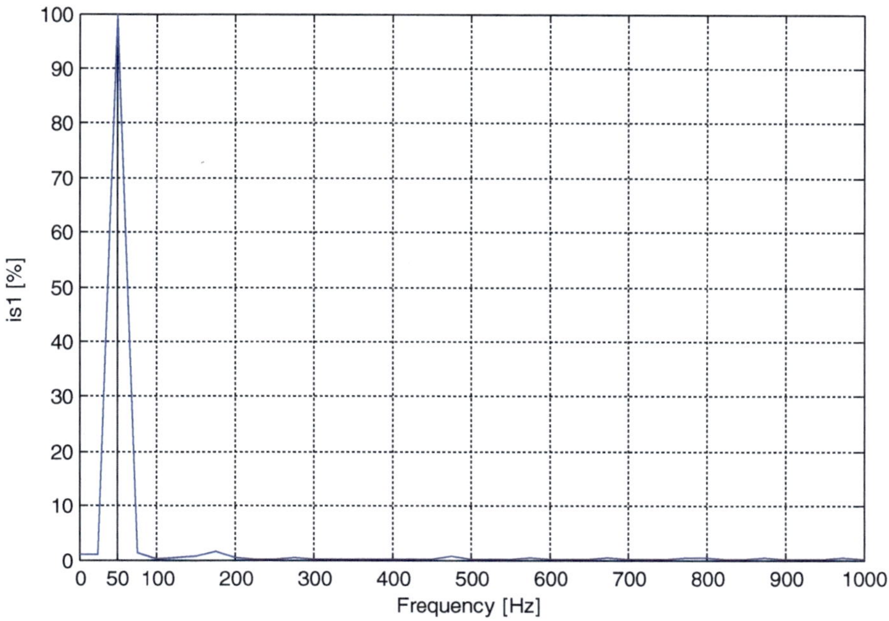

Fig. 7 Phase load current i_{s1} spectrum using the proposed model implemented in simulink

The multilevel converter provides a voltage staircase like waveform with low harmonic distortion as can be qualitatively seen in 6 and 7.

The presented model is valid independently of the power flow direction (DC-AC or AC-DC), or even with active balanced loads with line to line electromotive force peak value higher then DC bus voltage.

In order to prove these model features the following figures present simulation results of DC bus supplying current i_{L0} obtained, respectively, using the proposed model, Fig. 8. The results were obtained considering an active balanced load (*RLE*), with the previously presented *RL* values and applying a step in the electromotive force rms value, at $t = 0{,}02$ s, from 0 to 400 Vrms.

DC bus capacitor voltage divider is assumed to be connected in parallel with a non-ideal voltage source with internal resistance of 0.1 Ω.

The time evolution of i_{Lo} DC bus supplying current, presented in Fig. 8, shows power flow inversion at DC voltage source terminals, due to a rise in line-to-line voltage from 0 to a value higher then the DC bus voltage.

5 Conclusions

This work proposes new generalized state-space modelling for *m* level diodeclamped multilevel converters. The *m* generalized modelling was applied to a five level converter and compared to existing modelling approaches. The proposed generalized

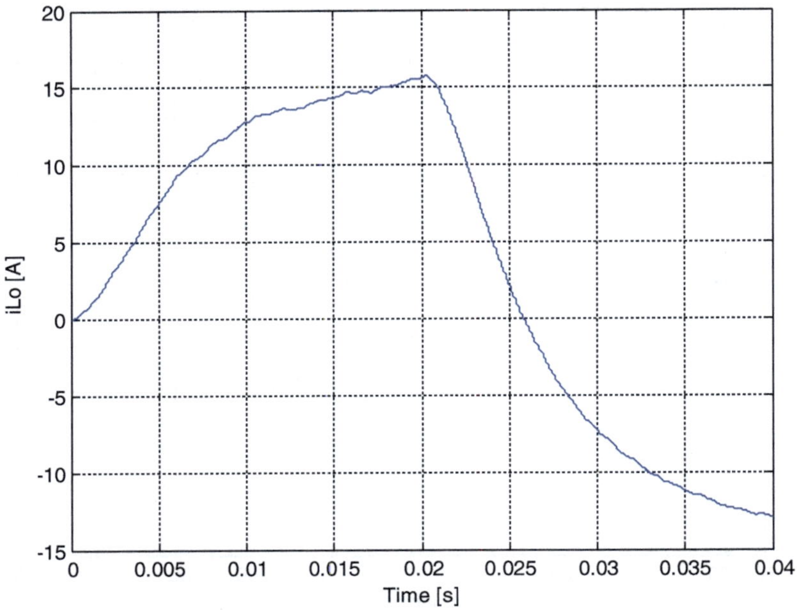

Fig. 8 Simulation of DC bus supplying current i_{Lo} using the proposed model implemented in simulink

approach is systematic and advantageous for converters with high number of levels. It is also suited for computer simulation.

Comparing obtained simulation results using developed models and the corresponding ones using SimPowerSystems toolbox it is concluded that the presented models show well the detailed system behaviour and are accurate enough.

The proposed modelling approaches are valid for passive and active loads. In the case of active loads, the proposed models are valid for both power flow directions and peak line-to line voltages lower or higher than the DC voltage value.

Advantages of multilevel converters were also confirmed by simulation waveform results, such as voltage staircase waveform, which means low load phase voltage and low current distortion as shown.

References

1. Nabae A, Takahashi I, Akagi H (Sept/Oct 1981) A new-neutral-point-clamped PWM inverter. IEEE Trans Ind Appl IA-17(5):518–523
2. Khomfi S, Tolbert M (2006) Multilevel power converters. In: Rashid MH (ed) Power electronics handbook. Elsevier, Burlington
3. Holmes DG, Lipo TA (2003) Pulsewidth modulation for power converters. Wiley, New York
4. Rodríguez J, Lai JS, Peng FZ (Aug 2002) Multilevel inverters: a survey of topologies, controls, and applications. IEEE Trans Ind Electron 49(4):724–738

5. Rodríguez FJ et al (2008) Control electronic platform based on floating-point DSP and FPGA for a NPC multilevel back-to-back converter. Electr Power Syst Res 78(9):1597–1609. doi:10.1016/j.epsr.2008.02.003 (Elsevier)
6. Akagi H, Fujita H, Yonetani S, Kondo Y (Oct 2005) A 6.6-kV transformerless STACOM based on a five-level diode-clamped PWM converter: system design and experimentation of a 200-V, 10-kVA laboratory model. Proc IEEE-IAS 1(4):557–564
7. Pou J, Pindado R, Boroyevich D, Rodriguez P (June 2004) Voltage balance strategies for diode-clamped multilevel converters. IEEE-PESC, vol 5:3988–3993, Aachen, Germany
8. da Silva JFA, Pires FV, Pinto SF, Barros JD (2003) Advanced control methods for power electronics systems. IMACS Math Comput Simul 63(3–5):281–295
9. Barros JD, Silva JF (2007) Optimal predictive control of three-phase NPC multilevel inverter: comparison to robust sliding mode controller. IEEE Annual Power Electronics Specialists Conference PESC07, Orlando

Modelling Codfish Drying: Comparison Between Artificial Neural Network, Diffusive and Semi-Empirical Models

CN Boeri, FJ Neto da Silva and JAF Ferreira

Abstract Convective drying is of prime importance in the food conservation industry and has been constantly studied and improved to obtain products with higher quality and lower processing time. In this work, three different models were used to perform the codfish drying simulation: artificial neural network (ANN), diffusive and semi-empirical models. The simulation results were compared for the following experimental conditions: drying air temperature of 20 °C, air velocities of 2 and 3 m/s and drying air relative humidities comprise between 55 and 65 %. The simulations showed good results for the semi-empirical and ANN models, requiring improvements to the diffusion model.

Keywords Artificial neural network · Diffusion law · Semi-empirical models · Codfish drying

1 Introduction

As all fish, fresh codfish is susceptible to deterioration by fast destructive action of enzymes, oxidation of lipids, high pH, high water activity and accentuated contents of non-protein nitrogen substances. Accordingly, it is of critical importance to adopt measures ensuring perfect conservation immediately after capture and during distribution and commercialization.

There are several techniques for processing fish (drying, salting, smoking, etc.) which can be used to meet the above requirements. In this work, the emphasis will be allocated to the drying conservation process.

The relevance of drying operations within a vast range of industrial processes is unquestionable: drying is present in the chemical, agricultural, wood processing,

CN Boeri (✉) · FJ Neto da Silva · JAF Ferreira
Department of Mechanical Engineering, University of Aveiro, Aveiro, Portugal
e-mail: camilaboeri@ua.pt

FJ Neto da Silva
e-mail: fneto@ua.pt

JAF Ferreira
e-mail: jaff@ua.pt

N. M. Fonseca Ferreira, J. A. Tenreiro Machado (eds.), *Mathematical Methods in Engineering*, DOI 10.1007/978-94-007-7183-3_8,

food, ceramics, and pulp and paper industries, among others. It is estimated that drying operations are responsible for 10–25 % of the national energy consumption in developed countries [1]. The correct definition of drying procedures of a vast range of products is crucial in what concerns energy minimization and minimal time of kiln residence without compromising the final product quality. Drying can change the sensory characteristics and nutritional value of foods, and the intensity of these changes depends upon the conditions used in the drying process and the specific characteristics of each product.

Several parameters influence the time required to reduce the moisture content of the product. The principal external factors to consider are air temperature, relative humidity and velocity. Whole salted codfish drying prevents the utilization of temperatures above 20–22 °C, since they lead to an inevitable deterioration of the product. Hence convective drying must be conducted without exceeding the required temperature limits. The limitation seriously conditions drying time and energy consumption.

Mathematical modelling of food drying processes represents an adequate and straightforward manner to predict drying behaviour of a given material in response to given drying conditions or to a change in these conditions. When a drying model is integrated with a proper control algorithm aiming at energy reduction and increased drying speeds, manual control can be replaced by automatic operation which may result in significant reductions in drying costs without compromises in product quality.

Hence the purposes of this work are the determination of the codfish drying kinetics for the following experimental conditions: drying air temperature of 20 °C, initial moisture contents comprise between 56 and 62 % (wet basis), air velocities of 2 and 3 m/s and drying air relative humidities comprise between 55 and 65 %, which replicate the conditions found in the whole salted codfish processing industry; and to compare the accuracy of several modelling techniques namely thin layer, diffusive and neural network models.

2 Materials and Methods

2.1 *Product and Dryer*

The experimental data used in this work were obtained from drying experiments carried out with the use of four salted whole green codfishes, with an initial mass which varied between 1,060 and 2,000 g. The codfishes were cut in samples with an approximate mass of 100 g each, and kept until utilization tightly wrapped in plastic bags in a refrigerator at a constant temperature of 5 °C. The oven drying method was used to determine the initial and the final moisture contents of the samples. The method implied extraction of parts of the sample and their drying in a controlled temperature environment at 105 ± 2 °C during, at least, 48 h.

Table 1 Drying air and product conditions for drying experiments

Experiment	T (°C)	v (m/s)	RH (%)	X_0 (%)
1	20	3	55	60.76
2	20	3	60	59.43
3	20	3	65	56.56
4	20	2	60	62.70

At the beginning of each experiment, a codfish sample was placed in the drying chamber over a digital balance. The mass of the sample was continuously determined by the monitoring system and the readings were recorded every 5 min. The equilibrium moisture content of the codfish was determined experimentally in a hygrometric chamber. During convective drying the sample mass was acquired at each time interval, which allowed for determination of the instantaneous moisture content using (1).

$$X = \frac{(X_e - X_0) \cdot (m - m_0)}{(m_f - m_0)} + X_0 \tag{1}$$

where X_e is the equilibrium moisture content, X_0 is the initial moisture content, m is the instantaneous mass, m_0 is the initial mass and m_f is the final mass.

The values of initial, equilibrium and instantaneous moisture content of the codfish sample were used to obtain the values of the dimensionless moisture content:

$$X_{dim} = \frac{X - X_e}{X_0 - X_e} \tag{2}$$

where X is the instantaneous moisture content of the product.

Four different drying situations were used in the experiments described in Table 1:

2.2 Drying Models

In the literature one can found several models to simulate the drying process. Here, experimental data were collected to adjust parameters on the thin layer and on the diffusive drying models and to train the neural network model. Modelling of the drying experiments was performed by resorting to the Page and Thompson semi-empirical models (thin layer drying model), to a Fick diffusion law (diffusive model) and to a neural network model.

The neural networks are a problem solving concept which represents an alternative to conventional algorithmic methods [2]. The concept was used herewith to build a neural model able to predict the behaviour of drying curves. The neural network used in this work was a multi-layer "feed-forward", consisting of four input layers, one hidden layer and one output layer with a convergence criterion for training purposes. In this network, each neuron of one layer is connected to all neurons in adjacent layers. The training of feed-forward networks is of the supervised type: this type of training requires a set of data for training, i.e., a series of pairs of inputs and desired outputs. The inputs are presented to the network and their weights are changed so that the output approximates the desired output. Therefore, experimental data of codfish drying was used in the steps of training and validation. Input variables were drying time, temperature, relative humidity and air velocity, with the output variable, the codfish moisture content. As convergence criterion, was used the mean square error of 1.0×10^{-5} for the network training.

The diffusive model is based on Fick's law, which states that the mass flow per unit area is proportional to the concentration water gradient. Crank [3] determined a large number of solutions of the diffusion equation for varied initial and boundary conditions. However, these solutions apply to simple geometric shapes and when the diffusivity is constant or varies linearly or exponentially with the concentration of water.

Considering a flat plaque:

$$\frac{\partial X}{\partial t} = \frac{\partial}{\partial z}\left(D_{ef}\frac{\partial X}{\partial z}\right), \quad 0 < z < L \tag{3}$$

Assuming the following initial and boundary conditions and the definition (7):

$$X(z, 0) = X_0 \tag{4}$$

$$\left.\frac{\partial X}{\partial z}\right|_{z=0} = 0 \tag{5}$$

$$X(L, t) = X_e \tag{6}$$

$$X = \frac{1}{L}\int_0^L X(z,\ t)dz \tag{7}$$

the analytical solution to Fick's diffusion equation is given by:

$$\frac{X - X_e}{X_0 - X_e} = \frac{8}{\pi^2}\sum_{i=0}^{\infty}\frac{1}{(2i+1)^2}\exp\left[-(2i+1)^2\pi^2 D_{ef}\frac{t}{4L^2}\right] \tag{8}$$

where t is the drying time, D_{ef} is the effective diffusivity, L is the sample half thickness and z is the direction.

The previous solution was adapted to the present drying situation by considering the codfish slab as a flat plaque for which the boundary located in contact with the digital balance plate was considered as impermeable to moisture transfer.

Semi-empirical models are known as exponential laws of drying. These models rely, in general, in an analogy with Newton's law of cooling, considering that the drying rate is proportional to the difference between the current moisture content and moisture content equilibrium. $\frac{\partial x}{\partial t}\alpha = (X - X_e)$. Recent applications of such models can be found in abundance in the literature [4–7]. Here, the drying semi-empirical

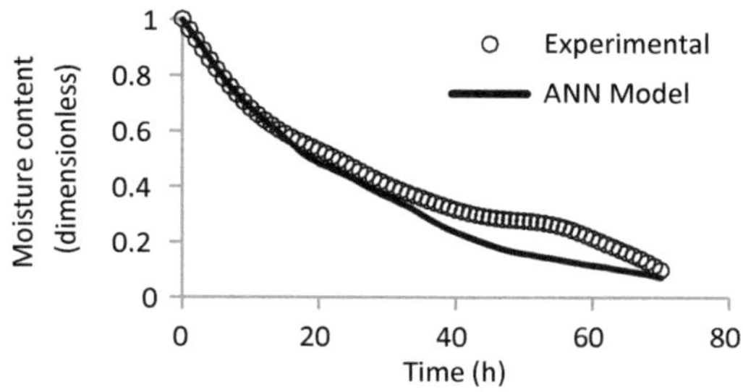

Fig. 1 Comparison between experimental data and neural network model—Experiment 1

models proposed by Page [8] and Thompson [9] were used. The Page equation was adjusted incorporating the air velocity and the drying air temperature as a model parameter by following the methodology proposed in the literature [10]:

$$X_{dim} = e^{\left[-v \cdot A \cdot X_0{}^{B} \cdot e^{(C \cdot T_S)} \cdot t^{D}\right]} \tag{9}$$

where v is the air velocity, Ts is the temperature of drying air, t is the drying time, A, B, C and D are constants related with the drying process and with the product.

Thompson's model is given by:

$$X_{dim} = e^{\left[\left(-A-\left(A^2+4 \cdot B \cdot t\right)^{0,5}\right)/2B\right]} \tag{10}$$

where t is the drying time, A and B are constants that depend of the drying process and the type of product.

3 Results and Discussion

In the modelling performed by artificial neural networks, several models were trained using the set of experimental data, through the use of Matlab® 7.0.1 software. The most satisfactory performance for the training phase was reached with the Levenberg-Marquardt's algorithm, where the best error was obtained using 50 neurons in the hidden layer, with 62 training epochs. The transfer functions used were the Sigmoid Function (Tansig) for the hidden layer and the linear function (Purelin) for the output layer. Were used 11 drying experiments in the training phase and 6 experiments in the validation phase of the network.

Results from the validation of the neural network are shown in the Figs. 1, 2, 3 and 4:

In what concerns the diffusive model for each drying conditions, the diffusion coefficients were calculated using five terms of the series, by nonlinear estimation, with the use of estimation method Quasi-Newton.

The values obtained for the diffusion coefficients in each experiment are shown in Table 2. The obtained diffusivity coefficients are in accordance with the values found in literature [11].

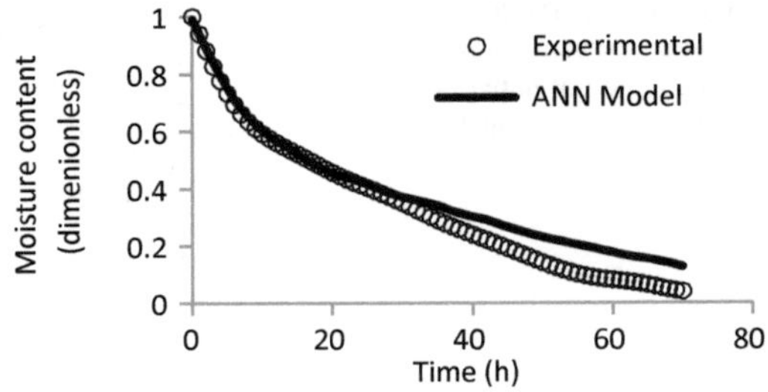

Fig. 2 Comparison between experimental data and neural network model—Experiment 2

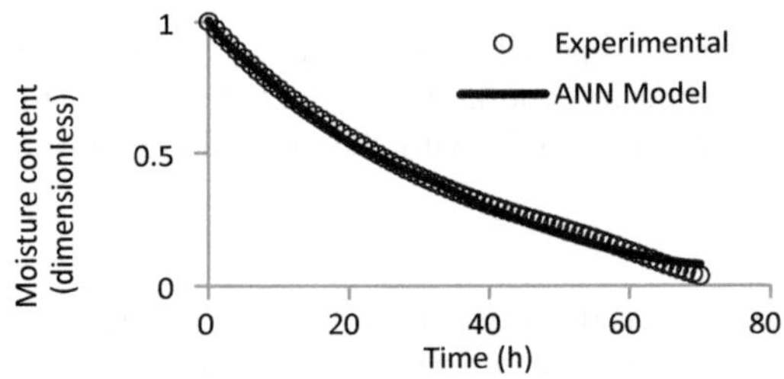

Fig. 3 Comparison between experimental data and neural network model—Experiment 3

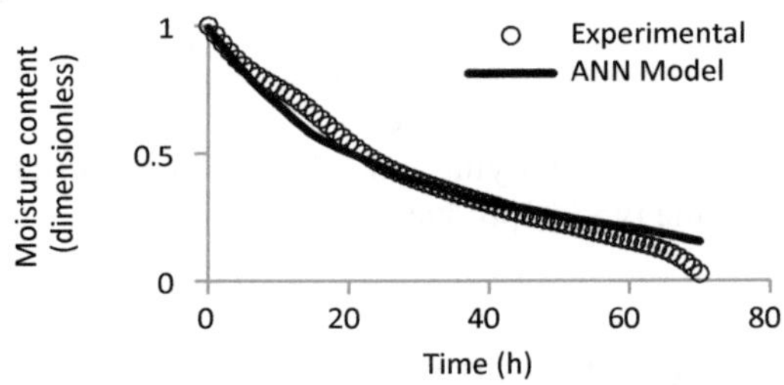

Fig. 4 Comparison between experimental data and neural network model—Experiment 4

Table 2 Values obtained for diffusion coefficients

Experiment	Diffusion coefficients (m^2/s)
1	3.13226×10^{-10}
2	4.39182×10^{-10}
3	2.70456×10^{-10}
4	2.72722×10^{-10}

The results obtained for the simulations accomplished with diffusive model, together with the experimental data, are shown in Figs. 5, 6, 7 and 8:

The relative inaccuracy of the diffusive type models reflect the constraints imposed both by the choice of the boundary condition at the exposed surface and by adopting constant diffusion coefficients; in fact it is quite doubtful that equilibrium moisture content at the surface may be attained immediately; future work will also be conducted on the adoption of a moisture dependent diffusion coefficient.

The parameters of Page's and Thompson's semi-empirical models were estimated by using the simplex algorithm, implemented in the *fminsearch* function of the MatLab® optimization toolbox and the data obtained from each experiment (Table 3):

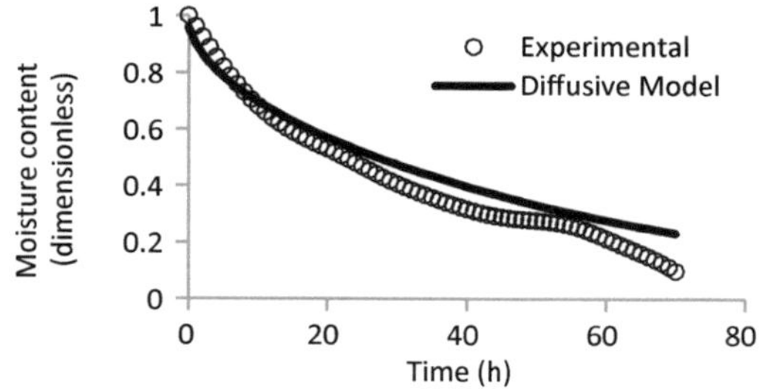

Fig. 5 Comparison between experimental data and diffusive model—Experiment 1

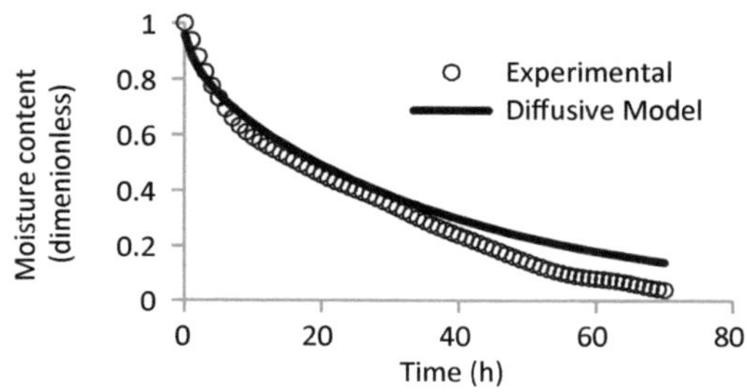

Fig. 6 Comparison between experimental data and diffusive model—Experiment 2

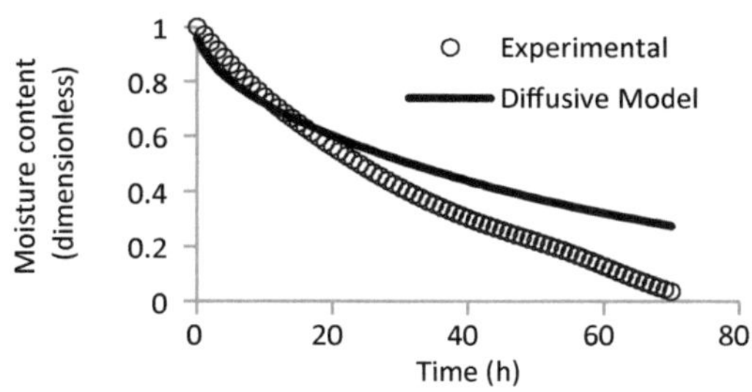

Fig. 7 Comparison between experimental data and diffusive model—Experiment 3

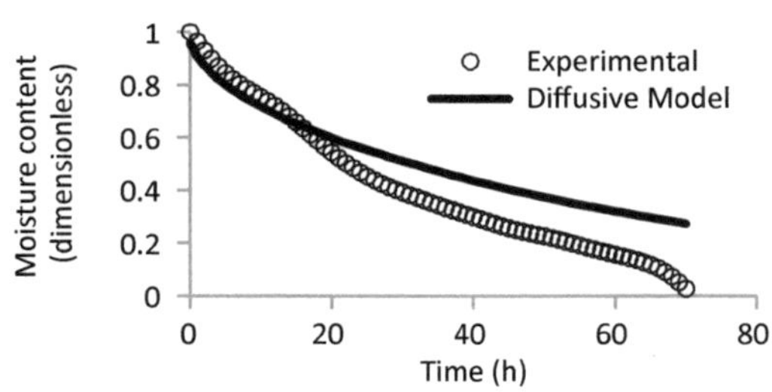

Fig. 8 Comparison between experimental data and diffusive model—Experiment 4

Table 3 Parameters found for Page's and Thompson's models

Parameters	Page	Thompson
A	0.00027295	– 33.102
B	5.2444	1.0506
C	0.022755	–
D	0.95259	–

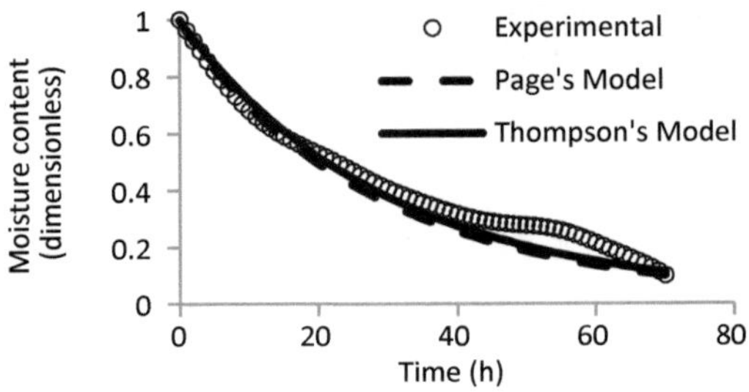

Fig. 9 Comparison between experimental data and semi-empirical models—Experiment 1

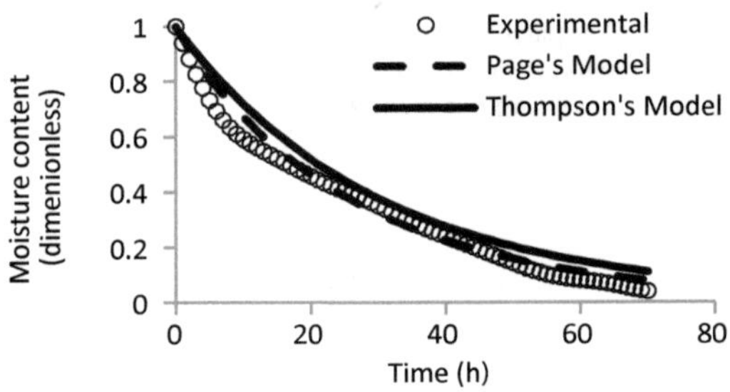

Fig. 10 Comparison between experimental data and semi-empirical models—Experiment 2

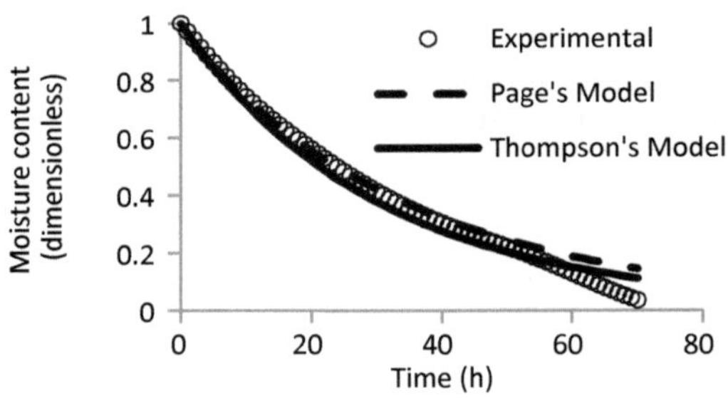

Fig. 11 Comparison between experimental data and semi-empirical models—Experiment 3

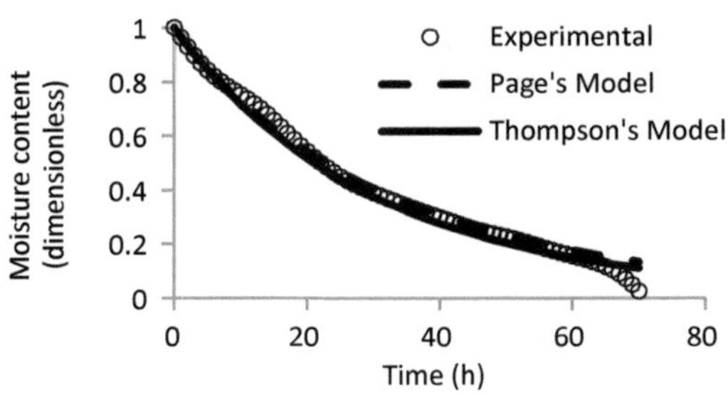

Fig. 12 Comparison between experimental data and semi-empirical models—Experiment 4

The results obtained for the simulations accomplished with Page's and Thompson's model, together with the experimental data, are shown in Figs. 9, 10, 11 and 12:

The semi-empirical models used do not take into account the influence of the relative humidity of the drying fluid. In future work, it is intended to overcome this limitation by including this parameter into the model, in order to improve its accuracy.

Table 4 Statistical analysis

Analysis	Mean absolute error (%)				Pearson's correlation coefficient			
Experiment	Page	Thompson	Diffusive	ANN	Page	Thompson	Diffusive	ANN
1	4.783	3.765	5.987	5.706	0.995	0.995	0.956	0.995
2	3.005	6.359	6.134	4.927	0.992	0.993	0.959	0.996
3	2.529	2.894	11.480	1.172	0.992	0.995	0.864	0.999
4	1.856	2.059	11.264	3.587	0.997	0.996	0.869	0.992

The models' accuracy was assessed using the mean absolute error between the simulated and the experimental values of the codfish moisture content and the Pearson's correlation coefficient. Table 4 shows the result of statistical analysis for each model, for all drying situations studied.

The results show good agreement between experimental and simulation data for the semi-empirical and artificial neural network models. However, for the diffusive model, the results obtained do not reveal good agreement. This can also be confirmed through the statistical analysis that was performed for each curve. Regarding the Pearson's correlation coefficient, a high linear correlation between the experimental and simulation values is verified by Page's, Thompson's and neural network models. In the diffusion model, the values obtained for the correlation coefficient are lower. The same is verified when analyzing the values obtained for the errors between the experimental and simulated curves. The smaller errors were found for the Page's model and the bigger errors for the diffusive model.

4 Conclusions

Simulations with the artificial neural network model have shown good results. The error ranged between 1.17 and 5.7 % and showed a correlation coefficient between 0.992 and 0.999. Despite the fact that the neural network model has shown promising potential, new experimental data is being collected in order to improve the accuracy of the model and new training will be conducted with other network settings.

For the diffusion model the obtained diffusion coefficients varied from 2.7×10^{-10} to 4.4×10^{-10} m^2/s which are in agreement with those found in the literature. However, the diffusion model did not show good results since statistical correlation of the coefficients for all the set of experiments ranged from 0.864 to 0.959. The semi-empirical models provided a good representation of the drying curve; the statistical analysis performed on the drying curve has shown a Pearson's correlation coefficient ranging from 0.992 to 0.997 for Page's model, and from 0.993 to 0.996 for Thompson's model. However, these models could be improved if the air relative humidity was included in Page's model.

References

1. Dufour P (2006) Control engineering in drying technology: review and trends. Dry Technol 24:889–904
2. Kalogirou SA (2000) Applications of artificial neural-networks for energy systems. Appl Energy 67:17–35
3. Crank J (1975) The mathematics of diffusion, 2nd edn. Oxford Science Publications, Oxford
4. Aktas M et al (2009) Determination of drying characteristics of apples in a heat pump and solar dryer. Desalination 239:266–275
5. St. George SD, Cenkowski S (2009) Modeling of thin-layer drying of an inert sphere. Dry Technol 27:770–781
6. Doymaz I (2008) The kinetics of forced convective air-drying of pumpkin slices. J Food Eng 79:243–248
7. Ng AB, Deng S (2008) A new termination control method for a clothes drying process in a clothes dryer. Appl Energy 85:818–829
8. Page GE (1949) Factors influencing the maximum rates of air drying shelled corn in thin-layer. Dissertation (M.Sc.), Purdue University
9. Thompson TL, Feert RM, Foster GH (1968) Mathematical simulation of corn drying. Trans ASAE 11(4):582–586
10. Soares JB (1986) Curvas de secagem em camada fina e propriedades físicas de soja (Glicine max L.). Dissertação Mestrado, Universidade Federal de Viçosa
11. Park KJ (1998) Diffusional model with and without shrinkage during salted fish muscle drying. Dry Technol 16:889–905

Stability of Matrix Differential Equations with Commuting Matrix Constant Coefficients

Fernando Martins, Edgar Pereira, M. A. Facas Vicente and José Vitória

Abstract Sufficient conditions for the asymptotic stability of systems of first order linear differential equations with commuting matrix constant coefficients is studied. Stability criterion in terms of blocks is presented. Inertia of a block circulant matrix is obtained.

Keywords Block companion matrix · Block Hermite matrix · Block Hurwitz matrix · Block Routh matrix · Block Schwarz matrix · Matrix polynomials · Matrix differential equations · Lyapunov matrix equation · Block circulant matrix · Inertia

1 Introduction

Stability questions arise in many problems in Science, Engineering and Economics; in particular, when the control of discrete or continuous systems is the main preoccupation. Matrices partitioned into blocks play a significant role, when studying

---

2010 *Mathematics Subject Classification.* 15A27, 34D99, 93D20.

F. Martins (✉)
Instituto de Telecomunicações (Covilhã), Coimbra College of Education,
Polytechnic Institute of Coimbra, Praça Heróis do Ultramar, Solum,
3030-329 Coimbra, Portugal
e-mail: fmlmartins@esec.pt

E. Pereira
Instituto de Telecomunicações (Covilhã), Department of Informatics,
University of Beira Interior, Rua Marquês d'Ávila e Bolama, 6201-001 Covilhã, Portugal
e-mail: edgar@di.ubi.pt

M. A. Facas Vicente · J. Vitória
Department of Mathematics, University of Coimbra, Apartado 3008,
3001-454 Coimbra, Portugal
e-mail: vicente@mat.uc.pt

INESC-C—Instituto de Engenharia de Sistemas e Computadores-Coimbra,
Rua Antero de Quental, 199, 3000-033, Coimbra, Portugal

J. Vitória
e-mail: jvitoria@mat.uc.pt

N. M. Fonseca Ferreira, J. A. Tenreiro Machado (eds.), *Mathematical Methods in Engineering,* DOI 10.1007/978-94-007-7183-3_9,

control problems which are modeled by means of matrix differential equations or matrix difference equations. In the present paper, we deal with differential equations whose coefficients are matrices commuting into pairs.

Let $P_n(\mathcal{R})$ denote a set of $n-$order square real matrices that commute into pairs and let $M_{p,q}(P_n(\mathcal{R}))$ stand for the set of real matrices partitioned into $p \times q$ blocks each belonging to $P_n(\mathcal{R})$, and if $p = q = m$, we simply write $M_m(P_n(\mathcal{R}))$.

We consider systems of first order linear differential equations with matrix constant coefficients that can be written in a matrix form

$$y'(t) = A_b y(t) \tag{1}$$

where $y(t) \in \mathcal{R}^{mn}$ and $A_b \in M_m(P_n(\mathcal{R}))$ is a block matrix.

The **block determinant** of a matrix $A_b \in M_m(P_n(\mathcal{R}))$ is the matrix (block) obtained by developing the determinant of A_b, by considering the blocks as elements; it is denoted as $det_b(A_b)$ ([18]). The **block trace** of a matrix $A_b \in M_m(P_n(\mathcal{R}))$ is the matrix (block) that is the sum of the blocks A_{ii}, $i = 1, \ldots, m$, and is denoted as $tr_b(A_b)$ ([20]). Moreover, $A \otimes B$ is the Kronecker product of the matrices A and B.

In the following, we present some known results and definitions that we use in this paper.

Definition 1.1 ([19]) *The matrix polynomial $det_b(I_m \otimes \Lambda - A_b)$ is called the* ***characteristic matrix polynomial*** *of the matrix $A_b \in M_m(P_n(\mathcal{R}))$, where the indeterminate Λ belongs to $P_n(\mathcal{R})$.*

Proposition 1.1 ([20]) *Let $A_b \in M_m(P_n(\mathcal{R}))$ be a block matrix. Then*

$$P(X) = X^m + (-E_1)X^{m-1} + \cdots + (-E_{m-1})X + (-E_m), E_i \in P_n(\mathcal{R}),$$

is the characteristic matrix polynomial of A_b, where the matrix coefficients E_i are constructed by the following algorithm

$$\begin{array}{lll}
D_1 = A_b \to & E_1 = tr_b(D_1) \to & B_1 = D_1 - (I_m \otimes E_1) \to \\
D_2 = A_b B_1 \to & E_2 = \frac{1}{2} tr_b(D_2) \to & B_2 = D_2 - (I_m \otimes E_2) \to \\
\ldots & \ldots & \ldots \\
D_{m-1} = A_b B_{m-2} \to & E_{m-1} = \frac{1}{m-1} tr_b(D_{m-1}) \to & B_{m-1} = D_{m-1} - (I_m \otimes E_{m-1}) \to \\
D_m = A_b B_{m-1} \to & E_m = \frac{1}{m} tr_b(D_m) \to & B_m = D_m - (I_m \otimes E_m) = 0_{mn}.
\end{array}$$

Definition 1.2 ([17]) *A matrix Γ of order n is a* ***(right) solvent*** *of the matrix polynomial $P(X)$ if $P(\Gamma) = 0_n$.*

Definition 1.3 ([9], p. 72) *Let A_b be a block matrix of order mn. If*

$$A_b X_1 = X_1 \Lambda, \tag{2}$$

where Λ is a block (a matrix of order n) and the block vector X_1 (a matrix of dimension $mn \times n$) is full rank, then Λ is called a ***(right) block eigenvalue*** *of A_b and X_1 is the corresponding* ***(right) block eigenvector***.

Theorem 1.1 ([6], p. 265) *Any solvent of the characteristic matrix polynomial of the matrix $A_b \in M_m(P_n(\mathcal{R}))$ is a block eigenvalue of A_b.*

Definition 1.4 *Let* $P(X) = X^m + A_1 X^{m-1} + \cdots + A_{m-1}X + A_m$, $A_i \in P_n(\mathcal{R})$, *be a matrix polynomial. The matrix* C_b, *of order* mn, *partitioned into blocks of order* n, *given by*

$$C_b = \begin{bmatrix} 0_n & I_n & 0_n & \cdots & 0_n \\ \vdots & \ddots & \ddots & \ddots & \vdots \\ \vdots & & \ddots & \ddots & 0_n \\ 0_n & \cdots & \cdots & 0_n & I_n \\ -A_m & -A_{m-1} & \cdots & -A_2 & -A_1 \end{bmatrix}$$

is said to be a ***block companion matrix*** *associated to matrix polynomial* $P(X)$.

Theorem 1.2 ([9], p. 79) *If the matrix* Γ_1, *of order* n, *is a solvent of the matrix polynomial* $P(X)$, *then*

$$C_b X_1 = X_1 \Gamma_1,$$

where C_b *is the block companion matrix of* $P(X)$ *and*

$$X_1 = \begin{bmatrix} I_n \\ \Gamma_1 \\ \vdots \\ \Gamma_1^{m-1} \end{bmatrix}.$$

Theorem 1.3 ([9], p. 85) *Let* C_b *be a block companion matrix associated with the matrix polynomial* $P(X)$ *and let* Λ_1 *be a block eigenvalue of* C_b *associated with the full rank block eigenvector* X_1, *i.e.,*

$$C_b X_1 = X_1 \Lambda_1.$$

Under these conditions, if the first block X_{11}, *of* X_1, *is nonsingular, then* $\Gamma_1 = X_{11}\Lambda_1 X_{11}^{-1}$ *is a solvent of* $P(X)$.

Definition 1.5 ([16]) *Let* A_b *be a block matrix of order* mn *and let* $\Lambda_1, \Lambda_2, \ldots, \Lambda_m$ *be a set of block eigenvalues of* A_b. *This set is said to be a* ***complete set of block eigenvalues*** *when all the eigenvalues, and respective partial multiplicities, of these block eigenvalues are the eigenvalues, with the same partial multiplicities, of the matrix* A_b.

2 Sufficient Conditions for the Stability

In this section, we obtain sufficient conditions for the asymptotic stability of the equilibrium of the matrix differential Eq. (1). We use block versions of companion, Schwarz and Hermite matrices.

Definition 2.1 *Let Λ_1 be a block eigenvalue of the block matrix A_b. If*

$$Re(\lambda_i) < 0, \tag{3}$$

for all $\lambda_i \in \sigma(\Lambda_1), i = 1, \ldots, n$, then Λ_1 is said to be stable.

The stability criterion in terms of blocks for the equilibrium of the matrix differential Eq. (1) is stated next.

Proposition 2.1 *Let $\Lambda_1, \Lambda_2, \ldots, \Lambda_m$ be a complete set of block eigenvalues of the block matrix A_b. If all block eigenvalues, Λ_j, are stable, then the equilibrium of the matrix differential Eq. (1) is asymptotically stable.*

Furthermore, we will need the following two definitions and two theorems.

Definition 2.2 ([4], p. 67) *A matrix $N \in \mathcal{R}^{n\times n}$ is said to be* ***symmetrizable*** *if there exists a matrix $R = R^T \in \mathcal{R}^{n\times n}$ positive definite such* that $N^T R = RN$.

Definition 2.3 ([1]) *Two matrices $N_1, N_2 \in \mathcal{R}^{n\times n}$ are said to be* ***simultaneously symmetrizable*** *if there exists a matrix $R = R^T \in \mathcal{R}^{n\times n}$ positive definite such that*

$$N_1^T R = RN_1 \text{ and } N_2^T R = RN_2. \tag{4}$$

It follows directly from Definition 2.3 that N_1+N_2 and N_1-N_2 are simultaneously symmetrizable.

Theorem 2.1 ([12], p. 50) *Let $N_1, N_2 \in \mathcal{C}^{n\times n}$ be diagonalizable. Then, N_1 and N_2 comute if and only if they are simultaneously diagonalizable.*

Theorem 2.2 ([5]) *A set of matrices simultaneously diagonalizable is also a set of matrices simultaneously symmetrizable.*

Next, we are able to prove that, under certain conditions, the Lyapunov matrix equation $C_b^T V+VC_b = W$ has a unique solution V, where C_b is the block companion matrix associated to the matrix polynomial $P(X)$.

Proposition 2.2 *Let $C_b \in M_m(P_n(\mathcal{R}))$ be the block companion matrix associated with the matrix polynomial $P(X)$. If*

(i) $\sigma(C_b) \cap \sigma(-C_b) = \emptyset$;
(ii) $A_\alpha A_\beta \in P_n(\mathcal{R})$, *with* $0 \le \alpha < \beta \le m$, *are diagonalizable;*
(iii) $W = W^T = [W_{ij}] \in \mathcal{R}^{mn\times mn}$ *where*

$$W_{ij} = W_{ji} = \begin{cases} 2RA_{m+1-j}A_{m+1-i} & \text{if } m+i \text{ and } m+j \text{ are even}, \quad i \le j \\ 0_n & \text{if } m+i \text{ or } m+j \text{ is odd} \end{cases} \tag{5}$$

$(i, j = 1, 2, \ldots, m)$ and $R = R^T \in \mathcal{R}^{n\times n}$ is positive definite,

then the Lyapunov matrix equation

$$C_b^T V + VC_b = -W \tag{6}$$

has a unique solution $V = V^T = [V_{ij}] \in \mathcal{R}^{mn\times mn}$, *where*

$$V_{ij} = V_{ji} = \begin{cases} \sum_{k=0}^{i-1} (-1)^{k+i-1} R A_{m-i-j+k+1} A_{m-k} & \text{if } i+j \text{ is even}, \quad i \le j \\ 0_n & \text{if } i+j \text{ is odd} \end{cases}. \quad (7)$$

From the above proposition, we obtain two important results on the stability of a matrix differential equation, using a block companion matrix.

Corollary 2.1 *If the Lyapunov matrix equation*

$$C_b^T V + V C_b = -W, \quad (8)$$

has a unique symmetric solution V, *such that:*

(i) V *is positive definite;*
(ii) W *is positive semi-definite;*
(iii) $\begin{bmatrix} W^{1/2} \\ W^{1/2} C_b \\ \vdots \\ W^{1/2} C_b^{m-1} \end{bmatrix}$ *is of full rank;*
(iv) $\Lambda_1, \Lambda_2, \ldots, \Lambda_m$ *are a complete set of block eigenvalues* A_b,

then all block eigenvalues, Λ_j, *are stable.*

Corollary 2.2 *The equilibrium of the matrix differential equation*

$$y'(t) = A_b y(t)$$

is asymptotically stable.

Thereafter, we present the block versions of the Hermite, Hurwitz, Routh and Schwarz matrices.

Definition 2.4 ([15]) *Let* $P(X) = A_0 X^m + A_1 X^{m-1} + \cdots + A_{m-1} X + A_m$, $A_i \in P_n(\mathcal{R})$, *be a matrix polynomial. The matrix* $H_b^e = \left[H_{b(ij)}^e\right] \in M_m(P_n(\mathcal{R}))$, *where*

$$H_{b(ij)}^e = H_{b(ji)}^e = \begin{cases} \sum_{k=0}^{i-1} (-1)^{k+i-1} A_k A_{i+j-k-1} & \text{if } i+j \text{ is even}, i \le j \\ 0_n & \text{if } i+j \text{ is odd} \end{cases} \quad (9)$$

$(i, j = 1, 2, \ldots, m)$ *is said to be the* ***block Hermite matrix*** *of* $P(X)$.

Definition 2.5 ([6]) *Let* $P(X) = A_0 X^m + A_1 X^{m-1} + \cdots + A_{m-1} X + A_m$, $A_i \in P_n(\mathcal{R})$, *be a matrix polynomial. And let be the matrix* $H_b = \left[H_{b(ij)}\right] \in$

$M_m(P_n(\mathcal{R}))$, where $H_{b(ij)} = A_{2j-i}$ with $A_r = 0_n$ if $r < 0$ or $r > m$ $(i, j = 1, 2, \ldots, m)$. To this matrix H_b, given by

$$H_b = \begin{bmatrix} A_1 & A_3 & A_5 & \cdots & \cdots & \cdots & A_{2m-1} \\ A_0 & A_2 & A_4 & \cdots & \cdots & \cdots & A_{2m-2} \\ 0_n & A_1 & A_3 & \cdots & \cdots & \cdots & A_{2m-3} \\ 0_n & A_0 & A_2 & \ddots & \cdots & \cdots & A_{2m-4} \\ 0_n & 0_n & A_1 & \cdots & \ddots & \cdots & \vdots \\ \vdots & \cdots & \cdots & \cdots & \cdots & \ddots & \vdots \\ 0_n & 0_n & 0_n & \cdots & \cdots & \cdots & A_m \end{bmatrix},$$

*we call the **block Hurwitz matrix**.*

Definition 2.6 ([15]) *If we perform the block elimination of the block Hurwitz matrix, H_b, we obtain the matrix*

$$\begin{bmatrix} C_{21} & C_{22} & C_{23} & \cdots & \cdots & \cdots \\ 0_n & C_{31} & C_{32} & C_{33} & \cdots & \cdots \\ 0_n & 0_n & C_{41} & C_{42} & \cdots & \cdots \\ 0_n & 0_n & 0_n & C_{51} & C_{52} & \cdots \\ \vdots & & & \ddots & \ddots & \ddots \\ 0_n & 0_n & 0_n & \cdots & 0_n & C_{(m+1)1} \end{bmatrix} \in M_m(P_n(\mathcal{R})), \tag{10}$$

*which we call the **block Routh matrix**.*

Definition 2.7 ([6]) *Let $S_b = \left[S_{b(ij)}\right] \in M_m(P_n(\mathcal{R}))$, where*

$$S_{b(ij)} = \begin{cases} I_n & if\, j - i = 1 \\ -S_k & if\, i - j = 1 \quad (k = 2, \ldots, m) \\ -S_1 & if\, j = i = m \\ 0_n & otherwise \end{cases}.$$

The matrix

$$S_b = \begin{bmatrix} 0_n & I_n & 0_n & \cdots & 0_n \\ -S_m & 0_n & I_n & \ddots & \vdots \\ 0_n & \ddots & \ddots & \ddots & 0_n \\ \vdots & \ddots & -S_3 & 0_n & I_n \\ 0_n & \cdots & 0_n & -S_2 & -S_1 \end{bmatrix}, \tag{11}$$

*is named the **block Schwarz matrix**.*

A strong relationship between the block companion matrix and the block Schwarz matrix is that they are similar, i.e., $S_b = TC_bT^{-1}$ with

$$T = \begin{bmatrix} I_n & . & 0_n & 0_n & 0_n & 0_n & 0_n & 0_n & 0_n \\ . & . & . & . & . & . & . & . & . \\ C_{(m-1)1}^{-1}C_{(m-1)2} & . & I_n & 0_n & 0_n & 0_n & 0_n & 0_n & 0_n \\ 0_n & . & 0_n & I_n & 0_n & 0_n & 0_n & 0_n & 0_n \\ C_{(m-3)1}^{-1}C_{(m-1)3} & . & C_{61}^{-1}C_{62} & 0_n & I_n & 0_n & 0_n & 0_n & 0_n \\ 0_n & . & 0_n & C_{51}^{-1}C_{52} & 0_n & I_n & 0_n & 0_n & 0_n \\ C_{(m-5)1}^{-1}C_{(m-5)4} & . & C_{41}^{-1}C_{43} & 0_n & C_{41}^{-1}C_{42} & 0_n & I_n & 0_n & 0_n \\ 0_n & . & 0_n & C_{31}^{-1}C_{33} & 0_n & C_{31}^{-1}C_{32} & 0_n & I_n & 0_n \\ . & . & C_{21}^{-1}C_{24} & 0_n & C_{21}^{-1}C_{23} & 0_n & C_{21}^{-1}C_{22} & 0_n & I_n \end{bmatrix}, \tag{12}$$

where C_{ij} are the elements of the block Routh matrix, with $C_{i1} \in P_n(\mathcal{R})$, $i = 2, \cdots, m$, being nonsingular ([15]).

Next, we state conditions for the stability of a matrix differential equation through the block Schwarz matrix S_b.

Proposition 2.3 *Let $S_b \in M_m(P_n(\mathcal{R}))$ be a block Schwarz matrix and let the matrices $S_1, S_1S_2, S_1S_2S_3, \ldots, S_1S_2S_3\cdots S_{m-1}S_m \in P_n(\mathcal{R})$ be diagonalizable and positive definite. Then, for a symmetric and positive semi-definite matrix Q, there exists a unique solution M, symmetric and positive definite, of the Lyapunov matrix equation $S_b^TM + MS_b = -Q$.*

Corollary 2.3 *If $\Lambda_1, \Lambda_2, \ldots, \Lambda_m$ are a complete set of block eigenvalues A_b, then they are stable.*

Corollary 2.4 *The equilibrium of the matrix differential equation*

$$y'(t) = A_b y(t)$$

is asymptotically stable.

In the following result, we study the stability of a matrix differential equation, by using a block Hermite matrix.

Proposition 2.4 *Let H_b^e be a block Hermite matrix of $P(X)$. If H_b^e is positive definite and if $A_\alpha A_\beta \in P_n(\mathcal{R})$ are diagonalizable, $0 \leq \alpha \leq \beta \leq m$, and if A_0 is nonsingular, then the equilibrium of the matrix differential equation $y'(t) = A_b y(t)$ is asymptotically stable.*

3 Block Circulants Matrices with Commuting Blocks

In this section, we deal with questions of stability and localization. We shall treat eigenvalues location in terms of inertia.

We consider now the equation

$$y'(t) = A_b y(t), \tag{13}$$

where $y(t) \in \mathcal{R}^{mn}$ and the matrix A_b has a block circulant structure.

Let consider a real $mn \times mn$ **block circulant matrix**

$$A_b = bcirc(A_1, A_2, \dots, A_m) = \begin{bmatrix} A_1 & A_2 & \cdots & A_m \\ A_m & A_1 & \cdots & A_{m-1} \\ \cdots & \cdots & \cdots & \cdots \\ A_2 & \cdots & A_m & A_1 \end{bmatrix},$$

where $A_1, A_2, \dots, A_m \in P_n(\mathcal{R})$.

In the next result it is stated, essentially, that the eigenvalues of A_b are given in terms of the eigenvalues of the blocks of a block diagonal matrix.

Theorem 3.1 ([8], p. 181) Let $F_p^* = \frac{1}{\sqrt{p}} \underbrace{V(1, w, w^2, \dots, w^{p-1})}_{\text{Vandermonde Matrix}}$, with $w = e^{\left(\frac{i2\pi}{p}\right)}$, and let $B_i = F_n A_i F_n^*, \quad i = 1, 2, \dots, m$. Then, we have:

(i)

$$\begin{bmatrix} M_1 \\ M_2 \\ M_3 \\ \vdots \\ M_m \end{bmatrix} = (\sqrt{m} F_m^* \otimes I_n) \begin{bmatrix} B_1 \\ B_2 \\ B_3 \\ \vdots \\ B_m \end{bmatrix};$$

(ii) $(F_m \otimes F_n) A_b (F_m \otimes F_n)^* = diag(M_1, M_2, \dots, M_m)$;

(iii) The eigenvalues of $M_1, M_2, \dots, M_m$ are eigenvalues of A_b.

We notice that the matrices $M_1, M_2, \dots, M_m$ are the Block Eigenvalues of A_b.

Proposition 3.1 *If $M_1, M_2, \dots, M_m$ are stable then the equilibrium of the matrix differential equation*

$$y'(t) = A_b y(t)$$

is asymptotically stable.

In Engineering and Applied Sciences [2, 10, 13, 14] inertia results are used, this mea-ning that the number of eigenvalues with positive, negative and null real parts is taken into account. Stability and eigenvalues location are strongly related. Stability of systems is studied through the Lyapunov matrix equation. By using a Lyapunov matrix equation, we get information on the number of eigenvalues having negative real parts.

So, calculating the inertia of the matrix A_b, we obtain some useful information about the eigenvalues having negative real parts. The structure of the block circulant matrices allows us to have some knowledge about their inertia and consequently concerning their stability. In the present case of block circulant matrices, we are able to exhibit some interesting formulas.

We approach the inertia of A_b, by considering two cases for m:

(1) m is even.
Let $\lambda_{k_1}, \lambda_{k_2}, \dots, \lambda_{k_n}$ be eigenvalues of $M_k, k = 1, \dots, \left[\frac{m}{2}\right] + 1$, such that

$$M_k = (\sqrt{m} F_m^* \otimes I_n)_k \begin{bmatrix} B_1 \\ B_2 \\ B_3 \\ \vdots \\ B_m \end{bmatrix},$$

where $(\sqrt{m} F_m^* \otimes I_n)_k$ is the kth block row of type n by mn of the block matrix $(\sqrt{m} F_m^* \otimes I_n)$.
Thus,

(i) if $x_{k_p} = Re(\lambda_{k_p}), p = 1, \dots, n,\ N_-\left(x_{k_1}, x_{k_2}, \dots, x_{k_n}\right) = l_k$, then the matrix A_b has $l_1 + 2\left(l_2 + \dots + l_{[\frac{m}{2}]}\right) + l_{[\frac{m}{2}]+1}$ eigenvalues with negative real parts;
(ii) if $x_{k_p} = Re(\lambda_{k_p}), p = 1, \dots, n,\ N_+\left(x_{k_1}, x_{k_2}, \dots, x_{k_n}\right) = r_k$, then the matrix A_b has $r_1 + 2\left(r_2 + \dots + r_{[\frac{m}{2}]}\right) + r_{[\frac{m}{2}]+1}$ eigenvalues with positive real parts;
(iii) if $x_{k_p} = Re(\lambda_{k_p}), p = 1, \dots, n, N_0\left(x_{k_1}, x_{k_2}, \dots, x_{k_n}\right) = u_k$, then the matrix A_b has $u_1 + 2\left(u_2 + \dots + u_{[\frac{m}{2}]}\right) + u_{[\frac{m}{2}]+1}$ eigenvalues with null real parts.

(2) m is odd.
Let $\lambda_{k_1}, \lambda_{k_2}, \dots, \lambda_{k_n}$ be eigenvalues of $M_k, k = 1, \dots, \left[\frac{m}{2}\right] + 1$, such that

$$M_k = (\sqrt{m} F_m^* \otimes I_n)_k \begin{bmatrix} B_1 \\ B_2 \\ B_3 \\ \vdots \\ B_m \end{bmatrix},$$

where $(\sqrt{m} F_m^* \otimes I_n)_k$ is the kth block row of type n by mn of the block matrix $(\sqrt{m} F_m^* \otimes I_n)$.
Thus,

(i) if $x_{k_p} = Re(\lambda_{k_p}), p = 1, \dots, n, N_-\left(x_{k_1}, x_{k_2}, \dots, x_{k_n}\right) = l_k$, then the matrix A_b has $l_1 + 2\left(l_2 + \dots + l_{[\frac{m}{2}]+1}\right)$ eigenvalues with negative real parts;
(ii) if $x_{k_p} = Re(\lambda_{k_p}), p = 1, \dots, n,\ N_+\left(x_{k_1}, x_{k_2}, \dots, x_{k_n}\right) = r_k$, then the matrix A_b has $r_1 + 2\left(r_2 + \dots + r_{[\frac{m}{2}]+1}\right)$ eigenvalues with positive real parts;
(iii) if $x_{k_p} = Re(\lambda_{k_p}), p = 1, \dots, n,\ N_0\left(x_{k_1}, x_{k_2}, \dots, x_{k_n}\right) = u_k$, then the matrix A_b has $u_1 + 2\left(u_2 + \dots + u_{[\frac{m}{2}]+1}\right)$ eigenvalues with null real parts.

The results in this section remain valid when the blocks of the block circulant matrices are neither necessarily commuting into pairs nor necessarily real.

4 Final Remarks and Conclusions

The Propositions 2.2, 2.3 and 2.4 correspond to results stated in a more general context, when the blocks of the considered block matrices are not necessarily commuting into pairs [15]. We omit the proofs of the above mentioned results. These proofs follow from [15] by using the Theorems 2.1 and 2.2.

Datta [7], when referring to Barnett [3], stated the following: "once a matrix has been transformed into Schwarz form the stability problem is solved immediately". We hope, in the future, to be able to assert something similar about the block Schwarz matrix.

We have seen that, in the case when a block matrix has a block circulant structure, we got interesting formulas for solving the stability problem.

In future work, we can consider the inertia of the block Schwarz matrix S_b and the inertia of the block circulant matrix

$$bcirc(A_1, A_2, \ldots, A_m) = \begin{bmatrix} A_1 & A_2 & \cdots & A_m \\ A_m & A_1 & \cdots & A_{m-1} \\ \cdots & \cdots & \cdots & \cdots \\ A_2 & \cdots & A_m & A_1 \end{bmatrix},$$

where $A_1, A_2, \ldots, A_m$ are the matrix constant coefficients of the matrix polynomial $P(X)$, in order to study the respective stability.

The quest for root localization of polynomials—albeit a quite old activity of mathematicians, applied scientists and engineers and despite the rich history of the subject in the case of univariate polynomials—keeps vivid [11].

Readers with some acquaintance with block structured matrices and who are using them for dealing with root localization may feel far from comfortable, when wondering how long and hard should be the path ahead, in order to achieve the degree of maturity displayed in [11], for the case structured matrices having scalar entries.

References

1. Adhikari S (2000) On symmetrizable systems of second kind. J Appl Mech 67:797–802
2. Barnett S (1983) Polynomials and linear control systems. Marcel Dekker, New York
3. Barnett S, Storey C (1970) Matrix methods in stability theory. Thomas Nelson, London
4. Bellman R (1960) Introduction to matrix analysis. McGraw-Hill, New York
5. Bhaskar A (2001) Taussky's theorem, symmetrizability and modal analysis revisited. Proc R Soc Lond A457:2455–2480
6. Costa C (2001) José Vicente Gonçalves: Matemático... porque Professor! Centro de Estudos de História do Atlântico, Funchal. (ISBN:972-8263-33-3)
7. Datta BN (1974) A constructive method for finding the Schwarz form of a Hessenberg matrix. IEEE Trans Automat Contr 19:616–617
8. Davis PJ (1979) Circulant matrices. Wiley, New York
9. Dennis E, Traub JF, Weber RP (1971) On the matrix polynomial, lambda-matrix and block eigenvalue problems. Computer Science Department, Technical Report, Cornell University, Ithaca, New York and Carnegie-Mellon University, Pittsburgh, Pennsylvania. (Disponível em http://cs-tr.cs.cornell.edu)

10. Gantmacher FR (1974) The theory of matrices, vol II. Chelsea, New York
11. Holtz O, Tyaglov M (28 February 2010) Structured matrices, continued fractions, and root localization of polynomials, 78 p, arXiv:0912.4703v2 [math.CA]
12. Horn RA, Johnson CR (1985) Matrix analysis. Cambridge University, New York
13. Lancaster P, Tismenetsky M (1985) The theory of matrices. Academic, London
14. Lehnigk SH (1966) Stability theorems for linear motions with an introduction to Lyapunov's direct method. Prentice-Hall, Englewood Cliffs
15. Martins F, Pereira E (2007) Block matrices and stability theory. Tatra Mt Math Publ 38:147–162
16. Pereira E (2003) Block eigenvalues and solutions of differential matrix equations. Mathematical Notes (Miskolc) 4:45–51
17. Pereira E, Vitória J (2001) Deflation for block eigenvalues of block partitioned matrices with an application to matrix polynomials of commuting matrices. Comput Math Appl 42:1177–1188
18. Vitória J (1982) Block eigenvalues of block compound matrices. Linear Algebra Appl 47:23–34
19. Vitória J (1982) A block Cayley Hamilton theorem. Bull Mathemátique (Roumanie) 26:93–97
20. Vitória J (1988) Some questions of numerical algebra related to differential equations. Numerical Methods (Miskolc, 1986), 127–140, Greenspan D, Rózsa P (eds) Colloquia mathematica cocietatis János Bolyai, vol 50, North-Holland

Identification of Material Thermophysical Parameters with Regard to Optimum Location of Sensors

Ewa Majchrzak and Jerzy Mendakiewicz

Abstract As a practical example illustrating the considerations presented in the paper the thermal processes proceeding in a system casting-mould are considered.

The casting is made from Fe-C alloy (cast iron) and the austenite and eutectic latent heats of this material should be identified. To estimate these parameters the knowledge of temperature history at the points selected from the domain considered is necessary. The location of sensors should assure the best conditions of identification process.

So, the algorithm of optimum location of sensors basing on the D-optimality criterion is presented, while the inverse problem is solved using the gradient method.

Keywords Heat transfer · Solidification process · Inverse problem

1 Formulation of Problem

A system casting-mould-environment is considered. Temperature field in casting domain is described by equation [1, 2]

$$x \in \Omega : C(T)\frac{\partial T(x,t)}{\partial t} = \lambda \nabla^2 T\,(x,t) \tag{1}$$

where $C(T)$ is the substitute thermal capacity of cast iron, λ is the thermal conductivity, T, x, t denote the temperature, geometrical co-ordinates and time.

The following approximation of substitute thermal capacity is taken into account

$$C\,(T) = \begin{cases} c_L, & T > T_L \\ \dfrac{c_L + c_S}{2} + \dfrac{Q_{aus}}{T_L - T_E}, & T_E < T \le T_L \\ \dfrac{c_L + c_S}{2} + \dfrac{Q_{eu}}{T_E - T_S}, & T_S < T \le T_E \\ c_S, & T \le T_S \end{cases} \tag{2}$$

E. Majchrzak (✉) · J. Mendakiewicz
Silesian University of Technology, Gliwice, Poland
e-mail: ewa.majchrzak@polsl.pl

N. M. Fonseca Ferreira, J. A. Tenreiro Machado (eds.), *Mathematical Methods in Engineering,* DOI 10.1007/978-94-007-7183-3_10,

where T_L, T_S are the liquidus and solidus temperatures, respectively, T_E is the temperature corresponding to the beginning of eutectic crystallization, Q_{aus}, Q_{eu} are the latent heats connected with the austenite and eutectic phases evolution, c_L, c_S are constant volumetric specific heats of molten metal and solid one, respectively.

The temperature field in mould sub-domain is described by equation

$$x \in \Omega_m : c_m \frac{\partial T_m(x,t)}{\partial t} = \lambda_m \nabla^2 T_m(x,t) \tag{3}$$

where c_m is the mould volumetric specific heat, λ_m is the mould thermal conductivity.

On the contact surface between casting and mould the continuity condition in the form

$$x \in \Gamma_c : \begin{cases} -\lambda \mathbf{n} \cdot \nabla T(x,t) = -\lambda_m \mathbf{n} \cdot \nabla T_m(x,t) \\ T(x,t) = T_m(x,t) \end{cases} \tag{4}$$

can be accepted.

On the external surface of the system the Robin condition

$$x \in \Gamma_0 : -\lambda_m \mathbf{n} \cdot \nabla T_m(x,t) = \alpha \left[T_m(x,t) - T_a \right] \tag{5}$$

is given (α is the heat transfer coefficient, T_a is the ambient temperature).

For time $t = 0$ the initial condition

$$t = 0 : T(x,0) = T_0(x), T_m(x,0) = T_{m0}(x) \tag{6}$$

is also known.

It is assumed that the aim of experiments is to determine the latent heats Q_{aus}, Q_{eu} of casting material and in order to find the optimum location of sensors the D-optimality criterion [3, 4] is taken into account.

2 Optimal Sensors Location

The model above formulated contains two unknown parameters Q_{aus}, Q_{eu} which will be reconstructed on the basis of observations. Let us assume that the approximate values of Q_{aus}^0, Q_{eu}^0 are available e.g. from preliminary experiments. Our goal is to determine the optimum sensors location in order to maximize the expected accuracy of parameters identification which will be found using the data generated in new experiments.

In Fig. 1 the domain considered and its discretization is shown. Let $x^1 = (x_1^1, x_2^1)$, $x^2 = (x_1^2, x_2^2)$, …, $x^M = (x_1^M, x_2^M)$ are the points from the casting sub-domain which are taken into account as the possible sensors location (Fig. 1). The design problem consists in the selection of the best positions of sensors under the assumption that only two sensors will be taken into account (it corresponds to the number of estimated parameters).

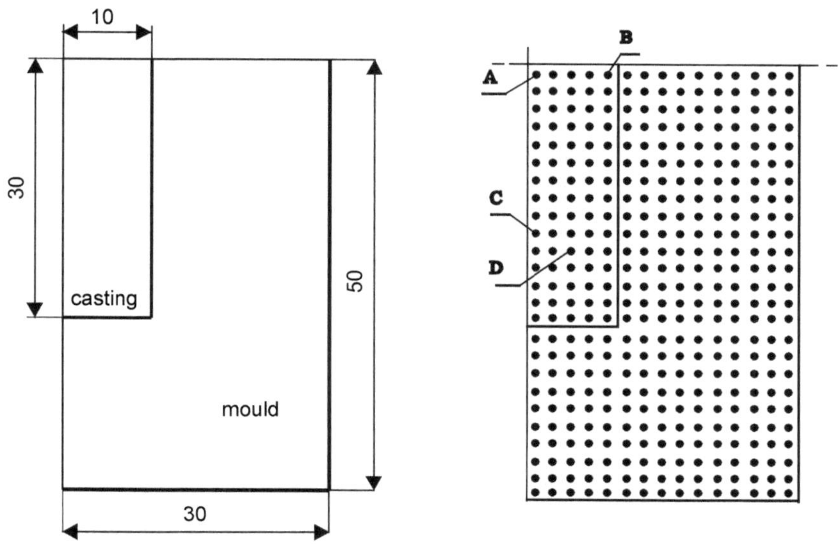

Fig. 1 Domain considered and its discretization

Let us introduce the sensitivity coefficients

$$Z_{l1}^{f} = \frac{\partial T\left(x^{l}, t^{f}, Q_{aus}^{0}, Q_{eu}^{0}\right)}{\partial Q_{aus}^{0}}, \quad Z_{l2}^{f} = \frac{\partial T\left(x^{l}, t^{f}, Q_{aus}^{0}, Q_{eu}^{0}\right)}{\partial Q_{eu}^{0}} \tag{7}$$

where l denotes number of node ($l = 1, 2, \ldots, M$) and t^{f} is the moment of time ($f = 1, 2, \ldots, F$). On the basis of (7) the sensitivity matrix is constructed

$$\mathbf{Z}\left(x^{i}, x^{j}\right) = \begin{bmatrix} Z_{i1}^{1} & Z_{i2}^{1} \\ \cdots & \cdots \\ Z_{i1}^{F} & Z_{i2}^{F} \\ Z_{j1}^{1} & Z_{j2}^{1} \\ \cdots & \cdots \\ Z_{j1}^{F} & Z_{j2}^{F} \end{bmatrix} \tag{8}$$

where x^{i}, x^{j} are the pair of nodes, at the same time $i, j = 1, 2, \ldots, M$, $i \neq j$.

D-optimality criterion used in the design of sensors location is the following [3, 4]

$$\det \mathbf{M}\left(x^{i*}, x^{j*}\right) = \max_{(x^{i}, x^{j})} \det \mathbf{M}\left(x^{i}, x^{j}\right) \tag{9}$$

where

$$\mathbf{M}\left(x^{i}, x^{j}\right) = \mathbf{Z}^{T}\left(x^{i}, x^{j}\right) \mathbf{Z}\left(x^{i}, x^{j}\right) \tag{10}$$

It is easy to check that

$$\mathbf{M}\left(x^i, x^j\right) = \begin{bmatrix} \sum\limits_{f=1}^{F}\left(Z_{i1}^f\right)^2 + \left(Z_{j1}^f\right)^2 & \sum\limits_{f=1}^{F} Z_{i1}^f Z_{i2}^f + Z_{j1}^f Z_{j2}^f \\ \sum\limits_{f=1}^{F} Z_{i1}^f Z_{i2}^f + Z_{j1}^f Z_{j2}^f & \sum\limits_{f=1}^{F}\left(Z_{i2}^f\right)^2 + \left(Z_{j2}^f\right)^2 \end{bmatrix} \tag{11}$$

The nodes (x^{i^*}, x^{j^*}) being the solution of optimum problem (9) correspond to the best sensors location in the case of simultaneous identification of parameters Q_{aus} and Q_{eu}.

3 Sensitivity Coefficients

Elementary step in the design of optimal sensor locations is to use an effective procedure for the computations of sensitivity coefficients. One of the method is the direct differentiation of governing equations with respect to the identified parameters [2, 5], [6–8]. So, differentiation of Eqs. (1–6) with respect to $p_1 = Q_{aus}$, $p_2 = Q_{eu}$ leads to the following additional boundary-initial problems

$$\begin{aligned}
&x \in \Omega : \frac{\partial C(T)}{\partial p_e}\frac{\partial T(x,t)}{\partial t} + C(T)\frac{\partial}{\partial p_e}\left(\frac{\partial T(x,t)}{\partial t}\right) = \lambda \frac{\partial \nabla^2 T(x,t)}{\partial p_e} \\
&x \in \Omega_m : c_m \frac{\partial}{\partial p_e}\left(\frac{\partial T_m(x,t)}{\partial t}\right) = \lambda_m \frac{\partial \nabla^2 T_m(x,t)}{\partial p_e} \\
&x \in \Gamma_c : \begin{cases} -\lambda \mathbf{n} \cdot \dfrac{\partial \nabla T(x,t)}{\partial p_e} = -\lambda_m \mathbf{n} \cdot \dfrac{\partial \nabla T_m(x,t)}{\partial p_e} \\ \dfrac{\partial T(x,t)}{\partial p_e} = \dfrac{\partial T_m(x,t)}{\partial p_e} \end{cases} \\
&x \in \Gamma_0 : -\lambda_m \mathbf{n} \cdot \frac{\partial \nabla T_m(x,t)}{\partial p_e} = \alpha \frac{\partial T_m(x,t)}{\partial p_e} \\
&t = 0 : \frac{\partial T(x,0)}{\partial p_e} = 0, \frac{\partial T_m(x,0)}{\partial p_e} = 0
\end{aligned} \tag{12}$$

or

$$\begin{aligned}
&x \in \Omega : C(T)\frac{\partial Z_e(x,t)}{\partial t} = \lambda \nabla^2 Z_e(x,t) - \frac{\partial C(T)}{\partial p_e}\frac{\partial T(x,t)}{\partial t} \\
&x \in \Omega_m : c_m \frac{\partial Z_{me}(x,t)}{\partial t} = \lambda_m \nabla^2 Z_{me}(x,t) \\
&x \in \Gamma_c : \begin{cases} -\lambda \mathbf{n} \cdot \nabla Z_e(x,t) = -\lambda_m \mathbf{n} \cdot \nabla Z_{me}(x,t) \\ Z_e(x,t) = Z_{me}(x,t) \end{cases} \\
&x \in \Gamma_0 : -\lambda_m \mathbf{n} \cdot \nabla Z_{me}(x,t) = \alpha Z_{me}(x,t) \\
&t = 0 : Z_e(x,0) = 0, Z_{m\,e}(x,0) = 0
\end{aligned} \tag{13}$$

where

$$Z_e(x,t) = \frac{\partial T(x,t)}{\partial p_e}, Z_{me}(x,t) = \frac{\partial T_m(x,t)}{\partial p_e} \tag{14}$$

Summing up, in order to construct the sensitivity matrix (8) the basic problem described by Eqs. (1–6) and the sensitivity problems (13) should be solved.

4 Inverse Problem Solution

As it was mentioned above, the parameters appearing in the mathematical model of casting solidification are known except the latent heats Q_{aus} and Q_{eu}. It is assumed that the temperature values T_{dl}^f at the points $x^1 = x^{i*}$, $x^2 = x^{j*}$ (c.f. Eq. (9)) located in the casting sub-domain for times t^f are known

$$T_{dl}^f = T_d\left(x^l, t^f\right), \quad l = 1, 2, \quad f = 1, 2, \ldots, F \tag{15}$$

To solve the inverse problem the least squares criterion is applied

$$S\left(Q_{aus}, Q_{eu}\right) = \frac{1}{2F}\sum_{l=1}^{2}\sum_{f=1}^{F}\left(T_l^f - T_{dl}^f\right)^2 \tag{16}$$

where T_{dl}^f and $T_l^f = T\ (x^l,\ t^f)$ are the measured and estimated temperatures, respectively. The estimated temperatures are obtained from the solution of the direct problem (c.f. Chap. 1) by using the current available estimate of these parameters e.g. from preliminary experiments.

In the case of typical gradient method application [2, 5, 6] the criterion (16) is differentiated with respect to the unknown parameters Q_{aus}, Q_{eu} and next the necessary condition of optimum is used.

So, one obtains the following system of equations

$$\begin{cases} \dfrac{\partial S}{\partial Q_{aus}} = \dfrac{1}{F}\sum\limits_{l=1}^{2}\sum\limits_{f=1}^{F}\left(T_l^f - T_{dl}^f\right)\left(Z_{l1}^f\right)^k = 0 \\ \dfrac{\partial S}{\partial Q_{eu}} = \dfrac{1}{F}\sum\limits_{l=1}^{2}\sum\limits_{f=1}^{F}\left(T_l^f - T_{dl}^f\right)\left(Z_{l2}^f\right)^k = 0 \end{cases} \tag{17}$$

where k is the number of iteration.

Function T_l^f is expanded in a Taylor series about known values of $Q_{aus}{}^k$, $Q_{eu}{}^k$, this means

$$T_l^f = \left(T_l^f\right)^k + \left(Z_{l1}^f\right)^k\left(Q_{aus}^{k+1} - Q_{aus}^k\right) + \left(Z_{l2}^f\right)^k\left(Q_{eu}^{k+1} - Q_{eu}^k\right) \tag{18}$$

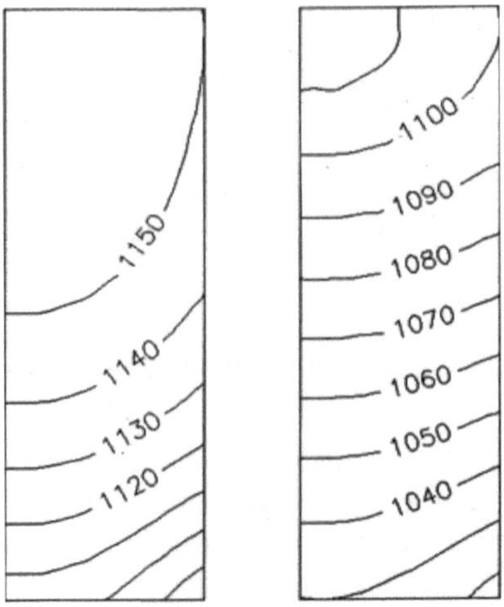

Fig. 2 Temperature field in casting sub-domain (t = 90 s and t = 180 s)

Putting (18) into (17) one obtains

$$\begin{cases} \sum_{l=1}^{2}\sum_{f=1}^{F}\left[\left(Z_{l1}^{f}\right)^{k}\right]^{2} Q_{aus}^{k+1} + \sum_{l=1}^{2}\sum_{f=1}^{F}\left(Z_{l1}^{f}\right)^{k}\left(Z_{l2}^{f}\right)^{k} Q_{eu}^{k+1} = \\ \sum_{l=1}^{2}\sum_{f=1}^{F}\left[\left(Z_{l1}^{f}\right)^{k}\right]^{2} Q_{aus}^{k} + \sum_{l=1}^{2}\sum_{f=1}^{F}\left(Z_{l1}^{f}\right)^{k}\left(Z_{l2}^{f}\right)^{k} Q_{eu}^{k} + \\ \sum_{l=1}^{2}\sum_{f=1}^{F}\left[T_{dl}^{f} - \left(T_{l}^{f}\right)^{k}\right]\left(Z_{l1}^{f}\right)^{k} \\ \sum_{l=1}^{2}\sum_{f=1}^{F}\left(Z_{l1}^{f}\right)^{k}\left(Z_{l2}^{f}\right)^{k} Q_{aus}^{k+1} + \sum_{l=1}^{2}\sum_{f=1}^{F}\left[\left(Z_{l2}^{f}\right)^{k}\right]^{2} Q_{eu}^{k+1} = \\ \sum_{l=1}^{2}\sum_{f=1}^{F}\left(Z_{l1}^{f}\right)^{k}\left(Z_{l2}^{f}\right)^{k} Q_{aus}^{k} + \sum_{l=1}^{2}\sum_{f=1}^{F}\left[\left(Z_{l2}^{f}\right)^{k}\right]^{2} Q_{eu}^{k} + \\ \sum_{l=1}^{2}\sum_{f=1}^{F}\left[T_{dl}^{f} - \left(T_{l}^{f}\right)^{k}\right]\left(Z_{l2}^{f}\right)^{k} \end{cases} \tag{19}$$

This system of equations allows one to find the values of ${Q_{aus}}^{k+1}$ and ${Q_{eu}}^{k+1}$. The iteration process is stopped when the assumed number of iterations K is achieved.

5 Example of Computations

The casting–mould system shown in Fig. 1 has been considered. The basic problem and additional problems connected with the sensitivity functions have been solved using the explicit scheme of FDM [1]. The regular mesh created by 25 × 15 nodes with constant step $h = 0.002$ [m] has been introduced, time step $\Delta t = 0.1$ [s]. The following input data have been assumed: $\lambda = 30$ [W/(mK)], $\lambda_m =$

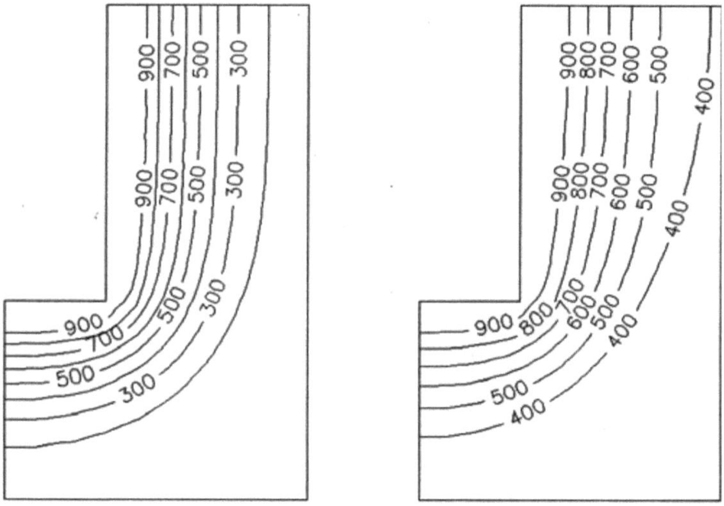

Fig. 3 Temperature field in mould sub-domain (t = 90 s and t = 180 s)

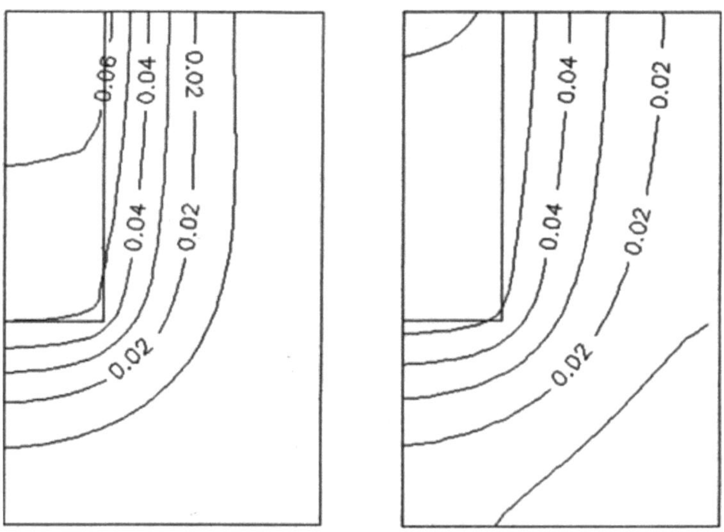

Fig. 4 Distribution of sensitivity function $\partial T/\partial Q_{aus}$ (t = 90 s and t = 180 s)

1 [W/(mK)], $c_L = 5.88$ [MJ/(m^{3}K)], $c_S = 5.4$ [MJ/(m^{3}K)], $c_m = 1.75$ [MJ/(m^{3}K)], pouring temperature $T_0 = 1300\,°\mathrm{C}$, liquidus temperature $T_L = 1250\,°\mathrm{C}$, border temperature $T_E = 1160\,°\mathrm{C}$, solidus temperature $T_S = 1110\,°\mathrm{C}$, initial mould temperature $T_{m0} = 20\,°\mathrm{C}$.

At first, the direct problem under the assumption that $Q_{aus} = 923$ [MJ/m^3], $Q_{eu} = 964$ [MJ/m^3] has been solved. In Figs. 2 and 3 the temperature field for $t = 90$ s

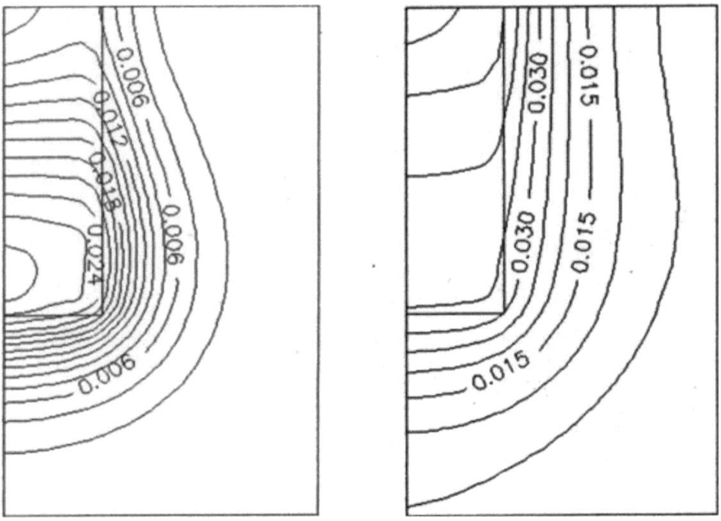

Fig. 5 Distribution of sensitivity function $\partial T/\partial Q_{eu}$ (t = 90 s and t = 180 s)

Fig. 6 Cooling curves

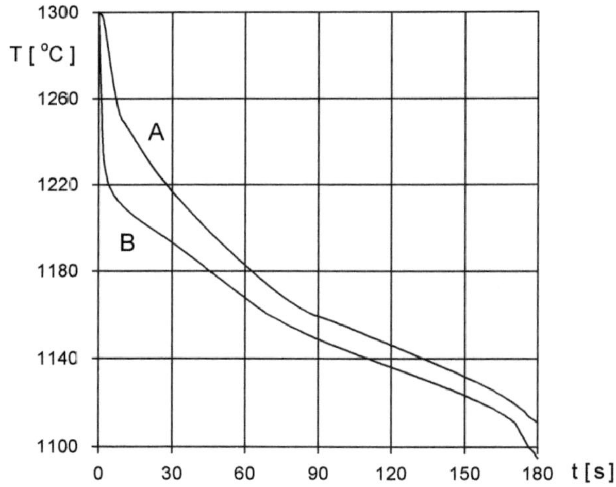

and $t = 180$ s in the casting and mould sub-domains is shown. Figs. 4 and 5 illustrate the distributions of sensitivity functions for $t = 90$ s and $t = 180$ s.

Next, the inverse problem has been considered. The problem of optimum sensors location has been solved under the assumption that $Q_{aus}^{\ 0} = 900$ [MJ/m^3], $Q_{eu}^{\ 0} = 1000$ [MJ/m^3]. The application of optimization procedure showed that the best sensors position corresponds to the nodes from casting domain marked by A and B in Fig. 1.

Figure 6 illustrates the cooling curves at the points A, B obtained for the real values of Q_{aus}, Q_{eu}($Q_{aus} = 923$ [MJ/m^3] $Q_{eu} = 994$ [MJ/m^3]). Using these curves

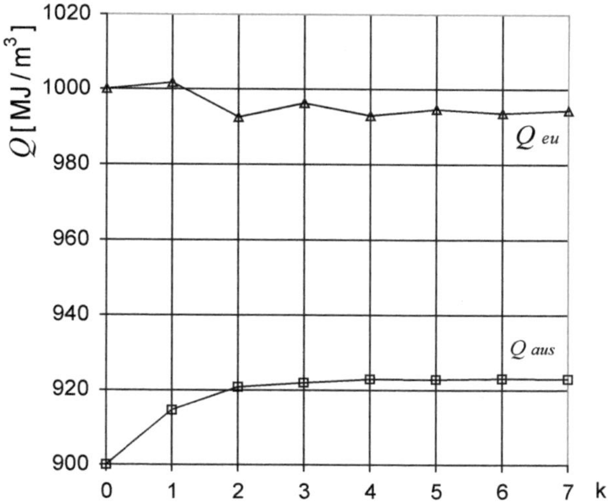

Fig. 7 Inverse problem solution

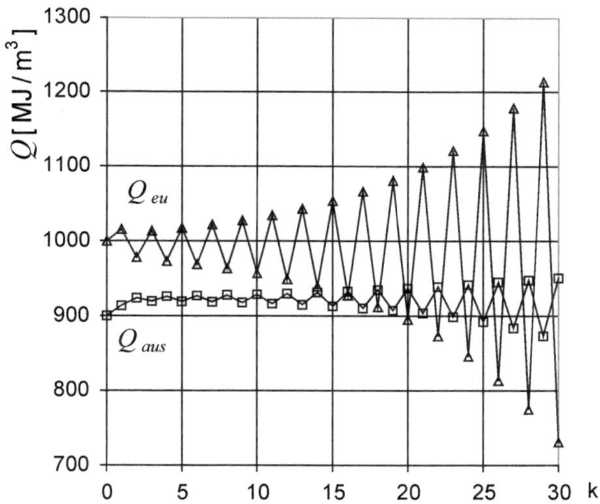

Fig. 8 Solution using points C and D

the inverse problem has been solved. The successive iterations of $Q_{aus}{}^k$, $Q_{eu}{}^k$ are shown in Fig. 7. It is visible that the iteration process is quickly convergent and the identified latent heats correspond to the previously assumed values.

The proper choice of sensors location seems to be very essential because it assures the effective and exact solution of identification problem. The good confirmation of this fact is the situation shown in Fig. 8. In this Figure the successive steps of iteration process (Eq. (19)) are marked. One can see that the solution is not convergent and the identification of unknown parameters is impossible. The results shown in Fig. 8 have been obtained for two randomly selected points (sensors) from casting domain (C and D in Fig. 1).

For the others positions of sensors location one can obtain the good values of searched parameters but the number of iterations will be probably greater than for optimal sensors position.

Acknowledgments This work was funded by Grant No N N507 3592 33.

References

1. Mochnacki B, Suchy JS (1995) Numerical methods in computations of foundry processes. PFTA, Cracow
2. Majchrzak E, Mendakiewicz J (2006) Identification of cast iron substitute thermal capacity. Arch Foundry 6(22):310–315
3. Patan M, Uciński D (2002) Optimal location of sensors for parameter estimation of static distributed systems. In: Wyrzykowski R et al (ed) PPAM 2001. LNCS, vol 2328. Springer, Berlin, pp 729–737
4. Dowding KJ, Blackwell BF (1999) Blackwell, Design experiments to estimate temperature dependent thermal properties, Inverse Problems in Engineering: Theory and Practice. ASME, U S, pp 509–518
5. Majchrzak E, Mochnacki B (2007) Identification of thermal properties of the system casting—mould. Mater Sci Forum 539–543:2491–2496
6. Mochnacki B, Majchrzak E, Szopa R, Suchy JS (2006) Inverse problems in the thermal theory of foundry, 5th edn, vol 1. Scientific Research of the Institute of Mathematics and Computer Science, Czestochowa University of Technology, pp 154–179
7. Kleiber M (1997) Parameter sensitivity. Wiley, Chichester
8. Mendakiewicz J (2008) Application of sensitivity analysis in experiments design, 7th edn, vol 1. Scientific Research of the Institute of Mathematics and Computer Science, Czestochowa University of Technology, pp 141–148

Mathematical Modeling of Heat and Mass Transfer in Domain of Solidifying Alloy

Bohdan Mochnacki and Ewa Majchrzak

Abstract In the paper the mathematical model, numerical algorithm and example of cylindrical casting solidification are presented. In particular the casting made from Cu-Zn alloy is considered. It is assumed that the temperature corresponding to the beginning of solidification is time-dependent and it is a function of temporary alloy component concentration. The course of macrosegregation has been modeled using the mass balances in the set of control volumes resulting from a domain discretization. The balances have been constructed in different ways, in particular under the assumption of instant equalization of alloy chemical constitution (a lever arm rule), next the Scheil model (e.g. Sczygiol 2000, Publ Czest Univ Techn Monographs, 71) has been used and finally the broken line model (Curran et al. 1980, Appl Math Modelling, 4, 398–400) has been taken into account. On a stage of numerical algorithm construction the boundary element method has been used in the variant called BEM using discretization in time (Curran et al. 1980, Appl Math Modelling, 4, 398–400; Sichert 1989, Technischen Fakultat der Universitat Erlangen; Szopa 1999, Publ. of the Silesian Univ. of Techn, 54) supplemented by the alternating phase truncation procedure (Majchrzak and B.Mochnacki 1995, Engineering Analysis with Boundary Elements, 16, 99–121; Lara 2003, Application of generalized FDM in numerical modelling of moving boundary problems, Doctoral Theses, Czestochowa).

Keywords Mathematical modeling of solidification process · Numerical methods

1 Governing Equations

In a casting domain, two changing with time sub-domains are distinguished. They correspond to liquid and solid phases. The moving boundary is identified by a temporary position of liquidus temperature $T^*(z_L)$, where z_L is a temporary concentration

B. Mochnacki (✉)
Higher School of Labour Safety Manasgement, Bankowa 8, 40-007 Katowice, Poland
e-mail: bohdan.mochnacki@im.pcz.pl

E. Majchrzak
Silesian University of Technology, Akademicka 2a, 44-100 Gliwice, Poland
e-mail: ewa.majchrzak@polsl.pl

N. M. Fonseca Ferreira, J. A. Tenreiro Machado (eds.), *Mathematical Methods in Engineering*, DOI 10.1007/978-94-007-7183-3_11,

of alloy component of liquid state near a border surface (in a case of lever arm and Scheil models z_L corresponds to concentration in the whole liquid part of casting domain). In the model proposed a presence of mushy zone is neglected and in a place of T^*one can introduce the so-called equivalent solidification point [7].

A transient temperature field in domain considered (taking into account the cylindrical geometry of casting—1D task) is determined by the following system of partial differential equations

$$\begin{aligned} c_L \rho_L \frac{\partial T_L(r,t)}{\partial t} &= \frac{\lambda_L}{r} \frac{\partial}{\partial r}\left[r \frac{\partial T_L(r,t)}{\partial r} \right], \\ c_S \rho_S \frac{\partial T_S(r,t)}{\partial t} &= \frac{\lambda_S}{r} \frac{\partial}{\partial r}\left[r \frac{\partial T_S(r,t)}{\partial r} \right] \end{aligned} \tag{1}$$

In Eq. (1) c, ρ, λ denote the specific heats, mass densities and thermal conductivities, T, r, t—are the temperature, geometrical co-ordinates and time.

On a border surface the Stefan condition is given:

$$r = \eta(t): \begin{cases} \lambda_S \dfrac{\partial T_S(r,t)}{\partial r} - \lambda_L \dfrac{\partial T_L(r,t)}{\partial r} = L_V \dfrac{\mathrm{d} r}{\mathrm{d} t} \\ T_S(r,t) = T_L(r,t) = T^*(z_L) \end{cases} \tag{2}$$

where L_V is a volumetric latent heat.

On an external surface the following continuity condition is assumed

$$r = R: \quad q(r,t) = \alpha\,[T(r,t) - T_a] \tag{3}$$

where α is a substitute heat transfer coefficient, T_a is an ambient temperature. At the moment $t=0$: $T_L(r,0) = T_0$, $z_L(r,0) = z_0$, at the same time T_0 is the pouring temperature, z_0—initial concentration of alloy component.

The algorithm of numerical simulation bases on the alternating phase truncation procedure. This approach requires the application of enthalpy approach on a stage of governing equations construction. So, we introduce the following definition of physical enthalpy

$$H(T) = \int_0^T c(\mu)\rho(\mu)\,\mathrm{d}\mu + L_V u(T) \tag{4}$$

where

$$u(T) = \begin{cases} 0 & T < T^*(z_L) \\ 1 & T \ge T^*(z_L) \end{cases} \tag{5}$$

The course of enthalpy function is shown in Fig. 1.

Fig. 1 Enthalpy diagram

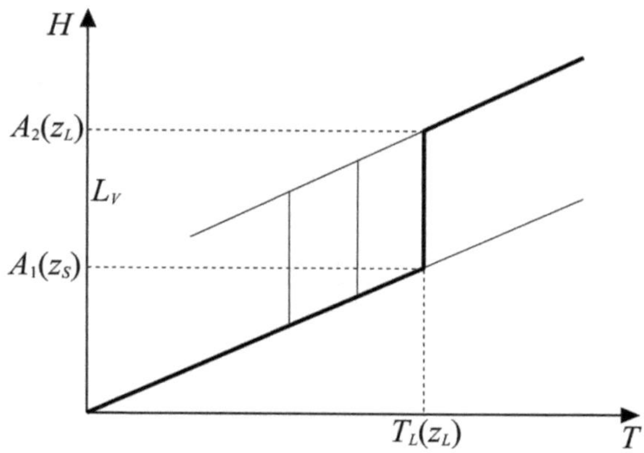

The system of Eq. (1) written using the enthalpy convention takes a form

$$\begin{aligned}\frac{\partial H_L(r,t)}{\partial t} &= \frac{a_L}{r}\frac{\partial}{\partial r}\left[r\frac{\partial H_L(r,t)}{\partial r}\right],\\ \frac{\partial H_S(r,t)}{\partial t} &= \frac{a_S}{r}\frac{\partial}{\partial r}\left[r\frac{\partial H_S(r,t)}{\partial r}\right]\end{aligned} \tag{6}$$

where a_L and a_S are the heat diffusion coefficients ($a = \lambda/c\rho$).

The Stefan boundary condition can be written as follows

$$r = \eta : \begin{cases} a_S\dfrac{\partial H_S(r,t)}{\partial r} - a_L\dfrac{\partial H_L(r,t)}{\partial r} = L_V\dfrac{\mathrm{d}r}{\mathrm{d}t} \\ A_L\,(z_L) = A_S\,(z_S) + L_V \end{cases} \tag{7}$$

where A_L and A_S are the right hand side and left hand side limits of enthalpy at the point T^* (see: Fig. 1).

The Robin boundary condition is of the form

$$r = R: \quad q(r,t) = \beta\,[H(r,t) - H_a] \tag{8}$$

where $\beta = a/c\rho$ is a substitute heat transfer coefficient written using the enthalpy convention, H_a is the enthalpy corresponding to T_a. The initial condition is also given:

$$t = 0: H\,(r,\,0) = H_0.$$

The adequate fragment of equilibrium diagram [8] of alloy considered (Zn < 30 %) one can approximate by two sectors starting from the same point. In a such situation the partition coefficient $k = \mathrm{const}$ and $z_S = k \cdot z_L$. The formula determining the liquidus line is of the form $T^* = T_m + m \cdot z_L$, where T_m is a solidification point of pure metal (Cu), while m—is a slope of straight line.

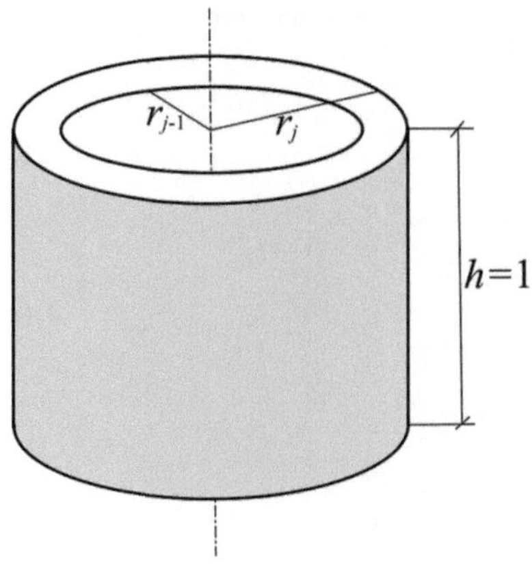

Fig. 2 Control volume V_j

2 Mass Balance Under the Assumption of Lever Arm Model

The mass balance of component alloy in domain of casting can be written in the form

$$m_0\, z_0 = m_S(t) z_S(t) + m_L(t) z_L(t) \tag{9}$$

where m_0 denotes a mass of component.

The domain considered is divided into control volumes (cylindrical rings) which altitude can be assumed in optional way (e.g. $h = 1$). Internal radius of element V_j is denoted by r_{j-1}, while an external one by r_j—Fig. 2.

Solid state fraction in volume V_j at time t equals $S_j\,(t)$. A mass of metal in solid and liquid state results from equations

$$m_{Sj} = S_j(t) V_j \rho_S, \quad m_{Lj} = \left[1 - S_j(t)\right] V_j \rho_L \tag{10}$$

Now, the time grid should be introduced

$$0 = t^0 < t^1 < \ldots < t^f < t^{f+1} < \ldots < t^F < \infty, \quad \Delta t = t^{f+1} - t^f \tag{11}$$

The local values of S_j result from the numerical model of solidification and they are defined in the following way

$$S_j\left(t^{f+1}\right) = \frac{A_L - H\left(r_j, t^{f+1}\right)}{A_L - A_S}, \quad A_S < H\left(r_j, t^{f+1}\right) < A_L \tag{12}$$

while for the others enthalpy values the function S_j equals 0 and 1, correspondingly.

Mass balance (9) written for time t^{f+1} leads to the equation

$$z_L\left(t^{f+1}\right) = \frac{m_0 z_0}{k m_S\left(t^{f+1}\right) + m_L\left(t^{f+1}\right)} \tag{13}$$

In the last equation the definition of partition coefficient k has been introduced. Finally

$$z_L\left(t^{f+1}\right)=\frac{R^2\rho_L z_0}{k\sum\limits_{j=1}^{n}V_j\rho_S S_j\left(t^{f+1}\right)+\sum\limits_{j=1}^{n}V_j\rho_L\left[1-S_j\left(t^{f+1}\right)\right]} \tag{14}$$

A temporary value of alloy component concentration determines a new value of solidification point T^* and border values A_L and A_S.

3 Mass Balance under the Assumption of Scheil Model

The Scheil model results from the assumption of limiting form of macrosegregation model determining the mass diffusion in a casting domain. Because the diffusion coefficient for solid state is essentially less than the same coefficient for molten metal and, from the other hand, the convection proceeding in a molten metal causes the equalization of chemical constitution in this domain, therefore it is assumed that $D_S=0$, while $D_L\rightarrow\infty$ (D is a diffusion coefficient). So, the mass balance resulting from Scheil's assumptions takes a form

$$\begin{aligned} m_0\, z_0 =& m_S\left(t^1\right)z_S\left(t^1\right)+m_S\left(t^2\right)z_S\left(t^2\right)+\ldots+\\ &+m_S\left(t^f\right)z_S\left(t^f\right)+m_S\left(t^{f+1}\right)z_S\left(t^{f+1}\right)+m_L\left(t^{f+1}\right)z_L\left(t^{f+1}\right) \end{aligned} \tag{15}$$

or

$$z_L\left(t^{f+1}\right)=\frac{m_0\, z_0-\left[m_S\left(t^1\right)z_S\left(t^1\right)+m_S\left(t^2\right)z_S\left(t^2\right)+\ldots+m_S\left(t^f\right)z_S\left(t^f\right)\right]}{k\, m_S\left(t^{f+1}\right)+m_L\left(t^{f+1}\right)} \tag{16}$$

After mathematical manipulations one obtains

$$z_L\left(t^{f+1}\right)=\frac{R^2\rho_L\, z_0-\sum\limits_{p=1}^{f}\sum\limits_{j=1}^{n}V_j\rho_S\, z_S\left(t^p\right)\left(S_j^p-S_j^{p-1}\right)}{k\sum\limits_{j=1}^{n}V_j\,\rho_S\left(S_j^{f+1}-S_j^f\right)+\sum\limits_{j=1}^{n}V_j\,\rho_L\left[1-S_j^{f+1}\right]} \tag{17}$$

where $S_j^p=S_j\left(t^p\right)$ etc.

Similarly, as in a case of previous model, the calculated value of $z_L(t^{f+1})$ determines a temporary temperature T^* and the border values A_L and A_S.

4 Broken Line Model

Macrosegregation process proceeding in the cylindrical casting domain is described by the system of diffusion equations in the form

$$\begin{aligned} P(r) \in \Omega_L &: \frac{\partial z_L(r,t)}{\partial t} = \frac{1}{r}\frac{\partial}{\partial r}\left(D_L r \frac{\partial z_L(r,t)}{\partial r}\right) \\ P(r) \in \Omega_S &: \frac{\partial z_S(r,t)}{\partial t} = \frac{1}{r}\frac{\partial}{\partial r}\left(D_S r \frac{\partial z_S(r,t)}{\partial r}\right) \end{aligned} \tag{18}$$

where z_L, z_S are the concentrations of alloy component for liquid and solid state sub-domains, D_L, D_S are the diffusion coefficients, r, t denote spatial co-ordinates and time. It is assumed that the diffusion coefficients of liquid and solid sub-domains are the constant values.

On the moving boundary between liquid and solid sub-domains the condition resulting from the mass balance is given [1, 5, 7]

$$r = \eta: \quad D_L \frac{\partial z_L(r,t)}{\partial r} - D_S \frac{\partial z_S(r,t)}{\partial r} = \frac{\mathrm{d}r}{\mathrm{d}t}\left[z_L(r,t) - z_S(r,t)\right] \tag{19}$$

Introducing the partition coefficient k one obtains the other form of condition (19)

$$r = \eta: \quad D_L \frac{\partial z_L(r,t)}{\partial r} - D_S \frac{\partial z_S(r,t)}{\partial r} = (1-k)\, z_L(r,t) \frac{dr}{dt} \tag{20}$$

If the mass transfer in the solid body is neglected ($D_S = 0$)

$$r = \eta: \quad D_L \frac{\partial z_L(r,t)}{\partial r} = (1-k) v z_L \tag{21}$$

where $v = \mathrm{d}r/\mathrm{d}t$ denotes the solidification rate.

On the outer surface of the system the no-flux condition should be assumed

$$r \in \Gamma_0: \quad \frac{\partial z_S(r,t)}{\partial r} = 0 \tag{22}$$

For time $t = 0$: $z_L(r, 0) = z_0$.

The idea of broken line model is the following. The concentration field in molten metal is assumed in the form of broken line. In particular the first segment corresponds to a certain layer δ while the second one to the other part of liquid state. The parameters of above distribution result from condition (20) and mass balance. The concentration in solid body results from partition coefficient k (Fig. 3). The details of this approach can be found in [9].

Fig. 3 The broken line model

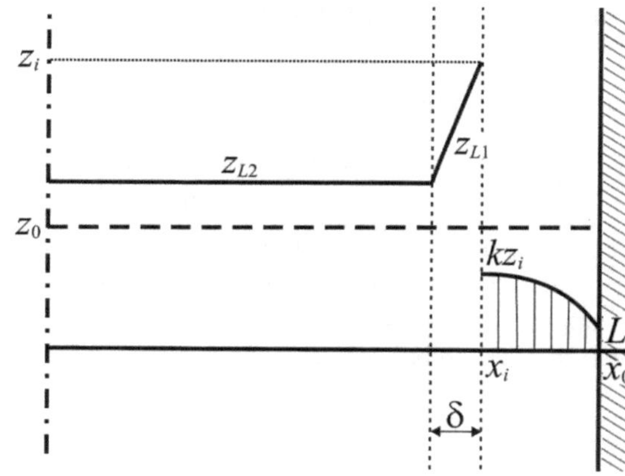

5 Alternating Phase Truncation Method

In this paper the classical variant of APTM presented by Rogers, Ciment and Berger (e.g. [6]) is used. Generalized form of the method can be found, among others, in [6, 7]. The algorithm of numerical solution of problem discussed (Eqs. (6), (7) and (8)), this means the computations concerning the transition from time t^f to time t^{f+1} is the following. Let us denote by H_j^f the discrete set of enthalpy values in the casting domain for timet^f at points r_j. In the first stage of computations the casting domain is conventionally treated as a liquid one. At the points r_j for which enthalpy H_j^f is less than A_L one assumes the local value of enthalpy equal to A_L, while for the others nodes the local value of enthalpy is invariable. So, the real enthalpy distribution is substituted by the following one

$$V_1(r_j,\ t^f) = \max\left\{H_j^f,\ A_L\,(z_L)\right\} \tag{23}$$

For homogeneous (molten metal) casting domain the enthalpy field for time t^{f+1} is calculated (using the optional numerical method). The solution obtained we denote as $V_1'(r_j, t^{f+1})$ (parameter a in Eq. (6) corresponds to a_L). The first stage of algorithm goes to the end by substraction of previously added enthalpy, this means

$$V_1\left(r_j,\ t^{f+1}\right) = V_1'\left(r_j,\ t^{f+1}\right) + H_j^f - V_1\left(r_j, t^f\right) \tag{24}$$

The second stage of computations concerning the transition $t^f \rightarrow t^{f+1}$, starts from the homogenization of casting domain to the solid state, in other words

$$V_2\left(r_j,\ t^f\right) = \min\left\{A_S\,(z_L), V_1\left(r_j,\ t^{f+1}\right)\right\} \tag{25}$$

The enthalpy field $V_2(r_j,\ t^f)$ is again calculated ($a = a_S$). The final solution concerning the time t^{f+1} results from the formula

$$H_j^{f+1} = V_2'\left(r_j,\ t^{f+1}\right) + V_1\left(r_j,\ t^{f+1}\right) - V_2\left(r_j, t^f\right) \tag{26}$$

6 Boundary Element Method

The numerical solution of equation

$$\frac{\partial H(r,t)}{\partial t} = a\frac{\partial^2 H(r,t)}{\partial r^2} + \frac{a}{r}\frac{\partial H(r,t)}{\partial r} \tag{27}$$

has been found using the boundary element method. Because the BEM algorithm for the objects oriented in cylindrical co-ordinate system is very complicated, the simpler approach is proposed.

Equation (22) can be written in the form

$$\frac{\partial H(r,t)}{\partial t} = a\frac{\partial^2 H(r,t)}{\partial r^2} + Q \tag{28}$$

where Q is the artificial source function and

$$Q(r,t) = \frac{a}{r}\frac{\partial H(r,t)}{\partial r} \tag{29}$$

In this way one obtains the energy equation corresponding to the objects oriented in cartesian co-ordinate system for which the BEM algorithm is simple and effective on a stage of numerical simulation.

In the case of variant called the BEM using discretization in time, the derivative $\partial H/\partial t$ for transition $t^f \to t^{f+1}$ is substituted by differential quotient and the Eq. (23) takes a form

$$\frac{H\left(r,t^{f+1}\right) - H\left(r,t^f\right)}{\Delta t} = a\frac{\partial^2 H(r,t)}{\partial r^2} + Q \tag{30}$$

A basic BEM equation for the problem (24) results from the weighted residual method application and then one obtains

$$\int_0^R \left[\frac{\partial^2 H\left(r,t^{f+1}\right)}{\partial r^2} - \frac{1}{a\Delta t}H\left(r,t^{f+1}\right) + \frac{1}{a\Delta t}H\left(r,t^f\right) + \frac{Q}{a}\right] H^*(\xi,r)\,\mathrm{d}r = 0 \tag{31}$$

where $H^*(\xi,r)$, $\xi \in (0,R)$ is the fundamental solution. In the case considered it is a function of the form [3, 5]

$$H^*(\xi,r) = \frac{\sqrt{a\Delta t}}{2}\exp\left(-\frac{|r-\xi|}{\sqrt{a\Delta t}}\right) \tag{32}$$

After mathematical manipulations, the Eq. (28) takes a form

$$H\left(\xi,t^{f+1}\right) + \left[\frac{1}{a}H^*(\xi,r)\,q\left(r,t^{f+1}\right)\right]_0^R = \frac{1}{a}\left[q^*(\xi,r)H\left(r,t^{f+1}\right)\right]_0^R + p(\xi) + z(\xi) \tag{33}$$

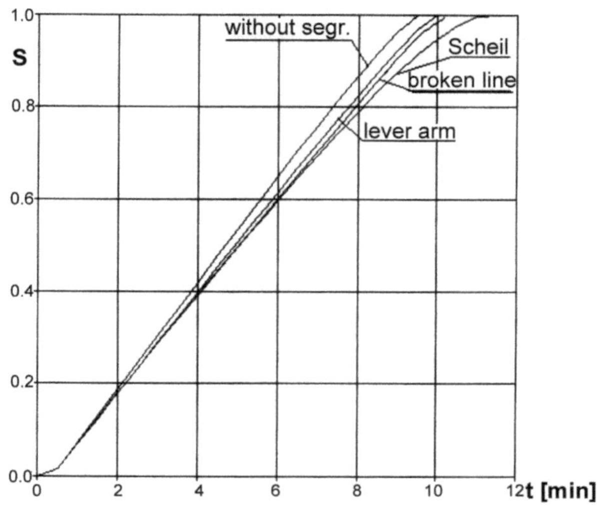

Fig. 4 Kinetics of solidification

where $q^*(\xi, r) = -a\partial H^*/\partial r$, while

$$p(\xi) = \frac{1}{a\Delta t}\int_0^R H^*(\xi, r)\, H\left(r, t^f\right) \mathrm{d}\, r \tag{34}$$

and

$$z(\xi) = \frac{1}{a}\int_0^R QH^*(\xi, r)\, \mathrm{d}\, r \tag{35}$$

For $\xi \to 0^+$ i $\xi \to L^-$ one has

$$\begin{bmatrix} g_{11} & g_{12} \\ g_{21} & g_{22} \end{bmatrix} \begin{bmatrix} q\left(0,\, t^{f+1}\right) \\ q\left(R,\, t^{f+1}\right) \end{bmatrix} = \begin{bmatrix} h_{11} & h_{12} \\ h_{21} & h_{22} \end{bmatrix} \begin{bmatrix} H\left(0,\, t^{f+1}\right) \\ H\left(R,\, t^{f+1}\right) \end{bmatrix} + \begin{bmatrix} p(0) \\ p(R) \end{bmatrix} + \begin{bmatrix} z(0) \\ z(R) \end{bmatrix} \tag{36}$$

Values of matrices **g**, **h**, **p**, **z** coefficients result from Eq. (33).

The capacity of artificial heat source can be found by substitution of $\partial H/\partial r$ by the adequate differential quotient [8].

7 Example of Computations

We consider the solidification of cylindrical casting ($R = 8$ cm) made from Cu-Zn alloy (10 % Zn). The following thermophysical parameters have been assumed $\lambda_L = \lambda_S = \lambda = 0.12$ kW/mK, $c_L = c_S = c = 3354$ kJ/m^{3}K, $\rho_L = \rho_S = \rho = 8600$kg/m^3,

Fig. 5 Cooling curves (axis and $r = 2\,\text{cm}$)

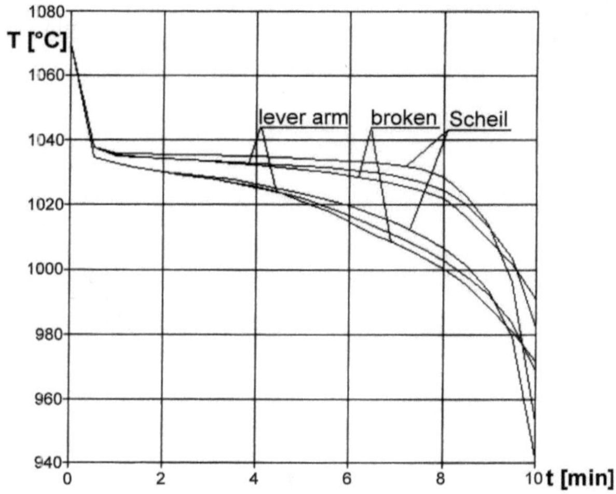

$L_V = 1.634 \cdot 10^6 \text{kJ/m}^3$, $k = 0.855$, $T^* = 1083 - 473.68 \cdot z_L$, $D_L = 3.5 \cdot 10^{-8} \text{m}^2/\text{s}$, $\delta = 1.5$ mm, initial temperature 1070°C. On the outer surface the Robin condition has been taken into account ($\alpha = 40 \text{W/m}^2\text{K}$, $T_a = 30$°C).

In Figs. 4 and 5 the kinetics of casting solidification is shown, at the same time the different models of macrosegregation have been considered. The next Figure shows the cooling curves at the points from casting domain

8 Final Remarks

The differences between solutions are non-drastic, but visible. It seems, that the numerical algorithm of solidification supplemented by the simple procedures taking into account the changes of alloy chemical composition are closer to the real physical conditions of the process and can be used on a stage of process modelling.

References

1. Sczygiol N (2000) Numerical modelling of thermo-mechanical phenomena in a solidifying casting and a mould, Series Monographs, No 71 (in Polish). Publishing House Czestochowa University of Technology, Czestochowa
2. Suchy JS, Mochnacki B (2003) Analysis of segregation process using the broken line model. Theor base Arch Foundry 3(10):229–234
3. Curran DAS, Cross M, Lewis BA (1980) Solution of parabolic differential equations by the BEM using discretization in time. Appl Math Modelling 4:398–400
4. Sichert W (1989) Berechnung von Instationaren thermischen Problemen mittels der Randelementmethode. Doctoral Theses, Technischen Fakultat der Universitat Erlangen

5. Szopa R (1999) Modelling of solidification and crystallization using combined boundary element method. Publ Silesian Univ Techn 54:1–177. (Gliwice)
6. Majchrzak E, Mochnacki B (1995) Application of the BEM in the thermal theory of foundry. Eng Anal Boundary Elem 16:99–121
7. Lara S (2003) Application of generalized FDM in numerical modelling of moving boundary problems. Doctoral Theses, Czestochowa. (in Polish)
8. Engineer's handbook (1988) Foundry engineering. WNT, Warsaw (in Polish)
9. Mochnacki B, Lara S, Pawlak E (2004) Analysis of segregation process using the broken line model for an infinite cylinder. Sci Res Inst Math Comput Sci 1(3):153–160

Total Variation Approach to Density Reconstruction from X-Ray Radiograph Tomography

Suhua Wei and Guiping Zhao

Abstract For cone beam x-ray radiographic tomography, density reconstruction for 1-dimensional objects can be performed by Abel transform inversion. For 2-dimensional cylindrical objects, we propose to divide the object into small blocks, in each block the density is viewed as a constant. The projection operator corresponds to a matrix. Density reconstruction leads to solve a linear algebraic equation system. To deal with its ill conditioning, we use Total Variation regularization. Numerical experiments show that TV regularization gives correct recovery of object edges, while the density contrast may be changed in some smooth parts.

Keywords Total variation regularization · Image reconstruction · Abel transform · Tomography

1 Introduction

The density reconstruction of objects from a few of radiographic views is an important tomography problem. Because of the complexity of x-ray-generation instrument, a single radiography view is widely used to detect the density of axially symmetric objects. For 1-dimensional object, typical works are done by T. J. Asaki [1, 2] and K. M. Hanson [3–6]. They approached regularization method for Abel transform inversion and Bayesian theory. For 2-dimensional objects with general cylindrical symmetry, if the projection view is not generated by parallel x-ray beam, the tomography can not be viewed as Abel transform inversion. In the paper, we consider density reconstruction for 2-dimensional case with a single projection view generated by cone beam x-ray. Since radiographs are transmission intensity maps, the signal is attenuated by the object, if we neglect the scatter, then the logarithm of the intensity of the received signal is proportional to the integral of the absorption alone the beam path. From the mathematical point of view, we can assume that function $\rho(\mathrm{r},\ \mathrm{z})$ has known integral value on the projection plane along each beam path. Our goal is to reconstruct $\rho(\mathrm{r},\ \mathrm{z})$. Our method is based on the discretization of the object and the

S. Wei (✉) · G. Zhao
Institute of Applied Physics and Computational Mathematics, Beijing, China
e-mail: wei_suhua@iapcm.ac.cn

N. M. Fonseca Ferreira, J. A. Tenreiro Machado (eds.), *Mathematical Methods in Engineering,* DOI 10.1007/978-94-007-7183-3_12,

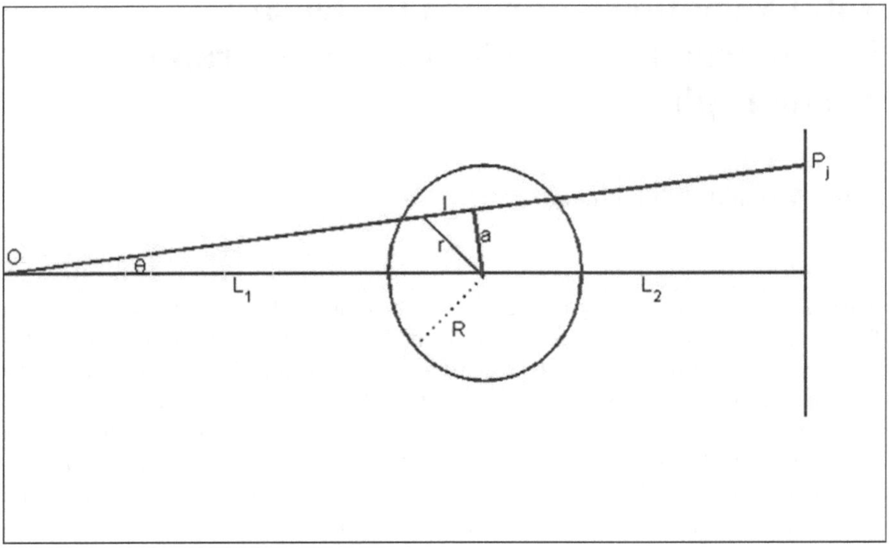

Fig. 1 High energy X-ray radiography schematic diagram

integral. After discretization, we need to solve a linear algebraic equation system. Because of the ill conditioning of the system, regularization is adopted. We compared Total Variation regularization with H^1 semi-norm regularization. TV has the advantage to reconstruct discontinuous functions, especially good for the restoration of edges. This is important for distinguishing materials with different densities. Our method is tested by 1- and 2-dimensional examples. The rest of the paper is organized as follows. In Sect. 2 we address our research background. In Sect. 3 we derive TV based density reconstruction model. In Sect. 4 we give numerical examples for 1-dimensional and 2-dimensional cases. Section 5 is our conclusion

2 Problem Description

In the case of this paper concerns, density reconstruction problem can be illustrated as in Fig. 1. The cone beam x-ray emits from source point which is indicated by letter "O". X-ray energy is attenuated after illuminating the object. In the recording plane we get a digitized image. For example, P_j is one pixel of the image. The pixel value is proportional to the integral of density function along x-ray path from O to P_j if we neglect the physics of interaction between the object and detectors with the radiation. Since function ρ represents volumetric density, then $\int \rho\,(r)^{dl}$ is areal density. Areal density can be known from the measured data of the digitized image. Our goal is to calculate ρ from $\int \rho^{\,dl}$.

Suppose the unknown object has limited volume and then density function has finite support, that is, the function ρ(r) is zero outside some radius R. In Fig. 1, letter “a” represents the distance from object centre to path line. “l”, “r” and “a” has the relationship $r^2 = l^2 + a^2$. Then $\int \rho(r)^{dl}$ can be rewritten as the Abel transform

$$\int \rho(r)\mathrm{d}1 = 2\int_{a}^{R} \frac{r\rho(r)}{\sqrt{r^2 - a^2}}\mathrm{d}r$$

The inversion of Abel transform has been discussed by Asaki in paper [1, 2]. We will focus on the 2-dimensional case. For simplicity, we address our method by starting from the discretization of object with 1-dimensional.

3 Total Variation Approach to Density Reconstruction

For 1-dimensional object with unknown density, we denote the density function as $\rho(r)$. r is the radius of the object. Suppose the function $\rho(r)$ has finite support, that is, $\rho(r) = 0$, when $r \geq R$.

We divide [0, R] by n intervals. Let $dr = \frac{R}{n}$, $r_i = i{*}dr$, $\rho_i = \rho(r_i)$, $i = 1, 2, \cdots, n$

In the projection plane every pixel of the image corresponds to an areal density value, all the values compose a matrix. It can be taken as a vector of m elements $b = (b_1, b_2, \cdots, b_m)^T$. The unknown density and the known areal density have the following relationship:

$$\int_{1_j} \rho(r)\mathrm{d}l = b_j, \quad j = 1, 2, \cdots, m. \tag{1}$$

Where 1_j denotes the trace line between source and the projection point corresponding to b_j.

To solve $\rho(r)$ from (1), we discretize every equation of the system and get a linear algebraic equation system:

$$A\rho = b \tag{2}$$

Where $\rho = (\rho_1, \rho_2, \cdots, \rho_n)^T$, $A = (a_{ij})$ is a m × n matrix. Every entry of A can be calculated according to the cross points of ray trace with blocks of object. For example, a_{ij} is equal to the length of ray l_j between sphere $r = r_{i-1}$ and $r = r_i$. Because of the ill-conditioning of matrix A and the noise contained in the measured data of right-hand side b, solving the least square problem $\min\|A\rho - b\|^2$ can not give the right answer of Eq. (2). Regularization is necessary to ensure that the small perturbation in right-hand side b will not cause high oscillation of numerical solution. In [7] Total Variation norm is proposed as regularization functional. TV does not penalize discontinuities in ρ and thus allows a better edge recovery. One purpose of density reconstruction is to get the edge position between different materials. Total Variation regularization is the proper choice for this goal. Therefore, density reconstruction problem can be written as:

$$\min\left\{\|A\rho - b\|_2^2 + \alpha\sum_{i=2}^{n}|\rho_i - \rho_{i-1}|\right\} \tag{3}$$

The minimization problem (3) can be solved by using fix-point iterative method [8].

For 2-dimensional axially symmetric objects, the density function has the form $\rho(r, z)$, where $r = \sqrt{x^2 + y^2}$. We discretize the object along r_1 and z_1, both of the two series have equal steps. Let $\rho_{i,j}$ denote the density value inside the block $r_{i-1} \leq r < r_i, z_{j-1} \leq z < z_j$. Suppose that $i = 1, 2, \cdots, n_1, j = 1, 2, \cdots, n_2$, the projection data is a matrix with size $m_1 \times m_2$. Similar to 1-dimensional case, we get the following equation

$$A\rho = b \tag{4}$$

Where A is a $m_1 \times m_2 \times n_1 \times n_2$ matrix. Its entry is equal to the length of ray trace inside blocks. b is a vector with size $m_1 \times m_2$. To solve Eq. (4), we use Total Variation regularization, the corresponding minimization problem is:

$$\min_{\rho} \left\{ \|A\rho - b\|_2^2 + \alpha \sum_{k=1}^{m_1 m_2 n_1 n_2} |\nabla \rho_k| \right\} \tag{5}$$

Where $|\nabla \rho_k|$ represents the gradient of function ρ at block k. Minimization problem (5) can be solved by Vogel and Oman's fast numerical method proposed in [9].

4 Numerical Examples

The first example is the reconstruction of an object of four nested density rings. Figure 2a is the exact density profile. The corresponding projection data (Fig. 2b) is a synthetically generated areal density with added Gaussian noise. Reconstructed density without using regularization is shown in Fig. 2c. Reconstruction with H^1 semi-norm regularization in Fig. 2d (The dotted line is our numerical solution, and the continuous line is the exact solution). Reconstruction with TV norm regularization in Fig. 2e. TV regularization can suppress noise and preserve density discontinuity.

In example 2, we tested density reconstruction for functions (6) to (8). First, we generate areal density by calculating integral values of these functions along cone beam x-ray traces. Secondly, we add Gaussian noise to the areal density. And then we use our method to reconstruct the density functions according to the synthetic areal density. Figure 3a, b and c are the numerical results of reconstructed density profiles. The corresponding exact solutions are functions (6) (7) (8). They are defined as follows.

The density function corresponding to Fig. 3a is (6).

$$\rho(r,z) = \begin{cases} 0.00129 & (r,z) \in C_1 \\ 18.25 & (r,z) \in C_2 - C_1 \\ 1.77 & (r,z) \in C_3 - C_2 \\ 2.64 & (r,z) \in C_4 - C_3 \\ 0 & (r,z) \notin C_4 \end{cases} \tag{6}$$

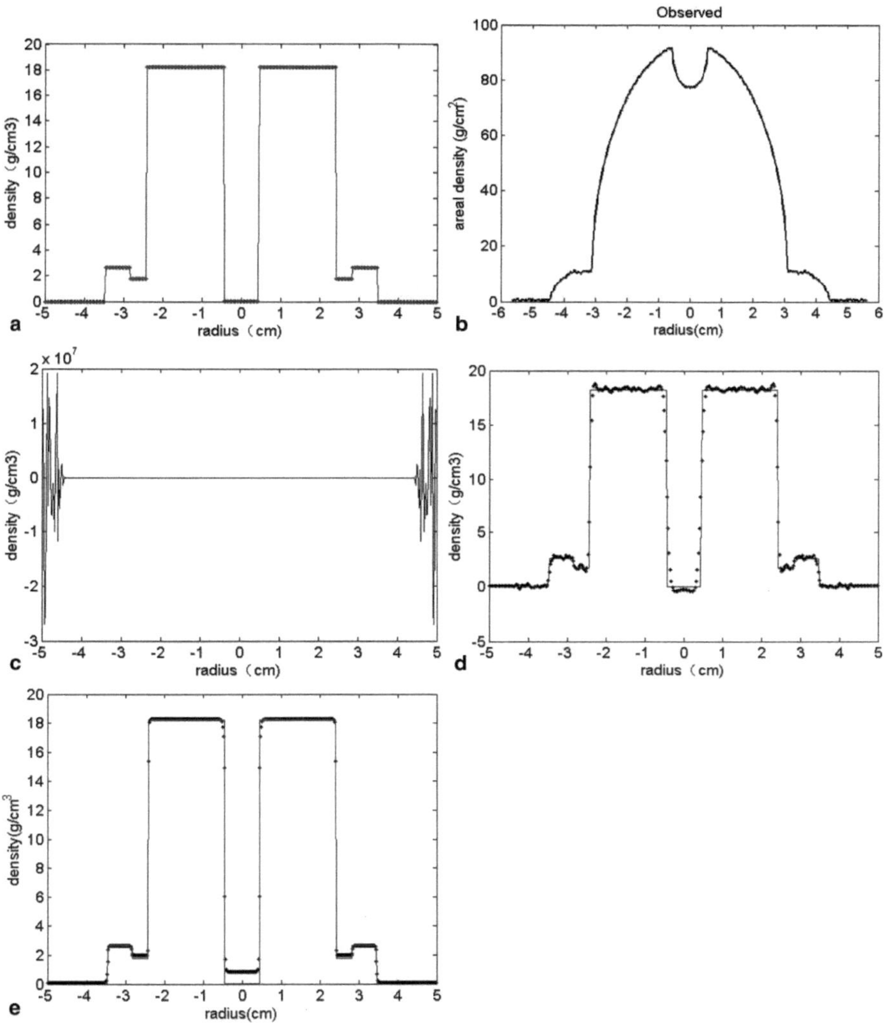

Fig. 2 **a** Exact solution. **b** Areal density. **c** Density reconstruction without regularization. **d** Density reconstruction with H^1 semi-norm regularization. **e** Density reconstruction with TV norm regularization

Where $C_1 = \{(r,z) : \frac{r^2}{0.5^2} + \frac{z^2}{1.0^2} \leq 1\}$, $C_2 = \{(r,z) : \frac{r^2}{2^2} + \frac{z^2}{3^2} \leq 1\}$, $C_3 = \{(r,z) : \frac{r^2}{2.3^2} + \frac{z^2}{3.3^2} \leq 1\}$, $C_4 = \{(r,z) : \frac{r^2}{2.8^2} + \frac{z^2}{3.8^2} \leq 1\}$.

The density function corresponding to Fig. 3b is (7).

$$\rho(r,z) = \begin{cases} 2.0 & (r,z) \in C_1 \\ 18.0 & (r,z) \in C_2 - C_1 \\ 0 & (r,z) \notin C_2 \end{cases} \tag{7}$$

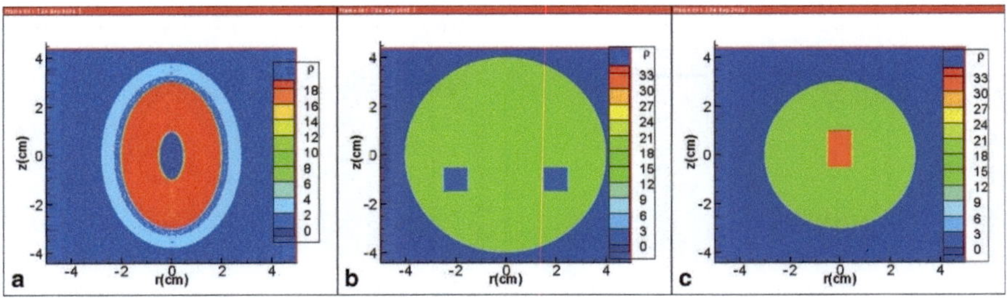

Fig. 3 Reconstructed density profile of axially symmetric objects. $r = 0$ is the synthetic axis

Where $c_1 = \{(r,z) : 1.5 \le r \le 2.5, -1.5 \le z \le -0.5\}, c_2 = \{(r, z) : r^2 + z^2 \le 4.0^2\}$.

The density function corresponding to Fig. 3c is (8).

$$\rho(r,z) = \begin{cases} 32.0 & (r,z) \in C_1 \\ 18.0 & (r,z) \in C_2 - C_1 \\ 0 & (r,z) \notin C_2 \end{cases} \quad (8)$$

Where $c_1 = \{(r,z) : 0 \le r \le 0.5, -0.5 \le z \le 1.0\}$, $c_2 = \{(r,z) : r^2 + z^2 \le 3.0^2\}$.

For different types of cylindrical objects, Total Variation regularization based density reconstruction gives correct recovery of object edges, while the density contrast may be changed in the process of recovery. See Fig. 2e. The recovered function is different the true solution in the region of r < 0.5.

5 Conclusions

The numerical method of density reconstruction from a single x-ray radiograph depends on how the beam is shaped. For parallel beam of x-ray, any cylindrical objects can be reconstructed by Abel transform inversion. For cone beam of x-ray, 1-dimensional case can be performed by Abel transform inversion, but two dimensional case can not be done like this. We propose to divide the object into small blocks, in each block the density is viewed as a constant. By exactly computing the trace length of x-ray inside each block we get a matrix, and the density reconstruction leads to solve a linear algebraic equation system. The ill-conditioning of the system can be overcome by adding TV regularization term. The obtained minimization problem can be solved using fix-point iterative method. Numerical experiments show that TV regularization gives correct recovery of object edges, while the density contrast may be changed in some smooth parts.

Acknowledgments This research is partially supported by NSFC (Project No.10971244).

References

1. Asaki TJ, Chartrand R, Vixie KR, Wohlberg B (2005) Abel inversion using total variation regularization: applications. LA-UR-05–2657. Los Alamos National Laboratory, USA
2. Asaki TJ, Chartrand R, Vixie KR, Wohlberg B (2005) Abel inversion using total variation regularization. Inverse Probl 21(6):1895–1903
3. Hanson KM, Cuningham GM, Jennings GR (1994) Tomographic reconstruction based on flexible geometric models. In: Proceedings of 1994 international conference on image processing, Austin, pp 145–147
4. Hanson KM (1989) A Bayesian approach to nonlinear inversion: Abel inversion from X-ray attenuation data, transport theory, invariant imbedding, and integral equations. Lecture notes in pure and applied mathematics, pp 363–378, edited by Nelson P et al, Marcel Dekker, New York
5. Hanson KM (1993) Special topics in test methodology: tomographic reconstruction of axially symmetric objects from a single dynamic radiograph. LA-UR-87-1670. Los Alamos National Laboratory, USA
6. Hanson KM (1984) Tomographic reconstruction of axially symmetric objects from a single radiograph. In: Procedings of the 16th international conference on high speed photography and photonics, Strasbourg
7. Rudin L, Osher S, Fatemi E (1992) Nonlinear total variation based noise removal algorithms. Physica D 60:259–268
8. Vogel CR, Oman ME (1996) Iterative methods for total variation denoising. SIAM J Sci Comput 17(1):227–238
9. Vogel CR, Oman ME (1995) Fast numerical methods for total variation minimization in image reconstruction. In: Luk FT (ed) Proceedings of SPIE 1995, advanced signal processing algorithms, vol 2563, San Diego

References

The Improvement of Total Variation Based Image Restoration Method and Its Application

Suhua Wei and Guiping Zhao

Abstract Total variation based image restoration method was first proposed by Rudin Osher and Fatermi in 1992. The images resulting from its application are usually piecewise constant, and have sometimes undesirable staircasing effect. To reduce this effect, we propose an improved model by combining the advantages of total variation and H^1 regularization. The new model substantially reduces the staircase effect, while preserving sharp edges. This model can be used in image reconstruction, it has advantages of keeping edges and recovering smooth region's value. We give 1D and 2D experimental results to show the efficiency of the proposed model.

Keywords Total variation regularization · Image restoration · Staircasing effect

1 Introduction

Image processing refers to the analysis and extraction of information from images, including restoration, compression and segmentation. Applications can be found in many areas like medical diagnosis, satellite surveying and computer techniques.

The aim of image restoration is to estimate the ideal true image from the recorded one. The direct problem is the computing of blurred image from a given image. The usual model for it is the convolution by a given kernel or point spread function. In many cases, the inverse problem of computing the true image from the observation is ill-posed. A general method to dealing with inverse problem is that of regularization. The choice of regularization will be essential for a satisfactory image restoration process. The solution of regularization based on least squares criteria is usually continuous, therefore, the image edges can not be well restored. To overcome this difficulty, a technique based on the minimization of total variation norm subject to some noise constraints is proposed by Rudin, Osher and Fatemi [1], that is, to seek solutions in BV space. The space of functions of bounded total variation plays an

S. Wei (✉) · G. Zhao
Institute of Applied Physics and Computational Mathematics, Beijing, China
e-mail: wei_suhua@iapcm.ac.cn

G. Zhao
e-mail: zhao_guiping@iapcm.ac.cn

N. M. Fonseca Ferreira, J. A. Tenreiro Machado (eds.), *Mathematical Methods in Engineering,* DOI 10.1007/978-94-007-7183-3_13,

important role when accurate estimation of discontinuities in solutions is required. The total variation (TV) denoising method preserves edges well, but has sometimes undesirable staircase effect, namely the transformation of smooth regions into piecewise constant regions (stairs), which implied that the finer details in the original image may not be recovered satisfactorily. To solve this problem, Chan, Marquina and Mulet [2] proposed an improved model, constructed by adding a nonlinear fourth order diffusive term to the Euler-Lagrange equations of the variational TV model. Marquina and Osher [3] preconditioned the right hand side of the parabolic equation with $|\nabla u|$ which had a staircase reducing effect. Another popular way to reduce staircasing is to introduce in some way higher order derivatives into the regularization term. Chambolle and Lions [4] do this by minimizing the inf-convolution of the TV norm and a second order functional. Instead of combing TV norm and second order derivatives within one regularization functional, Lysaker and Tai [5] use two regularization functionals. In [6], Blomgren, Chan and Mulet propose a "TV-H^1 interpolation" approach to address the staircase problem of the TV technique. The approach is performed by redefining the Total Variation functional $R(u)$ in view of the properties of TV-norm and H^1-seminorm. However, it is not completely clear how to choose a function Φ, which makes the regularizing functional $R(u)$ being convex. In this paper, we give a choice of function Φ, and the corresponding regularizing functional $R(u)$ verifies the sufficient conditions for convexity. This is mathematically desirable, for then the constrained optimization problem will have some kind of uniqueness.

The paper is organized as follows: in Sect. 2, we introduce the image restoration problem using the Total Variation norm as regularization functional. In Sect. 3, we describe the staircase effect caused by the TV model and briefly review some techniques proposed in literature to deal with it. In Sect. 4, we construct an improved regularizing functional to reduce the staircase effect. We then analysis our model and give its Euler-Lagrange equation as well as its discretization method. In Sect. 5,we give numerical examples to test the efficiency of our new model. The final part is our conclusion.

2 Total Variation Image Restoration

An image can be interpreted as either a real function defined on Ω, a bounded and open domain of R^2, or as a suitable discretization of this continuous image. Our aim is to restore an image which is contaminated with noise and blur. The restoration process includes the recovery of edges and smooth regions. Let us denote by z the observed image and u the real image. We assume that the degradation model is $Ku + n = z$, where K is a known linear blur operator, and n is a Gaussian white noise, i.e. the values n_i of n at pixels i are independent random variables, each with a Gaussian distribution of zero mean and variance σ^2. Our objective is to estimate u from given z. The inverse problem has many solutions and is ill-posed. If we impose a certain regularity condition on the solution u, then it may become well-posed [7].

In [1], it is proposed to use as regularization functional the so-called Total Variation norm or TV-norm:

$$TV(u) = \int_\Omega |\nabla u| dxdy = \int_\Omega \sqrt{u_x^2 + u_y^2} dxdy. \tag{1}$$

Since TV norm does not penalize discontinuities in u, thus we can recover the edges of the original image. The restoration problem can be written as:

$$\min_u \int_\Omega |\nabla u| dxdy, \tag{2a}$$

$$\text{subject to } ||Ku - z||_{L^2}^2 = |\Omega|\sigma^2. \tag{2b}$$

Using known techniques, the solution of problem (2) can be achieved by solving the equivalent unconstrained problem:

$$\min_u \int_\Omega \left(\alpha |\nabla u| + \frac{1}{2}(Ku - z)^2 \right) dxdy, \tag{3}$$

where α represents the tradeoff between smoothness and fidelity to the original data. Assuming homogeneous Neumann boundary conditions, the Euler-Lagrange equation of (3) is:

$$0 = -\alpha \nabla \cdot \left(\frac{\nabla u}{|\nabla u|} \right) + K^*(Ku - z). \tag{4}$$

The above Eq. (4) is not well defined at locations where $|\nabla u| = 0$, due to the presence of the term $1/|\nabla u|$. The common method to overcome this technical difficulty is to slightly perturb the total variation functional to become:

$$\int_\Omega \sqrt{|\nabla u|^2 + \beta} dxdy,$$

where β is a small positive number.

In [9] it is shown that the solutions of the perturbed problems

$$\min_u \int_\Omega \left(\alpha \sqrt{|\nabla u|^2 + \beta} + \frac{1}{2}(Ku - z)^2 \right) dxdy \tag{5}$$

converge to the solutions of (3) when $\beta \to 0$. The Euler-Lagrange equation of (5) is

$$0 = -\alpha \nabla \cdot \left(\frac{\nabla u}{\sqrt{|\nabla u|^2 + \beta}} \right) + K^*(Ku - z), \tag{6}$$

with homogeneous Neumann boundary conditions.

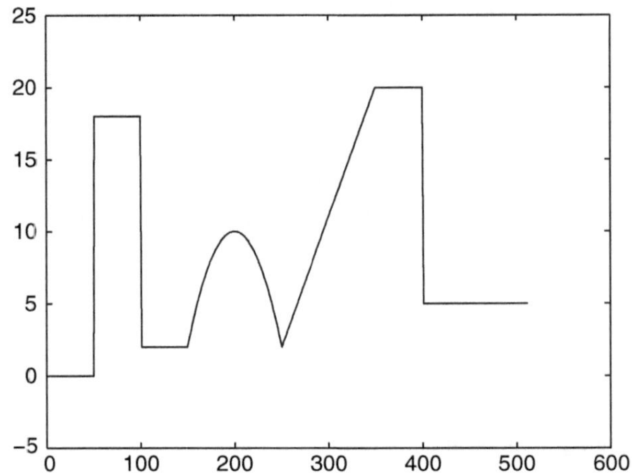

Fig. 1 Original image

3 The Staircase Effect

The image restoration model based on total variation regularization tends to yield piecewise constant images. This is 'staircasing effect'. Smooth regions in original image are recovered as piecewise smooth regions. In order to overcome this difficulty, some works focus on introducing higher order derivatives into the regularization term. Some starts from the parabolic equation and reform the right hand side of the equation to get reduced effect of staircasing. A popular approach to reducing staircasing is to combine the ability of TV denoising to preserve edges with the ability of H^1 to preserve smooth regions. Blomgren, Chan and Mulet [6] proposed to use as regularizing functionals the interpolation of TV-norm and H^1-seminorm, because staircase effect is partly due to the fact that the TV-norm is not biased against discontinuous nor continuous functions. On the other hand, the functional

$$H^1(u) = \int_\Omega |\nabla u|^2 dxdy,$$

has a strong bias against discontinuous functions.

Consider functionals of the type:

$$R(u) = \int_\Omega |\nabla u|^p dxdy, \tag{7}$$

where $p \in [1, 2]$. TV-norm and H^1 functionals can be obtained by Eq. (7) with $p = 1, 2$, respectively. In [6], numerical evidence show that sharp edges are obtained for $p = 1, 1.1$, and the staircase effect does exist. With the increasing of p, for instance $p = 1.5, 2$, those sharp edges are smeared, but the staircase effect is alleviated. In view of these results, the criterion of constructing regularization functionals should be that obtain TV behavior at sharp gradients (edges) and H^1 behavior away from

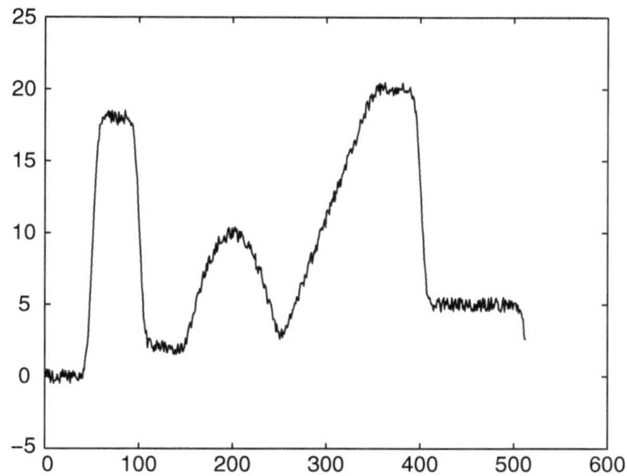

Fig. 2 Noisy and blurred image

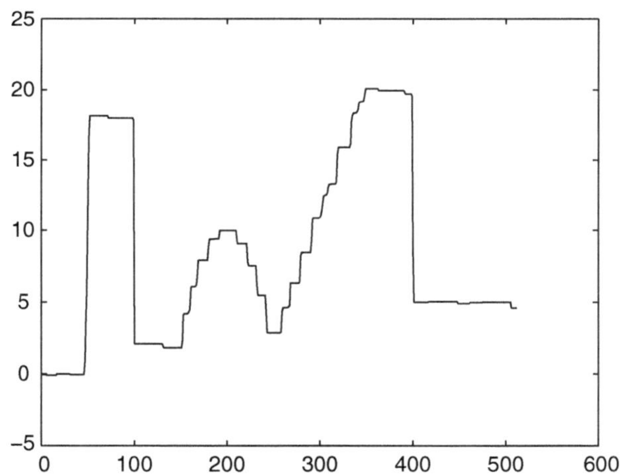

Fig. 3 Total Variation restoration

edges. The approach which is proposed by Blomgren, Chan and Mulet is to consider regularizing functionals of the type:

$$R(u) = \int_{\Omega} \Phi(|\nabla u|)\, dxdy, \tag{8}$$

$\Phi(|\nabla u|)$ could be a "convex combination" of x and x^2, with variable weight $\alpha(x) \in [0, 1]$:

$$\Phi(x) = \alpha(x)x + (1 - \alpha(x))\, x^2,$$

with $\alpha(x) \to 1$ when $x \to \infty$ and $\alpha(x) \to 0$ when $x \to 0$. That is, at edges where $|\nabla u|$ is very large, $\Phi(x)$ is close to x, the result of using functional $R(u)$ is approximately equal to that of TV-norm. At smoother region where $|\nabla u|$ is very

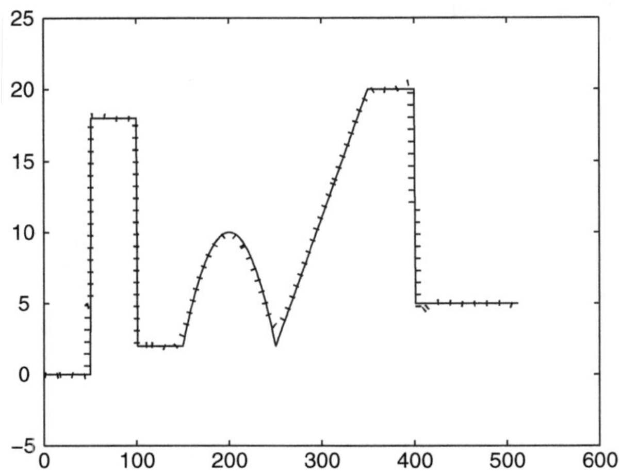

Fig. 4 BCM model restoration.Dotted line is reconstructed image, solid line is original image

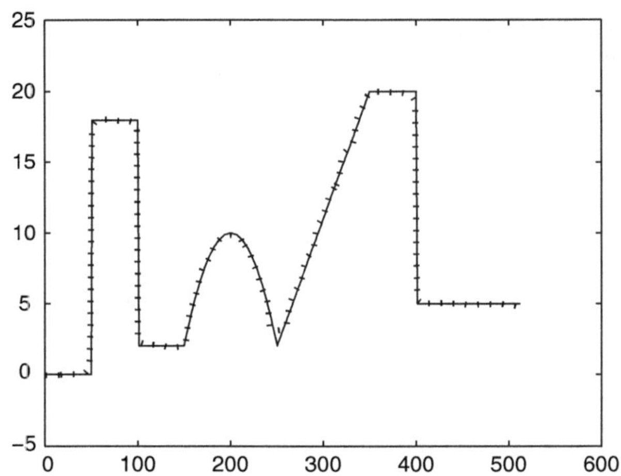

Fig. 5 Our proposed model restoration. Dotted line is reconstructed image, solid line is original image

small, $\Phi(x)$ is close to x^2, the result of using functional $R(u)$ is approximately equal to that of H^1-seminorm.

4 A Convex Regularizing Functional for Staircase Reduction

As stated in Sect. 3, we consider regularizing functional $R(u)$,

$$\begin{aligned} R(u) &= \int_\Omega \Phi(|\nabla u|)dxdy, \\ \Phi(x) &= \alpha(x)x + (1-\alpha(x))x^2 \end{aligned} \tag{9}$$

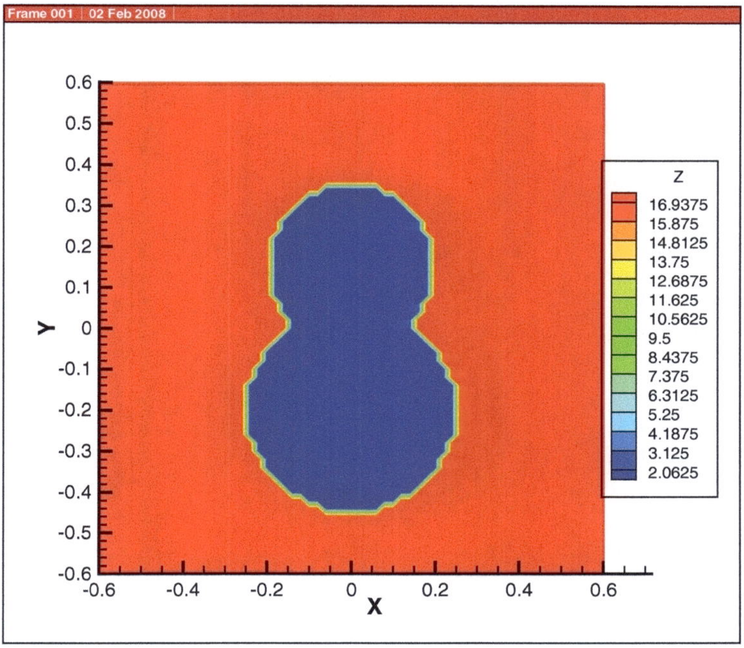

Fig. 6 Original image with value 1 inside and 18 outside

where $\alpha(x) = \frac{x}{1+x}$, which satisfies $\alpha(x) \to 1$ when $x \to \infty$ and $\alpha(x) \to 0$ when $x \to 0$. Thus we get regularizing functional

$$R(u) = \int_\Omega \frac{2|\nabla u|^2}{1+|\nabla u|}\, dxdy \tag{10}$$

Therefore, the new model for total variation denoising is

$$\min \alpha \int_\Omega \frac{2|\nabla u|^2}{1+|\nabla u|} + \frac{1}{2}\|Ku - z\|_{L^2}^2 \tag{11}$$

The Euler-Lagrange equation of (11) is

$$0 = -\nabla \cdot \left(\frac{2+|\nabla u|}{(1+|\nabla u|)^2} \nabla u \right) + \lambda K^*(Ku - z) \tag{12}$$

We calculate the derivatives of functional $R(u)$,

$$R'(u) = -\nabla \cdot \left(\frac{\Phi'(|\nabla u|)}{|\nabla u|} \nabla u \right) \tag{13}$$

$$R''(u)v = -\nabla \cdot \left(\frac{\Phi'(|\nabla u|)}{|\nabla u|} \left(\nabla v - \frac{(\nabla u, \nabla v)}{|\nabla u|^2} \nabla u \right) \right)$$

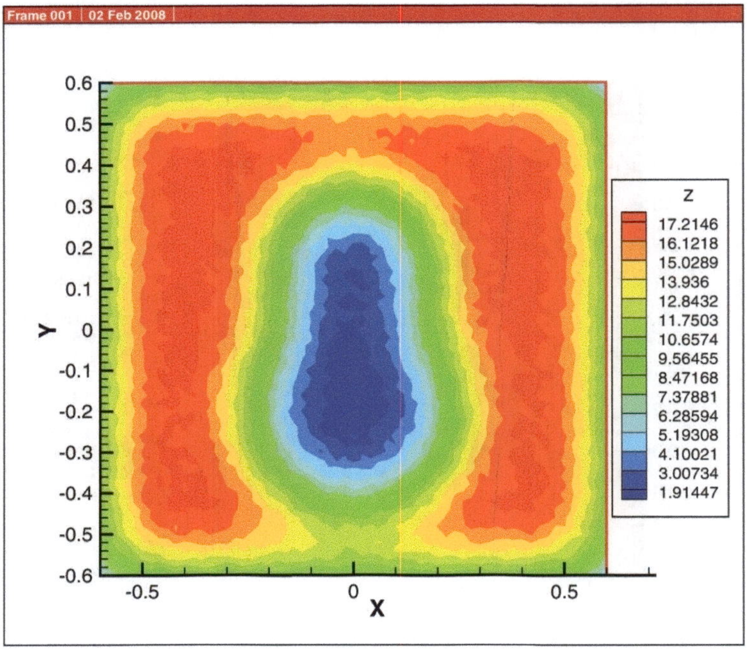

Fig. 7 Blurred image with random noise and gaussian kernel

$$+ \Phi'' (|\nabla u|) \frac{(\nabla u, \nabla v)}{|\nabla u|^2} \nabla u \Big) .$$

From (14) we deduce that:

$$(R''(u)v, v) = \int_\Omega \left(\frac{\Phi'(|\nabla u|)}{|\nabla u|} (|\nabla v|^2 - \frac{(\nabla u, \nabla v)^2}{|\nabla u|^2} \right) \tag{14}$$

$$+ \Phi''(|\nabla u|) \frac{(\nabla u, \nabla v)^2}{|\nabla u|^2}) dxdy. \tag{15}$$

The Cauchy-Schwartz inequality implies that

$$\nabla v|^2 - \frac{(\nabla u, \nabla v)^2}{|\nabla u|^2} \geq 0,$$

therefore $\Phi'(x) \geq 0$ and $\Phi''(x) \geq 0, x \geq 0$, that is, Φ is an increasing convex function in $[0, \infty)$, are sufficient conditions for the functional R of (10) being convex. It's easy to get the expression of $\Phi'(x)$ and $\Phi''(x)$:

$$\Phi'(x) = \frac{x(x+2)}{(1+x)^2}$$

$$\Phi''(x) = \frac{2}{(1+x)^3}$$

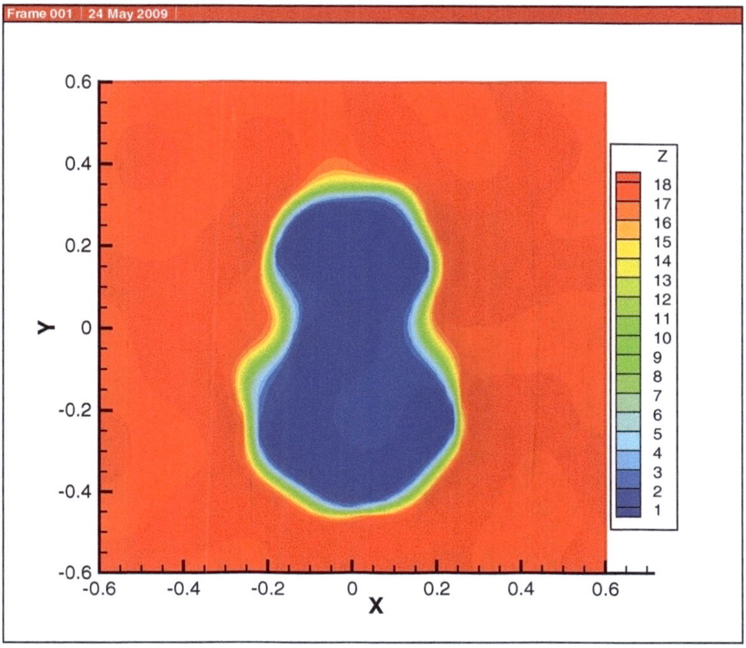

Fig. 8 Restored image by Total Variation regularization

Obviously, $\Phi'(x) \geq 0$, $\Phi''(x) > 0$ when $x \geq 0$. Therefore, the functional R of (10) is convex.

There are many methods to solve Euler-Lagrange Eq. (12). L. Rudin, S. Orsher and E. Fatemi [1] use a time marching scheme to reach a steady state of a parabolic equation; C. Vogel and M. Oman [8] propose the fixed point iteration method, which results in the lagged diffusivity fixed point algorithm. Chan and Mulet [9] give the convergence of the lagged diffusivity fixed point method. Considering the presence of highly nonlinear and non-differentiable term in Euler-Lagrange equation, Chan, Golub and Mulet proposed a nonlinear primal-dual method [10], Chan and Chen [11] intrduced the nonlinear multigrid method. Further works about fast total variation minimization method and algorithm can be seen in literature [12, 13]. In our computation, we referenced Vogel and Oman's fast, robust total variation-based image reconstruction method [14]. To solve Euler-Lagrange Eq. (12), fixed point iteration technique is adopted:

$u^0 = z$, solve for u^{k+1}:

$$-\nabla \cdot \left(\frac{2 + |\nabla u^k|}{(1 + |\nabla u^k|)^2} \nabla u^{k+1} \right) + \lambda K^* \left(K u^{k+1} - z \right) = 0. \tag{16}$$

The new model (11) has some advantages: First, because of the convexity of the regularizing functional $R(u)$, the solution to problem (11) has some kind of

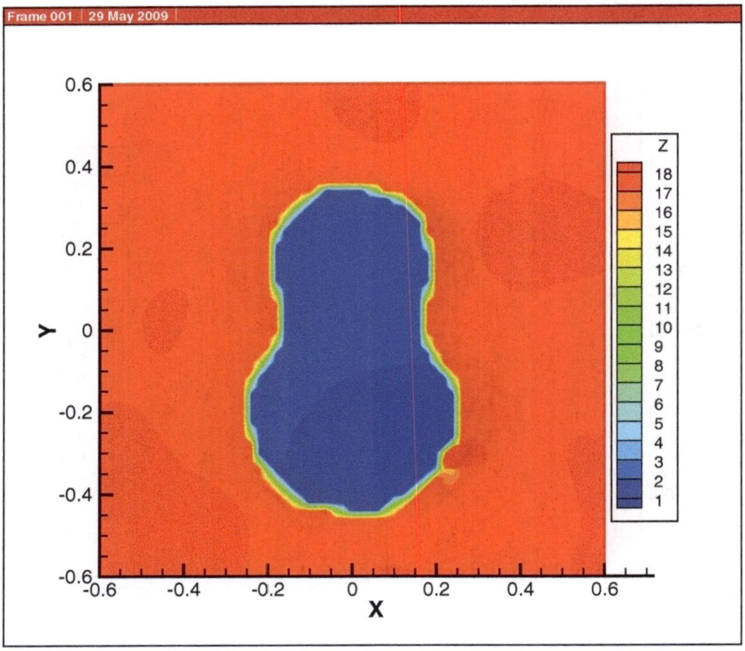

Fig. 9 Restored image by BCM model

uniqueness. Second, our model has no non-differentiable locations. It is not necessary to do numerical regularization, namely, to replace the term $|\nabla u|$ by $\sqrt{|\nabla u|^2 + \beta}$ for a small enough positive artificial parameter β. Third, the new model can efficiently reduce the staircase effect in smooth regions while keep sharp edges behaving like total variation based image restoration model.

5 Numerical Examples

In this section,we perform numerical experiments in 1D and 2D images. In the first experiment, we use a synthetic 1D image which includes piecewise constant, piecewise linear and piecewise parabolic regions. The original image, shown in Fig. 1, is added random noise and blurred by Gaussian kernel. The kernel is defined as

$$g(x) = \frac{1}{\sqrt{2\pi}\sigma} e^{\frac{-x^2}{2\sigma^2}}.$$

31 points of the discrete kernel with $\sigma = 4.5$ is used to get the contaminated image Fig. 2. From Figs. 3 to 5 we give three kinds of restoration of the corrupted image. Restoration by total variation regularization is shown in Fig. 3, restoration by BCM (Blomgren, Chan and Mulet) model [6] in Fig. 4 and our proposed model in Fig. 5.

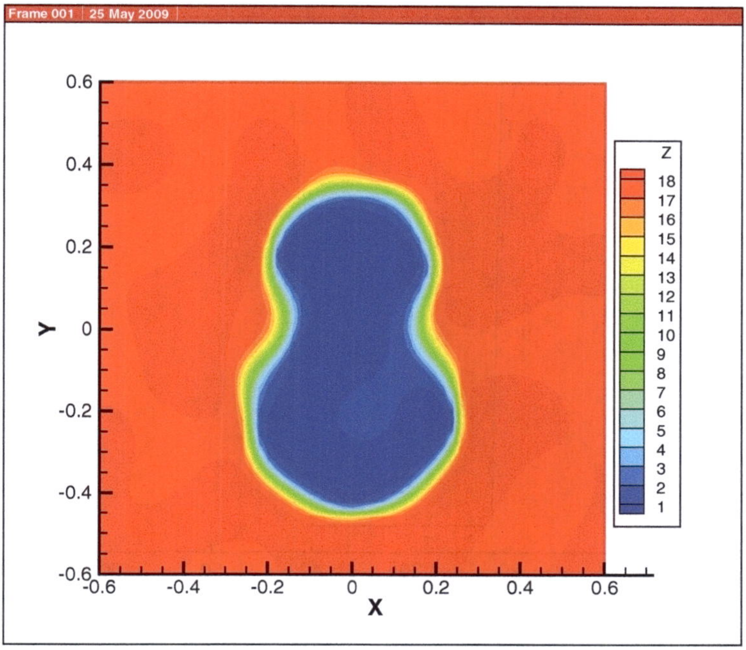

Fig. 10 Restored image by our proposed model

We observe that in Fig. 5 the staircase effect of smooth regions is improved and edges are correctly reserved. Figure 4 has better staircasing reduction in some region, but worse edge location retaining and the 'Gibbs' phenomenon exists. To evaluate the quality of restored images, we calculated signal to noise ratio (SNR) for each image. The SNR values of images in Figs. 4 and 5 are 14.5dB and 15.5dB respectively. Before restoration, the image SNR is 11.8dB. We can see that the reconstructed image with proposed model has better SNR value. In the computation, we found that the staircase effect depends on the choice of regularizing parameter λ in (16), therefore, we use the same value ($\lambda = 0.005$) when doing comparison.

In the second experiment, we perform 2D image restoration with different staircasing reduction models. The original image is created by a two values function, 1 inside and 18 outside (Fig. 6). We contaminate the image with random noise and Gaussian kernel (Fig. 7). The kernel takes the value of $\sigma = 4.5$. We discretize the gaussian function by step size $h = 0.02$ both in x and y directions. Similar to the 1D case, we use 31 by 31 points to blur the original image. In Fig. 8, the noisy and blurred image is restored using TV technique. Figures 9 and 10 are reconstructed images using BCM model and our model respectively. We can see that both TV and BCM model have problems on recovering curve edges. The resulted edges do not look so smooth as it should be. This is suffering from the staircasing effect. Using our model curve edges can be better recovered. Notice also that how the adjoint parts

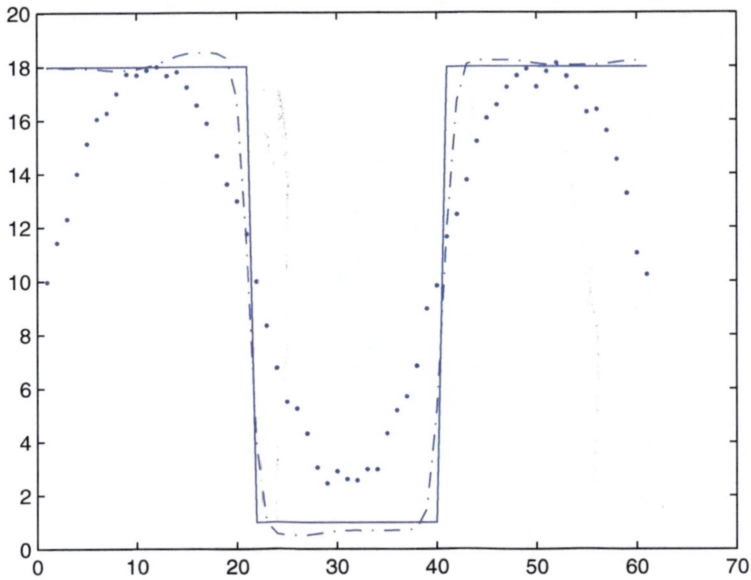

Fig. 11 Comparison by cross lines.The solid line, dotted line and dashed line respectively corresponds to the original image,blurred image and restored image by our model

of the two edge circles are recovered. Our model retains corner better than BCM model does. BCM model uses a third order polynomial interpolating between 0 and sg_{max}. g_{max} is the maximum reliable gradient on the discrete grid and $0 < s \leq 1$. The recovered image is sensitive to the choice of s. We have tried different s, Fig. 9 gives best image recovery among all other images we have obtained by BCM model. The SNR values corresponding to images from Figs. 8 to 10 are, respectively, 13.7dB, 14.4dB and 14.1dB. The contaminated image SNR value is 4.2dB. For the 2D image our model has very close SNR value improvement with BCM model, but the advantage of curve edge recovery is obvious. In Fig. 11 we plot three cross lines which respectively correspond to the original, the blurred and the restored images. We can observe that the proposed model is efficient in recovering image edges and pixel values. In application fields, it's necessary for both pixel values and edge locations be recovered.

6 Conclusions

Total Variation based image restoration method is widely used in image processing area. Its disadvantage is the staircase effect caused at smooth regions. We proposed an improved model which combines the advantage of TV and H^1. It can reduce the staircase effect and recover both the pixel values and correct edge locations. In the application field of nuclear physics, tomographic reconstruction of axially symmetric objects from a single dynamic radiograph is based on the inversion of Abel transform [15]. Abel inversion can be realized by using total variation regularization. It is better to use improved total variation based regularizing term proposed in this paper. We have tested it for density reconstruction from x-ray attenuation date generated by a single radiograph.

Acknowledgments This research is partially supported by NSFC (Project No.10971244).

References

1. Rudin L, Osher S, Fatemi E (1992) Nonlinear total variation based noise removal algorithms. Physica D 60:259–268
2. Chan T, Marquina A, Mulet P (2000) High order total variation-based image restoration. SIAM J Sci Comput 22(2):503–516
3. Marquina A, Osher S (2000) Explicit algorithms for a new time dependant model based on level set motion for nonlinear debluring and noise removal. SIAM J Sci Comput 22:387–405
4. Chambolle A, Lions PL (1997) Image recovery via total variation minimization and related problems. Numer Math 76:167–188
5. Lysaker OM, Tai X-C (2006) Iterative image restoration combining total variation minimization and a second-order functional. Int J Comp Vis 66:5–18
6. Blomgren P, Chan TF, Mulet P, Wong CK (1997) Total variation image restoration: numerical methods and extensions. In Proceedings of the 1997 IEEE international conference on image processing, San Diego, p 384–387
7. Fu H, Ng M, Nikolova M, Barlow J (2006) Efficient minimization methods of mixed ℓ_2-ℓ_1 and ℓ_1-ℓ_1 norms for image restoration. SIAM J Sci Comput 27:1881–1902
8. Vogel CR, Oman ME (1996) Iterative methods for total variation denoising. SIAM J Sci Statist Comput 17:227–238
9. Chan TF, Mulet P (1999) On the convergence of the lagged diffusivity fixed point method in image restoration. SIAM J Numer Anal 36(2):354–367
10. Chan TF, Golub GH, Mulet P (1999) A nonlinear primal-dual method for total variation-based image restoration. SIAM J Sci Comput 20:1964–1977
11. Chan TF, Chen K (2006) An Optimization-based multilevel algorithm for total variation image denoising. Multiscale Model Simul 5(2):615–645
12. Chambolle A (2004) An algorithm for total variation minimization and applications. J Math Imaging Vision 20:89–97
13. Huang Y, Ng MK, Wen Y (2008) A fast total variation minimization method for image restoration. SIAM J Multiscale Model Simul 7(2):774–795
14. Vogel CR, Oman ME (1998, Jun) Fast, robust total variation-based reconstruction of noisy, blured images. IEEE Trans Image Proc 7(6):813–824
15. Asaki TJ, Chartrand R, Vixie K R, Wohlberg B (2005) Abel inversion using total variation reg ularization . Inverse Prob 21(6):1895–1903

Adaptive System for Control of Active Ankle-Foot Orthosis and Gait Analysis

Ivanka Veneva and Nuno Ferreira

Abstract The main aim of this research is the development of an autonomous adaptive system for actuation, data acquisition and control of active ankle-foot orthosis. In this paper the design of a control unit composed by microcontroller, driver and sensor system, and its application to the actuation and position of the foot orthotic segment is presented. The research work combines hardware and software design of the intelligent control device with graphical interface for representation and analysis of the data acquired during human motion. The dynamic system simulation is done in Matlab Simulink and SimMechanics.

A laboratory model of the proposed system was implemented to demonstrate its autonomy and verify experimentally its functionality.

The proposed control device can be used in several applications involving human motion analysis and control of different types of orthoses or functional electrical stimulation used for gait correction.

Keywords Control · Active ankle-foot orthoses · ATmega128 microcontroller · Biomechanics · Rehabilitation robotics

1 Introduction

Ankle foot orthoses (AFO) are assistive devices for Drop foot pathology. Drop foot is the inability of an individual to lift their foot because of reduced or no muscle activity around their ankle. The major causes of drop foot are severing of the nerve, stroke, cerebral palsy and multiple sclerosis. The standard AFO is a rigid polypropylene structure that prevents any ankle motion. There are several commercial products currently on the market. The more widely used device is a dorsiflexion-assist spring ankle foot orthosis, produced by several manufacturers, i.e. Tamarack Joints, Becker

I. Veneva (✉)
Institute of Mechanics, Bulgarian Academy of Sciences, Sofia, Bulgaria
e-mail: veneva@imbm.bas.bg

N. Ferreira
Engineering Institute of Coimbra, Polytechnic Institute of Coimbra, Coimbra, Portugal
e-mail: nunomig@isec.pt

N. M. Fonseca Ferreira, J. A. Tenreiro Machado (eds.), *Mathematical Methods in Engineering*, DOI 10.1007/978-94-007-7183-3_14,

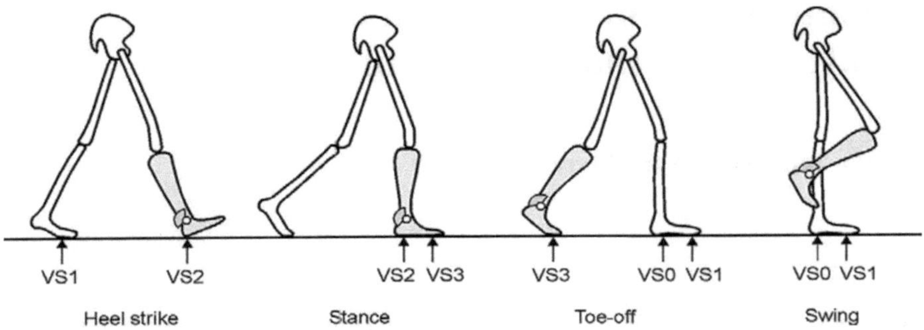

Fig. 1 States and transitional conditions

Orthoses. This AFO is able to help individuals during normal walking by lifting their toe during initial swing.

The idea of an actively powered orthotic device has been explored since the early 1980s using hydraulic and pneumatic device. More recently, compressed gas and DC motors have been researched to provide active assistance to the individuals with paraplegia [2, 3]. An active ankle-foot orthosis with a force-controllable series elastic actuator (SEA) was also designed [1] capable of controlling orthotic joint stiffness and damping for plantar and dorsiflexion ankle motions.

We propose an autonomous adaptive device for actuation, data acquisition and control of active ankle-foot orthosis. The aim of the work is to present an autonomous control and monitoring system for gait analysis using tactile sensors and active ankle-foot orthoses during normal level walking. The device is used to help or rehabilitate persons with control disorders and other weaknesses of ankle foot complex.

2 Methods

Within a given walking cycle, four distinct positions were used corresponding to the phases: *heel strike, stance, toe-off* and *swing*. During the swing phase, where the clearance of the toe is released, electro-mechanical system must actively adjust the flexion of the orthosis by actuator movement and keep this position till the heel strike appears. Thus the ankle torque has to be modulated from cycle-to-cycle throughout the duration of a particular gait phase (Fig. 1).

Active ankle-foot orthosis (AAFO) is a system with one degree of freedom which foot segment is connected to the shank segment by a rotational joint. A direct drive actuator is attached laterally to the AFO.

A feedback with Proportional-Integral-Derivative control was used to estimate the trajectory of the foot and position the actuated foot segment of AFO when the foot rotates about the ankle. Control signals are received in real time from two tactile sensor arrays incorporated in the foot part of AFO and in the insole of the healthy leg, which is the basement of the control algorithm. During each gait cycle

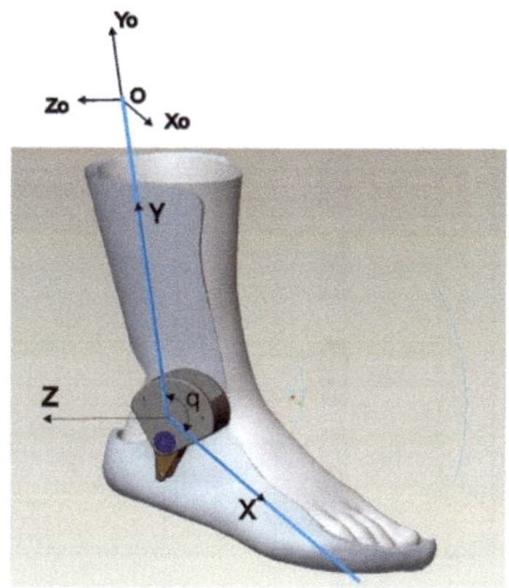

Fig. 2 Ankle-foot orthosis with direct drive actuator

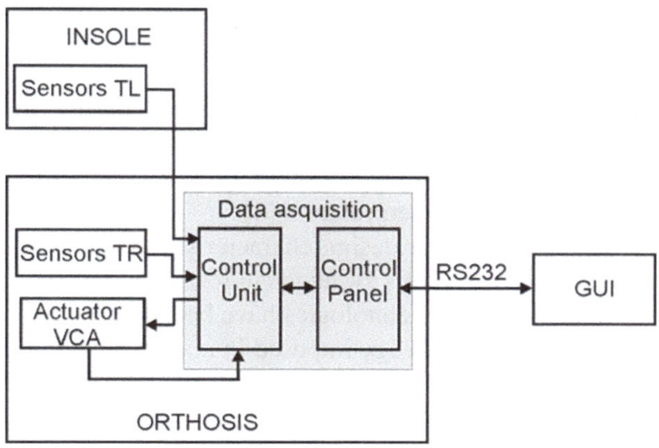

Fig. 3 Autonomous control and monitoring system with active ankle-foot orthoses

a microcontroller estimates forward speed and modulates swing phase flexion and extension in order to assure automatic adaptation of the joint torque [4, 5]. Realizing flexion/extension the actuator applies a torque adequate to the joint position of the human ankle during level ground walking (Fig. 2).

The used Voice Coil Actuator (VCA) has two build-in mechanical stops, which limit its range of motion to slightly less than 30°.

3 System Design

Active Ankle-foot orthoses is an electro-mechanical system controlled by a control module. The complete autonomous system consists of four primary components—sensing, data acquisition, communication and friendly oriented software for interpretation of the data (Fig. 3). The sensor system has mounted into two basic components:

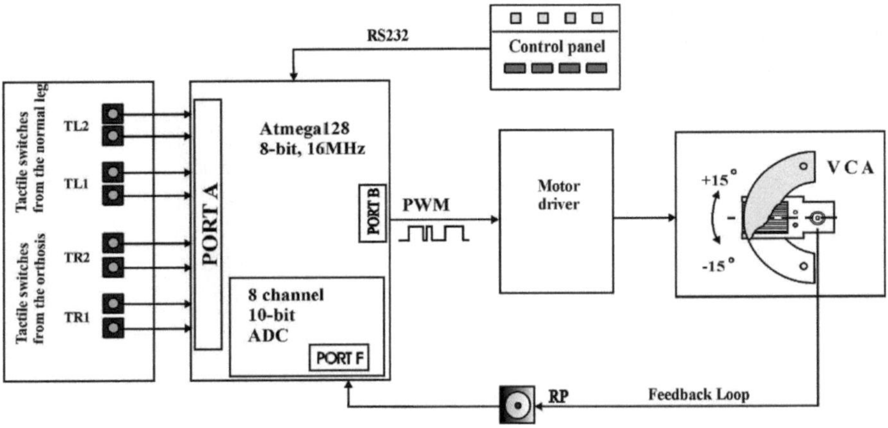

Fig. 4 ATmega128 controlling a VCA

insole for the healthy leg and ankle-foot orthoses. During the walking the acquisition unit gathers and digitizes the information from the sensors. In monitoring mode these data are transferred through the RS-232 lines to a graphical user interface for visualization and interpretation [4].

A very important design characteristic is the system power source. One of the main objectives of this system is to reach a significant level of autonomy.

Many batteries technologies have been lately developed. Nowadays Lithium Ion is the battery technology imposed in market having a good power density.

3.1 Control Module Prototype. Design and Implementation

The control module prototype (Figs. 4 and 5) has been realised using microcontroller ATmega128 (Atmel Co.), featuring basic hardware peripherals such as analog to digital converter (ADC), USART for RS232 communication and a timer with Pulse Width Modulation (PWM) output. The PWM channel is connected to the driver to control the direction and speed of the motor by varying the duty cycle of the PWM output. By varying the current flow thought the coil the speed and torque of the motor can be varied. The position control is handled by electronics according the outputs from a set of two sensor arrays—tactile sensors TR incorporated in the foot part of AFO and TL in the insole of the healthy leg. The sensors change their outputs throughout the stance and the swing phases of walking. Additional feedback element—angular position sensor RP is attached to the moving parts of the motor assemblies to sense the velocity and position.

Secondary functions for the electronics in control application is to ensure that the speed and overload detection are as desired by closed loop control.

Fig. 5 Control module prototype

3.2 Control Algorithm

The control algorithm is based on the biomechanical interpretation of the locomotion. During each gait cycle, by measuring the total time TL (for the left leg) and TR (for the right leg) when the foot remains in contact with the ground the microcontroller estimates the forward speed and modulates the swing phase flexion and extension in order to achieve quite normal lower limb dynamics (Fig. 6). Thus the joint torque of the actuator was automatically modulated to match patient specific gait requirements to permit smooth and natural heel strike to forefoot transition. The tactile sensors and a rotary potentiometer measure ankle joint position in real-time and send signals to the microcontroller. These sensors data are then used in every step of the control algorithm (Fig. 7) in order to optimize the heel to forefoot transition during the stance phase TL or swing phase TR of walking.

The microcontroller receives the diagnostic information about the system from the sensors and generates the torque command to the driver.

To get a more exact torque it is important to have a feedback and PID control in order to maintain stability when a foot load is applied [7] (Fig. 8).

The controller reads the system state y, by a rotational potentiometer, subtracts the measured angle from a desired reference y0, to generate the error value, *e*. The error will be managed in three terms—the proportional, T_p, the integral, T_i, and the derivative, T_d, terms are summed to calculate the output based on the PID algorithm:

$$u(t) = k_p \left[e(t) + \frac{1}{T_i} \int_0^t e(\tau) d\tau + T_d \frac{de(t)}{dt} \right] \tag{1}$$

Fig. 6 Signals from the external sensors

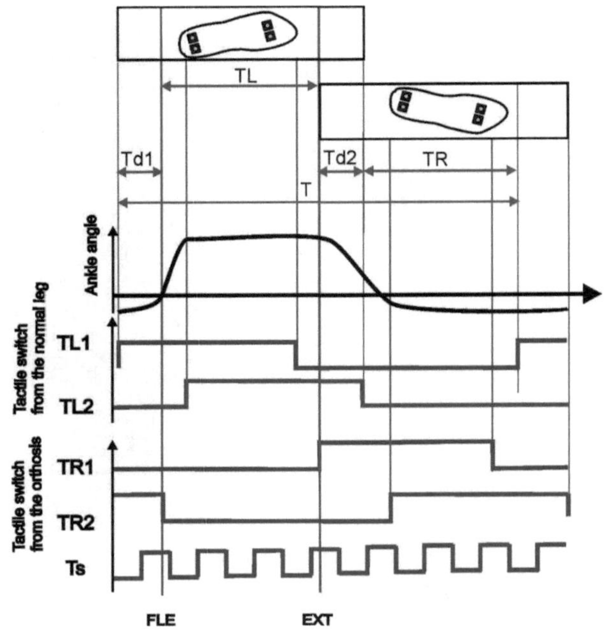

Approximating the integral and the derivative terms to get the discrete form this gives the controller:

$$u(n) = K_p e(n) + K_i \sum_{k=0}^{n} e(k) + K_d \left(y(n) - y(n-1)\right) \tag{2}$$

where n is the discrete step at time t.

4 System Analysis Models

In Matlab SimMechanics (Fig. 9) the ankle-foot orthosis is built of two segments connected by rotational joint with a single rotational degree of freedom: Body1 (shank) and Body2 (foot). We simulate the model in Inverse Dynamics mode to compute the joint torque required to rotate the foot in desired position. During the simulation the geometry of the orthosis is presented as a double pendulum. Once we know the computed torque, we can calculate the required dynamic motor torque and to decide which is the correct motor with appropriate parameters for joint actuation.

$$T_z = T_d - T_c - T_g, \tag{3}$$

$$T_d = \left(J_c + md^2\right)\ddot{q} + k\dot{q} + mgd\sin q, \tag{4}$$

Fig. 7 Control algorithm

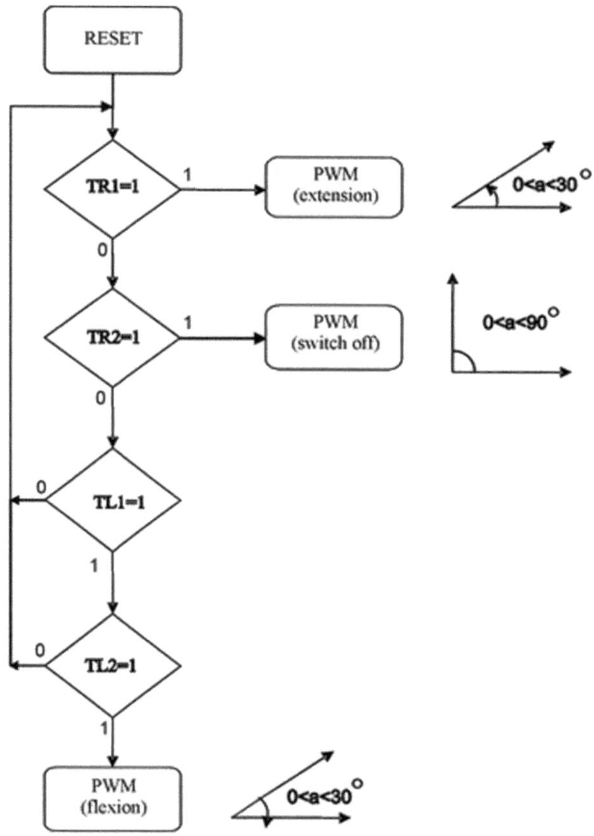

Fig. 8 PID controller with feedback

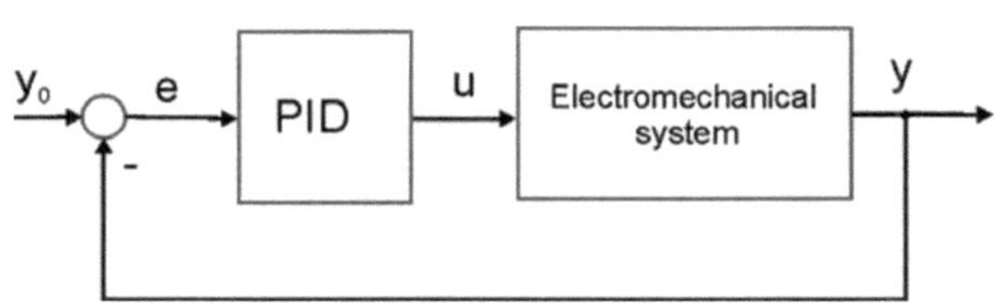

where T_d is the driving torque; T_c—the torque caused by the friction; T_g—torque caused by the gravity; J_c is the foot body inertia moment; q—generalized coordinate.

Using a model of a direct drive DC actuator driving an inertial load it is possible to develop differential equations that describe its behaviour:

$$\frac{di}{dt} = -\frac{R}{L} i\,(t)\,\frac{K_b}{L}\omega\,(t) + \frac{1}{L} u_{app}\,(t) \tag{5}$$

$$\frac{d\omega}{dt} = -\frac{1}{J} K_f \omega\,(t) + \frac{1}{J} K_m i\,(t), \quad \omega(t) = \frac{dq}{dt}. \tag{6}$$

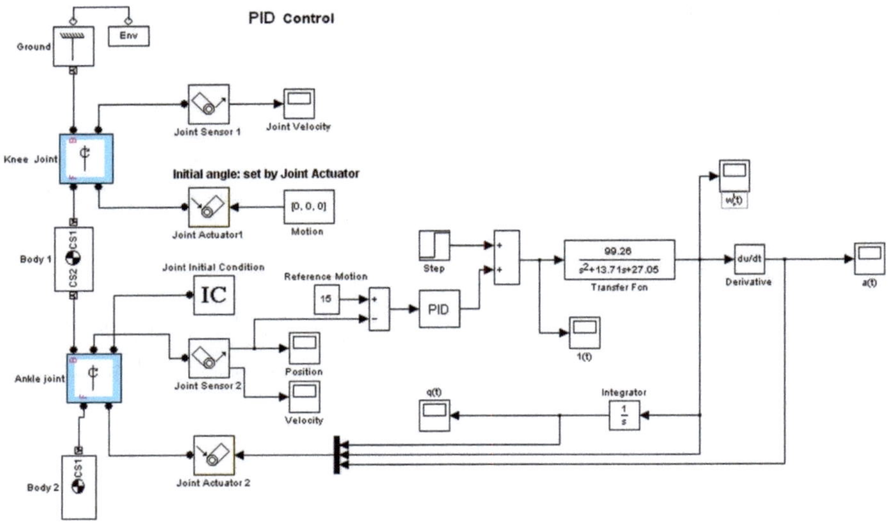

Fig. 9 Electromechanical system model with PID control

In this model the foot is the inertial load driven by the actuator. The model shows the angular velocity of the foot, *w(t)*, as the output and applied voltage, $u_{app}(t)$, as the input of the circuit. The resistance of the circuit is R and the self-inductance of the armature is L.

A state-space representation of the DC actuator as a dynamic system is developed in Matlab [6]. The current i and the angular velocity ω are the two states of the system. The applied voltage, u_{app}, is the input to the system, and the angular velocity ω is the output.

where

- J is the inertia of a body;
- K_m is the armature constant related to physical properties of the motor, such as magnetic field strength, the number of turns of wire around the conductor coil;
- K_b is the electromotive force constant;
- K_f is a linear approximation for viscous friction.

$$\frac{di}{dt}\begin{bmatrix} i \\ w \end{bmatrix} = \begin{bmatrix} -\frac{R}{L} & \frac{K_b}{L} \\ \frac{K_m}{J} & \frac{K_f}{J} \end{bmatrix} \cdot \begin{bmatrix} i \\ w \end{bmatrix} + \begin{bmatrix} \frac{1}{L} \\ 0 \end{bmatrix} .u_{app}(t), \tag{7}$$

$$y(t) = \begin{bmatrix} 0 & 1 \end{bmatrix} . \begin{bmatrix} i \\ w \end{bmatrix} + [0].u_{app}(t), \tag{8}$$

Giving the nominal values for parameters we can obtain the transfer function of the actuator. In Fig. 9, the actuator is represented by its transfer function.

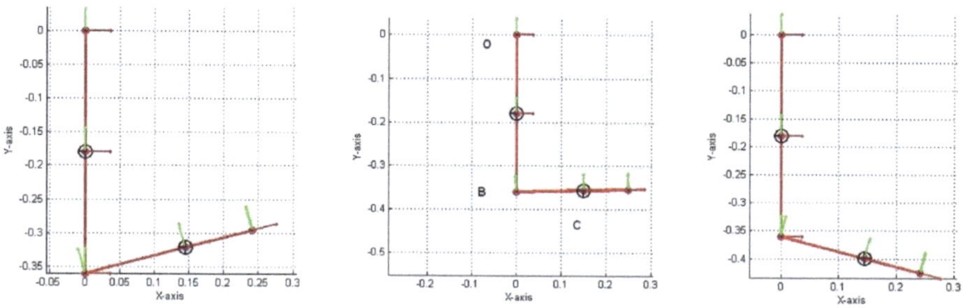

Fig. 10 The geometry of the orthosis is presented as a double pendulum during the simulation in MATLAB

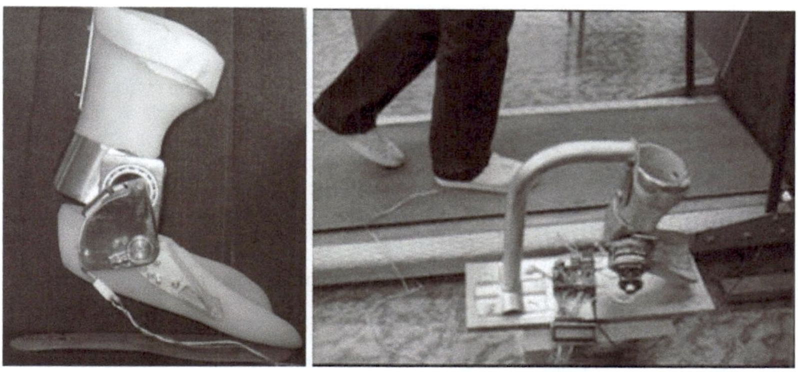

Fig. 11 AAFO with the hinge joint and attached laterally direct drive actuator

During the simulation the geometry of the orthosis is presented as a double pendulum. In Fig. 10, we can see the position oft he orthosis corresponding to the phases *swing, stance* and *toe-off*. During the swing phase, where the clearance of the toe is released, electro-mechanical system must actively adjust the flexion of the orthosis by actuator movement and keep this position till the heel strike appears.

Once we know the actuator parameters and computed torque, we can verify that this is the correct answer of the system simulation by analysing driven angular motion for the articulation of the ankle joint (foot) in Matlab Simulink.

5 Experimental Results

The proposed control module is designed and tested. In order to test the control algorithm and system functionalities a laboratory model of orthosis with hinge joint and attached laterally direct drive actuator was designed (Fig. 11). The orthoses is restricted in the ankle joint to $+/-$ 20°.

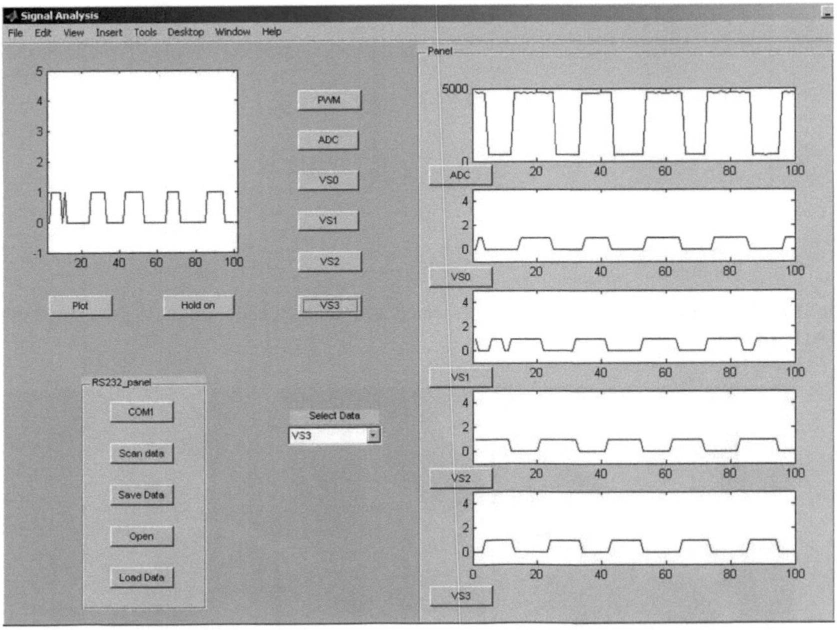

Fig. 12 Graphical program written in MATLAB for visualization of human motion data

A healthy subject equipped with the sensors mounted under the heel and the toes part of the insole (TR1, TR2 for the right leg and TL1, TL2 for the left leg), performs different trials of slow and normal level walking. During walking the motion of the orthosis is observed—if the time and phase parameters of the orthosis coincide with these of the right leg.

The sensors work together to detect walking over one given interval of time and to collect the following parameters: ankle joint angles, foot (heel and toe) contacts and foot velocities. Angle sensor (rotary potentiometer) detects joint position and provides a signal ADC during swing and stance phase. The tactile sensors acted like a switch, turning off when the foot was over the sensor and turning on when the leg moved away from it. The microcontroller collects sensor data in four VS buffers, ADC buffer and PWM duty-cycle buffer.

In monitoring mode the data are transmitted to the PC through the RS232 serial interface. A graphical program module written in MATLAB receives the data and visualizes it in its own window, giving us the representation of the signals (Fig. 12).

VS0 and VS1 are recorded digital signals from the tactile sensors mounted on the orthosis. Signals from the lower insole of the healthy leg are VS2 and VS3.

The features of normal walking are important for overall autonomous control and are not achievable with most mechanically passive orthoses. We assumed that normal gait is symmetrical and that deviation from a reference pattern is a sign of disability. To analyze asymmetry, the step time of the affected side (TR) was subtracted from the unaffected side (TL). The difference in stride lengths should be zero for symmetric gait.

6 Discussion

Autonomy of a system for use by a disabled is a crucial goal in mobility restoration. The percentage of persons suffering from muscular weakness of the lower limb can oscillate between 0.05 and 1 % of the total European population.

The presented device for control of active ankle-foot orthosis integrates biomechanics based algorithms with active control system. The autonomy of the developed system has been demonstrated presenting experimental data during walking. The system controls the orthosis functionalities, records the data received from sensors during the gait and transfers recorded data to graphical user interface for visualization and future analysis.

The developed device provides broad information for both control and gait analysis. The data from the sensors are used in every step from the control algorithm. The actuator joint torque is automatically modulated in order to optimize the heel-to-forefoot transition during the stance or the swing phase of walking.

The experimental data discussed in this paper can be used in cases of the drop foot treatment and lower limb rehabilitation to enhance the AAFO functional performance and to improve the patient gait.

Acknowledgments This work is supported by the European Social Fund and Bulgarian Ministry of Education, Youth and Science under Operative Program "Human Resources Development", Grant BG051PO001-3.3.04/40.

References

1. Blaya JA, Herr H (2004) Adaptive control of a variable-impedance ankle-foot orthosis to assist drop-foot gait. IEEE Trans Neural Syst Rehabil Eng 12(1):24–31
2. Kaneoko S, Yano H (1997) A new concept of dynamic orthosis for paraplegia: the weight bearing control (wbc) orthosis. Prosthet Orthot Int 221: 222–228
3. Veneva I (2007) Autonomous control system for gait analysis. In: Proceedings of the 9th international robotics conference (pp 230–236), Varna, Bulgaria
4. Veneva I (2008) PID Control for active ankle-foot orthosis positioning. Edition Machine Union, Year XIV, vol 6, ISSN1310-3946, pp 52–57
5. Veneva I, Boiadjiev G (2008) Mathematical model of electromechanical controlled module in orthopedics. In: Proceedings of the 6th international conference on bionics and prosthetics, biomechanics and mechanics, mechatronics and robotics (pp 22–26), Varna, Bulgaria
6. Veneva I, Toshev Y (2007) Control algorithm for ankle-foot orthosis and applications. In: Proceedings of the 9th international robotics conference (pp 225–230), Varna, Bulgaria
7. Wasylewski N, Ruthenberg B (1997) An experimental device for investigating the force and power requirements of a powered gait orthosis. J Rehabil Res Dev 34:203–213

Vector Ellipsoidal Harmonics Structure Peculiarities and Limitations

George Dassios

Abstract The theory of scalar ellipsoidal harmonics was introduced by Lamé in 1837, more than half a century after Laplace introduced his theory of spherical harmonics in 1782. It is amazing that the relative theory of vector spherical harmonics was demonstrated as late as 1935 by Hansen.The appearance of a corresponding theory for vector ellipsoidal harmonics was resisting until 2009, for the very simple reason that such a theory can not exist, at least not in such a nice form as the theory of their spherical counterparts. The intrinsic difficulties of the ellipsoidal coordinate system, the fine and symmetric structure that is encoded in the anisotropic character of the ellipsoidal geometry, as well as the necessary generalizations and limitations that are needed are discussed in this presentation. Furthermore, the new analytical techniques that are suggested through the introduction of vector ellipsoidal harmonics are also demonstrated via special examples.

Keywords Vector harmonics · Ellipsoidal harmonics · Potential theory

1 Introduction

Separation of variables led to spectral theory for linear operators with physical implications that can hardly be compared with any other mathematical method in Physics. A quick glance to the mathematical basis of Vibration Theory, or Quantum Mechanics will justify this statement. The method of eigenfunction expansions, which has been invented through separation of variables, provides the most effective way to solve analytically boundary value problems. But, nature is seldom very generous, and in order to obtain a manageable separable system we need linearity of the governing operators, simple geometry, and above all, a lot of luck!

In almost all cases of physical interest, the mathematical characteristics of the separated ordinary differential equations, such as convergence, orthogonality, completence of the eigensolutions and so on, are concluded from the general theory of Sturm–Liouville systems [10]. In the early 1950's, Eisenhart, Moon and Spencer attempted to find all coordinate systems that allow separation for the Laplace and

G. Dassios (✉)
Department of Chemical Engineering, University of Patras, Patras 265 04, Greece
e-mail: gdassios@chemeng.upatras.gr

N. M. Fonseca Ferreira, J. A. Tenreiro Machado (eds.), *Mathematical Methods in Engineering*, DOI 10.1007/978-94-007-7183-3_15,

the Helmholtz equations [7]. They found that, among all the orthogonal curvilinear coordinate systems with first and second degree coordinate surfaces, there are only 11 systems that allow for splitting of the partial differential equation to three ordinary differential equations. Among these coordinate systems, the simplest one is the Cartesian and the most complicated one is the ellipsoidal [4], [5].

If the partial differential operator acts on a vector field, which is decomposed in three scalar components with respect to some curvilinear system, then separation of variables seeks for the splitting of the vector equation into three scalar partial differential equations, each one involving only a single component of the decomposed field. Then, each one of these three scalar equations can be further separated into three ordinary differential equations and the theory of scalar equations is readily applied.

In the present work we will investigate the case of the ellipsoidal system, where such a program is not applicable and we will try to find how we can handle this situation. We will demonstrate a novel approach to the problem of evaluating coefficients of a not completely orthogonal system.

2 The Spherical Case

If the eigensolutions of a scalar problem are known then the naive approach to solve a vector problem, is to expand every Cartesian component in terms of the corresponding scalar eigenfunctions and look for solutions in the form

$$\boldsymbol{U}(\boldsymbol{r}) = \sum_{n}\sum_{m} \boldsymbol{C}_{nm} u_{nm}(\boldsymbol{r}) \tag{1}$$

where

$$\boldsymbol{C}_{nm} = \left(C^1_{nm}, C^2_{nm}, C^3_{nm}\right) \tag{2}$$

denotes the vector of the unknown coefficients. If we apply the boundary conditions to the above form of the solution we will find that the evaluation of the coefficients $\boldsymbol{CC}_{nm}$ is either very hard, or impossible. The reason for this difficulty is that we casted the vectorial character of the field $\boldsymbol{U}$ into the coefficients $\boldsymbol{CCC}_{nm}$ and not into the eigenfunctions, which carry information about the natural geometry of the problem. Therefore, the right approach is to introduce vector eigenfunctions that are compatible with the geometry of the boundaries [8, 9]. But this is not an easy task. In fact, vectorial eigenfunctions are known only for spherical and spheroidal geometries.

The vector spherical harmonics were introduced by Hansen [2]. The interior solid harmonics are given by

$$\boldsymbol{A}^m_{1n}(\boldsymbol{r}) = r^n \boldsymbol{r} Y^m_n(\hat{\boldsymbol{r}}) - \frac{1}{2n+3}\nabla\left(r^{n+2}Y^m_n(\hat{\boldsymbol{r}})\right) \tag{3}$$

$$\boldsymbol{A}^m_{2n}(\boldsymbol{r}) = \nabla \times \boldsymbol{A}^m_{1n}(\boldsymbol{r}) = \nabla \times \left(r^n \boldsymbol{r} Y^m_n(\hat{\boldsymbol{r}})\right) \tag{4}$$

$$\boldsymbol{A}^m_{3n}(\boldsymbol{r}) = \nabla \times \boldsymbol{A}^m_{2n}(\boldsymbol{r}) = \nabla\nabla \cdot \boldsymbol{A}^m_{1n}(\boldsymbol{r}) \tag{5}$$

and the exterior solid harmonics are given by

$$\boldsymbol{B}_{1n}^{m}(\boldsymbol{r}) = \boldsymbol{r}\frac{Y_n^m(\hat{\boldsymbol{r}})}{r^{n+1}} + \frac{1}{2n-1}\nabla\left(\frac{Y_n^m(\hat{\boldsymbol{r}})}{r^{n-1}}\right) \tag{6}$$

$$\boldsymbol{B}_{2n}^{m}(\boldsymbol{r}) = \nabla \times \boldsymbol{B}_{1n}^{m}(\boldsymbol{r}) = \nabla \times \left(\boldsymbol{r}\frac{Y_n^m(\hat{\boldsymbol{r}})}{r^{n+1}}\right) \tag{7}$$

$$\boldsymbol{B}_{3n}^{m}(\boldsymbol{r}) = \nabla \times \boldsymbol{B}_{2n}^{m}(\boldsymbol{r}) = \nabla\nabla \cdot \boldsymbol{B}_{1n}^{m}(\boldsymbol{r}) \tag{8}$$

where

$$Y_n^m(\hat{\boldsymbol{r}}) = C_n^m P_n^m(\cos\vartheta)e^{im\varphi} \tag{9}$$

denotes the complex normalized form of the scalar surface spherical harmonics, with C_n^m the normalization constants. The functions $\boldsymbol{A}_{in}^m$ and $\boldsymbol{B}_{in}^m$ solve the vector Laplace equation

$$\Delta \boldsymbol{F}(\boldsymbol{r}) = \nabla\nabla \cdot \boldsymbol{F}(\boldsymbol{r}) - \nabla \times \nabla \times \boldsymbol{F}(\boldsymbol{r}). \tag{10}$$

In applying boundary conditions, the important functions are the following surface vector harmonics

$$\boldsymbol{P}_n^m(\hat{\boldsymbol{r}}) = \hat{\boldsymbol{r}} Y_n^m(\hat{\boldsymbol{r}}) \tag{11}$$

$$\boldsymbol{B}_n^m(\hat{\boldsymbol{r}}) = \frac{1}{\sqrt{n(n+1)}}\boldsymbol{D} Y_n^m(\hat{\boldsymbol{r}}) \tag{12}$$

$$\boldsymbol{C}_n^m(\hat{\boldsymbol{r}}) = \frac{1}{\sqrt{n(n+1)}}(\boldsymbol{D} Y_n^m(\hat{\boldsymbol{r}})) \times \hat{\boldsymbol{r}} \tag{13}$$

where

$$\boldsymbol{D} = r\nabla = \hat{\boldsymbol{\vartheta}}\frac{\partial}{\partial\vartheta} + \frac{\hat{\boldsymbol{\varphi}}}{\sin\vartheta}\frac{\partial}{\partial\varphi}. \tag{14}$$

Observe that $\boldsymbol{P}_n^m$ are radial functions, while $\boldsymbol{B}_n^m$ and $\boldsymbol{C}_n^m$ are tangential functions. The set $\{\boldsymbol{P}_n^m, \boldsymbol{B}_n^m, \boldsymbol{C}_n^m\}_{n=0,m=-n}^{n=\infty,m=n}$ is a local orthogonal system.

Performing the indicated differentiations we obtain

$$\boldsymbol{A}_{1n}^{m}(\boldsymbol{r}) = \frac{\sqrt{n+1}}{2n+3}r^{n+1}\left[\sqrt{n+1}\boldsymbol{P}_n^m - \sqrt{n}\boldsymbol{B}_n^m\right] \tag{15}$$

$$\boldsymbol{A}_{2n}^{m}(\boldsymbol{r}) = \sqrt{n(n+1)}r^n \boldsymbol{C}_n^m \tag{16}$$

$$\boldsymbol{A}_{3n}^{m}(\boldsymbol{r}) = (n+1)\sqrt{n}r^{n-1}\left[\sqrt{n}\boldsymbol{P}_n^m + \sqrt{n+1}\boldsymbol{B}_n^m\right] \tag{17}$$

where $\boldsymbol{A}_{2n}^m$ are tangential for every n and m and $\boldsymbol{A}_{1n}^m$, $\boldsymbol{A}_{3n}^m$ are both perpendicular to $\boldsymbol{A}_{2n}^m$ for every n and m. Any vector harmonic function inside a sphere has the expansion

$$\boldsymbol{F}(\boldsymbol{r}) = \sum_{n,m} \left[c_{1n}^m \boldsymbol{A}_{1n}^m(\boldsymbol{r}) + c_{2n}^m \boldsymbol{A}_{2n}^m(\boldsymbol{r}) + c_{3n}^m \boldsymbol{A}_{3n}^m(\boldsymbol{r}) \right] \tag{18}$$

For Dirichlet data on the boundary $r = a$, the function $\boldsymbol{F}$ is then written as

$$\boldsymbol{F}(a\hat{\boldsymbol{r}}) = \sum_{n,m} \left[p_n^m \boldsymbol{P}_n^m(\hat{\boldsymbol{r}}) + b_n^m \boldsymbol{B}_n^m(\hat{\boldsymbol{r}}) + c_n^m \boldsymbol{C}_n^m(\hat{\boldsymbol{r}}) \right] \tag{19}$$

where p_n^m and b_n^m are linear combinations of c_{1n}^m and c_{3n}^m and c_n^m are proportional to c_{2n}^m. The orthogonality of $\boldsymbol{P}_n^m$, $\boldsymbol{B}_n^m$, $\boldsymbol{C}_n^m$ allows for the analytic evaluation of the unknown coefficients c_{1n}^m c_{2n}^m and c_{3n}^m. A similar approach solves the Neumann problem.

3 Vector Ellipsoidal Harmonics

The challenging question now is whether the above logic for the sphere can be extended to the case of the ellipsoidal system. We scrutinize the procedure that led to the definition of Vector Spherical Harmonics (VSH) for a general orthogonal curvilinear system (ξ_1, ξ_2, ξ_3) and a vector harmonic field

$$\boldsymbol{F}(\boldsymbol{\xi}) = f^1(\boldsymbol{\xi})\widehat{\boldsymbol{\xi}_1} + f^2(\boldsymbol{\xi})\widehat{\boldsymbol{\xi}_2} + f^3(\boldsymbol{\xi})\widehat{\boldsymbol{\xi}_3} \tag{20}$$

and we see that separability for the vector Laplacian is possible only when the following conditions among the metric coefficients h_{ξ_1}, h_{ξ_2}, h_{ξ_3} hold [9]

- $h_{\xi_1} = 1$
- $\frac{\partial}{\partial \xi_1} \frac{h_{\xi_2}}{h_{\xi_3}} = 0$
- $h_{\xi_2} h_{\xi_3}$ is proportional either to 1 or to ξ_1^2

It is easy to see that no one of these conditions holds for the ellipsoidal system. Hence, *we can not introduce Vector Ellipsoidal Harmonics* (VEH) which are as good as the VSH. But let us ignore this obstacle for the moment and proceed further, in order to find *what is the best we can do with the ellipsoidal system* [3, 6]

The ellipsoidal coordinates$(\rho,\ \mu,\ \nu)$ are connected to the Cartesian ones by the relations

$$x_1 = \frac{\rho\mu\nu}{h_2 h_3} \tag{21}$$

$$x_2 = \frac{\sqrt{\rho^2 - h_3^2}\sqrt{\mu^2 - h_3^2}\sqrt{h_3^2 - \nu^2}}{h_1 h_3} \tag{22}$$

$$x_3 = \frac{\sqrt{\rho^2 - h_2^2}\sqrt{h_2^2 - \mu^2}\sqrt{h_2^2 - \nu^2}}{h_1 h_2} \tag{23}$$

with

$$0 \leqslant \nu^2 \leqslant h_3^2 \leqslant \mu^2 \leqslant h_2^2 \leqslant \rho^2 < \infty \tag{24}$$

and

$$h_1^2 = a_2^2 - a_3^2, h_2^2 = a_1^2 - a_3^2, h_3^2 = a_1^2 - a_2^2 \tag{25}$$

are the semi-focal distances of the system. The ellipsoidal metric coefficients are

$$h_\rho^2 = \frac{(\rho^2 - \mu^2)(\rho^2 - \nu^2)}{(\rho^2 - h_3^2)(\rho^2 - h_2^2)} \tag{26}$$

$$h_\mu^2 = \frac{(\mu^2 - \rho^2)(\mu^2 - \nu^2)}{(\mu^2 - h_3^2)(\mu^2 - h_2^2)} \tag{27}$$

$$h_\nu^2 = \frac{(\nu^2 - \rho^2)(\nu^2 - \mu^2)}{(\nu^2 - h_3^2)(\nu^2 - h_2^2)} \tag{28}$$

from where it is obvious that none of the above conditions is satisfied, since all coefficients depend on all variables. Let

$$E_n^m(\rho, \mu, \nu) = E_n^m(\rho)E_n^m(\mu)E_n^m(\nu) \tag{29}$$

$$F_n^m(\rho, \mu, \nu) = F_n^m(\rho)E_n^m(\mu)E_n^m(\nu) \tag{30}$$

be the interior and exterior solid ellipsoidal harmonics, respectively, and let

$$S_n^m(\mu, \nu) = E_n^m(\mu)E_n^m(\nu) \tag{31}$$

be the corresponding surface ellipsoidal harmonics. As we mentioned earlier, the importance of the theory of vector spherical harmonics is that the the surface vector spherical harmonics $\boldsymbol{P}_n^m$, $\boldsymbol{B}_n^m$, $\boldsymbol{C}_n^m$ are complete and orthogonal over the unit sphere S^2. That allows for the calculation of the coefficients of an eigenexpansion in closed form.

For the spherical case, we choose one function in the normal direction $\hat{\boldsymbol{r}}$ and two tangential functions in the directions of ∇Y and $\hat{\boldsymbol{r}} \times \nabla Y$. Following a similar approach for the ellipsoid, we choose the normal direction $\hat{\boldsymbol{\rho}}$ and the two tangential directions ∇S and $\hat{\boldsymbol{\rho}} \times \nabla S$. Hence, we define the vector surface ellipsoidal harmonics as

$$\boldsymbol{R}_n^m(\mu, \nu) = d(\rho, \mu, \nu)\hat{\boldsymbol{\rho}} S_n^m(\mu, \nu) \tag{32}$$

$$\boldsymbol{D}_n^m(\mu, \nu) = f(\rho, \mu, \nu)\nabla S_n^m(\mu, \nu) \tag{33}$$

$$\boldsymbol{T}_n^m(\mu, \nu) = g(\rho, \mu, \nu)\hat{\boldsymbol{\rho}} \times \nabla S_n^m(\mu, \nu) \tag{34}$$

and specify the functions d, f and g in such a way as to secure as much orthogonality as possible. We know that complete orthogonality among these functions is not possible. So, let us try for the best possible, i.e. let us try to maximize the set of orthogonal relations among the functions (32)–(34).

The orthogonality relations we need to have are

$$\oint_{S_\rho} \boldsymbol{R}_n^m \cdot \boldsymbol{D}_{n'}^{m'} l_\rho ds_\rho = 0 \tag{35}$$

$$\oint_{S_\rho} \boldsymbol{D}_n^m \cdot \boldsymbol{T}_{n'}^{m'} l_\rho ds_\rho = 0 \tag{36}$$

$$\oint_{S_\rho} \boldsymbol{T}_n^m \cdot \boldsymbol{R}_{n'}^{m'} l_\rho ds_\rho = 0 \tag{37}$$

$$\oint_{S_\rho} \boldsymbol{R}_n^m \cdot \boldsymbol{R}_{n'}^{m'} l_\rho ds_\rho = R_n^m \delta_{nn'} \delta_{mm'} \tag{38}$$

$$\oint_{S_\rho} \boldsymbol{D}_n^m \cdot \boldsymbol{D}_{n'}^{m'} l_\rho ds_\rho = D_n^m \delta_{nn'} \delta_{mm'} \tag{39}$$

$$\oint_{S_\rho} \boldsymbol{T}_n^m \cdot \boldsymbol{T}_{n'}^{m'} l_\rho ds_\rho = T_n^m \delta_{nn'} \delta_{mm'}. \tag{40}$$

A long, detailed and tedious investigation leads to the choice

$$d(\rho, \mu, \nu) = 1 \tag{41}$$

$$f(\rho, \mu, \nu) = \rho \frac{\sqrt{\rho^2 - \mu^2}\sqrt{\rho^2 - \nu^2}}{\sqrt{\rho^2 - h_3^2}\sqrt{\rho^2 - h_2^2}} \tag{42}$$

$$g(\rho, \mu, \nu) = \rho \tag{43}$$

$$l_\rho(\mu, \nu) = \frac{1}{\sqrt{\rho^2 - \mu^2}\sqrt{\rho^2 - \nu^2}} \tag{44}$$

for which the orthogonality relations (35)–(39) are satisfied. Orthogonality (40) is managed if we replace the weighting function l_ρ by the function

$$\tilde{l}_\rho(\mu, \nu) = \frac{\sqrt{\rho^2 - \mu^2}\sqrt{\rho^2 - \nu^2}}{\left(\rho^2 - h_3^2\right)\left(\rho^2 - h_2^2\right)}. \tag{45}$$

Hence, complete orthogonality for the ellipsoidal system demands the use of two different inner products! For the relative proofs we refer to [1].

Note that, in contrast to the spherical system, where the surface harmonics $\{\boldsymbol{P}_n^m, \boldsymbol{B}_n^m, \boldsymbol{C}_n^m\}$ given by (11)–(13), are independent of the radial variable r, the ellipsoidal system $\{\boldsymbol{R}_n^m, \boldsymbol{D}_n^m, \boldsymbol{T}_n^m\}$ is $\rho-$dependent. In other words, in the case of the ellipsoid, the set of surface harmonics changes as we move from one ellipsoidal surface to another.

The completeness of the vector ellipsoidal harmonics, over the surface of any ellipsoid, is a consequence of the fact that every spherical harmonic is expandable in ellipsoidal harmonics and vice versa.

The expansion algorithm for a smooth vector field $\boldsymbol{F}$ defined on the ellipsoid $\rho =$ constant works as follows

$$\begin{aligned}\boldsymbol{F}(\mu,\nu;\rho) &= \frac{1}{h_\rho}\sum_{n=0}^{\infty}\sum_{m=1}^{2n+1} A_n^m(\rho)\boldsymbol{R}_n^m(\mu,\nu;\rho)\\ &+ \frac{1}{h_\rho}\sum_{n=1}^{\infty}\sum_{m=1}^{2n+1} B_n^m(\rho)\boldsymbol{D}_n^m(\mu,\nu;\rho)\\ &+ \frac{1}{h_\rho}\sum_{n=1}^{\infty}\sum_{m=1}^{2n+1} C_n^m(\rho)\boldsymbol{T}_n^m(\mu,\nu;\rho).\end{aligned} \tag{46}$$

Taking the inner product of (46) with $\boldsymbol{R}_{n'}^{m'}(\mu,\nu;\rho)$, with respect to the inner product defined by the weighting function $l_\rho(\mu,\nu)$ we obtain the coefficient $A_n^m(\rho)$. Taking the inner product of (46) with $\boldsymbol{D}_{n'}^{m'}(\mu,\nu;\rho)$, with respect to the inner product defined by the weighting function $\tilde{l}_\rho(\mu,\nu)$ we obtain the coefficient $B_n^m(\rho)$. Then the function

$$\begin{aligned}\boldsymbol{F}'(\mu,\nu;\rho) &= \boldsymbol{F}(\mu,\nu;\rho)\\ &- \frac{1}{h_\rho}\sum_{n=0}^{\infty}\sum_{m=1}^{2n+1} A_n^m(\rho)\boldsymbol{R}_n^m(\mu,\nu;\rho)\\ &- \frac{1}{h_\rho}\sum_{n=1}^{\infty}\sum_{m=1}^{2n+1} B_n^m(\rho)\boldsymbol{D}_n^m(\mu,\nu;\rho)\end{aligned} \tag{47}$$

becomes and has the expansion

$$\boldsymbol{F}'(\mu,\nu;\rho) = \frac{1}{h_\rho}\sum_{n=1}^{\infty}\sum_{m=1}^{2n+1} C_n^m(\rho)\boldsymbol{T}_n^m(\mu,\nu;\rho). \tag{48}$$

Finally, taking the second inner product with $\boldsymbol{T}_{n'}^{m'}(\mu,\nu;\rho)$ we obtain the coefficients $C_n^m(\rho)$. Hence, the above expansion can be completely calculated.

4 Conclusions

It is not possible to introduce vector ellipsoidal harmonics which behave as nice as the vector spherical harmonics. Perhaps, this is the reason why the vector ellipsoidal harmonics were lacking from the literature for 74 years after the vector spherical harmonics were introduced. Here we define a complete set of vector ellipsoidal harmonics, that depend on the ellipsoid on which they are living, and that are orthogonal with respect to two analytic structures defined by two inner products having different weighting functions. Using these two inner products it is easy to calculate the coefficients of any vectorial eigenfunction expansion. Just as it is with the case of spherical geometry, one set of vector surface ellipsoidal harmonics is locally normal to the particular ellipsoid and the other two sets are tangential to the surface of the ellipsoid at this point. The crucial difference though is that, in general, the directions of the position vector and the corresponding normal do not coincide, and this is the source of many difficulties with the ellipsoidal system.

The introduction of Vector Spherical Harmonics in 1935, provided a significant freedom in solving boundary value problems in spherical geometry. The present introduction of vector ellipsoidal harmonics identifies an area of classical applied mathematics that is open to many theoretical and real life problems. For example, we know that every vector boundary value problem in ellipsoidal geometry that has been solved up to now involves a tremendous amount of calculations, since it was done the "wrong"(but the only possible) way, i.e. using scalar ellipsoidal harmonics. Now, it can be solved the "right" way by using vector eigenfunctions. It is of interest to see how much easier a vector problem can be solved using vector instead of scalar ellipsoidal harmonics. Even more important it is to investigate which vector problems that were impossible with the scalar eigenfunctions are tractable with the corresponding vector eigenfunctions. We hope that these questions will find some answers in the near future.

References

1. Dassios G, Tsampas M (2009) Vector ellipsoidal harmonics and neuronal current decomposition in the brain. Inverse Probl Imaging 3:243–257
2. Hansen WW (1935) A new type of expansion in radiation problems. Phys Rev 47:139–143
3. Hobson EW (1931) The theory of spherical and ellipsoidal harmonics, 1st edn. Cambridge University Press, Cambridge
4. Lamé MG (1837) Sur les surfaces isothermes dans les corps solideshomogénes en équilibre de température. Journal de Mathématiques Pures et Appliquées 2:147–183
5. Lamé MG (1839) Sur l'équilibre des temperatures dans un ellipsoide à trois axes inégaux. Journal de Mathématiques Pures et Appliquées 4:126–163
6. MacMillan WD (1958) The theory of the potential, 1st edn. Dover, New York
7. Moon P, Spencer DE (1961) Field theory handbook. Springer-Verlag, Berlin
8. Morse PM, Feshbach H (1953) Methods of theoretical physics, vol I, 1st edn. McGraw-Hill, New York
9. Morse PM, Feshbach H (1953) Methods of theoretical physics, vol II, 1st edn. McGraw-Hill, New York
10. Whittaker ET, Watson GN (1920) A course of modern analysis, 3rd edn. Cambridge University Press, London

Casualties Distribution in Human and Natural Hazards

Carla M. A. Pinto, A. Mendes Lopes and J. A. Tenreiro Machado

Abstract Catastrophic events, such as wars and terrorist attacks, big tornadoes and hurricanes, huge earthquakes, tsunamis, floods, and landslides, are always accompanied by a large number of casualties. The size distribution of these casualties have separately been shown to follow approximate power law (PL) distributions. In this paper, we analyze the number of victims of catastrophic phenomena, in particular, terrorism, and find double PL behavior. This means that the data set is better approximated by two PLs instead of one. We have plotted the two PL parameters corresponding to all terrorist events occurred in every year, from 1980 to 2010. We observe an interesting pattern in the chart, where the lines, that connect each pair of points defining the double PLs, are roughly aligned to each other.

Keywords Casualties distribution · Power law behavior · Double power law

1 Introduction

Catastrophic events are characterized by a huge severity, usually defined by a large number of casualties. By catastrophic events, we mean wars, terrorist attacks, tornados, earthquakes, floods, and landslides. The distribution of the number of casualties in these events is proved to be a power law (PL) [4, 5, 6, 9, 12, 21, 23].

C. M. A. Pinto (✉)
Department of Mathematics, Institute of Engineering of Porto,
Rua Dr. António Bernardino de Almeida, 431, Porto, 4200-072, Portugal
e-mail: cap@isep.ipp.pt

Centro de Matemática da Universidade do Porto, Porto, Portugal

A. M. Lopes
UISPA, IDMEC - Polo FEUP Faculty of Engineering, University of Porto,
Rua Dr. Roberto Frias, Porto, 4200-465 Portugal
e-mail: aml@fe.up.pt

J. A. Tenreiro Machado
Department of Electrical Engineering, ISEP-Institute of Engineering
of Polytechnic of Porto, Rua Dr António Bernardino de Almeida, 431,
Porto, 4200-072 Portugal
e-mail: jtm@isep.ipp.pt

N. M. Fonseca Ferreira, J. A. Tenreiro Machado (eds.), *Mathematical Methods in Engineering*, DOI 10.1007/978-94-007-7183-3_16,

PL distributions were first mentioned in 1896, when Pareto described the distribution of income [19]. Pareto proved that the relative number of individuals with an annual income larger than a certain value x was proportional to a power of x. This has been known by Pareto distribution. After this work, Auerbach [1] demonstrated an analogous result for city size distributions. Ranking cities from 1 to n, with the city with bigger population ranked as 1, Auerbach demonstrated that the product of cities populations by their ranks was approximately constant, for a given territory. Estoup [10] and Zipf [25, 26] applied PLs to words frequencies in texts. They found that there are words that are used more often than others and that the distribution of words frequencies follows a PL. Zipf [26] described the distribution of city sizes by a Pareto distribution.

Often, to show that a certain data set follows a PL distribution, researchers depict a plot of the size and the frequency of the event studied. In logarithmic scales, they obtain a straight line with negative slope. In the case of the Pareto distribution, the behavior is exactly linear, and is given by

$$\ln\left(P[X \geq x]\right) = \ln C - \ln\tilde{\alpha} - \tilde{\alpha}\ln x \tag{1}$$

where X is a random variable following a PL distribution, $\tilde{\alpha} > 0$, $\tilde{C} = \frac{C}{\tilde{\alpha}} > 0$. In these distributions, the tail falls asymptotically according to the value of $\tilde{\alpha}$, translating in heavy tails, comparatively to other distributions. Zipf's law is a special case of the Pareto's law, with coefficient $\tilde{\alpha} = 1$. Relevant reviews on PL distributions can be found in [16, 20, 24].

This paper is organized as follows. In Sect. 2, we summarize results found in the literature concerning application of PL behavior to the number of casualties in natural or human-made disasters. In Sect. 3, we apply double PLs to data from real disasters. Finally, in Sect. 4, we enumerate the main results and conclusions of this paper.

2 Catastrophic Occurrences

Patterns seen in wars, terrorist attacks, tornadoes, earthquakes, landslides, floods, and other severe occurrences, have been at close attention by various researchers [4–6, 9, 12–15, 21–23]. Many attentive explanations have arisen in the literature. Nevertheless, a complete understanding of these patterns is a complicated task. Important and complex political, geographical, historical, and, even cultural, factors oppose to a better understanding. Predicting the number of casualties in natural or human-made disasters is extremely important in developing pre-disaster strategies. Aspects like rationalization of medical supplies and food, gathering emergency teams, organize shelter spaces, amongst others, have to be dealt with, in order to minimize the damage.

A PL behavior is indicative of a particular property of a system, it indicates that the size of an event is inversely proportional to its frequency. In this sense, large casualties are associated with low frequency phenomena, and more frequent events are less harmful in terms of preserving human lives [21, 22]. Examples of phenomena with low probability and huge casualties, are the two world wars (WW), high magnitude earthquakes, strong tornadoes, huge tsunamis, amongst others.

In 1948, Richardson [21], analyzed domestic and international cases of violence, in the period from 1820 to 1945. He distributed the cases, according to casualties measured in powers of ten, into five categories, being the two WWs in the highest category. In a later work [22], the same author showed that if the frequency of an occurrence decreased by a factor close to three, then the number of casualties increased by a power of ten.

Guzzetti [13] considers landslide events in specific periods in different countries, such as Italy, Canada, Alps, Hong Kong, Japan, and China. He shows that the plot of the cumulative distribution function of the number of landslide events vs the number of casualties is well approximated by a straight line. This result suggests a PL distribution of the data.

Cederman [6] followed Richardson's work [21, 22]. He used data from the Correlates of War (COW) Project [11], focusing on interstate wars. He computed the cumulative relative frequency of war size and showed that it obeyed a PL. The author proposed a self-organized critical dynamical system, that replicated the PL behavior seen in real data. Its model allowed conflict to spread and diffuse, potentially over long periods of time, due to the quasi-parallel execution.

In 2005, Jonkman [15] focused on the number of human deaths caused by three types of floods (river floods, flash floods and drainage issues), between January 1975 and June 2002. Highest average mortality was computed for flash floods. The author plotted of the global frequency of events with N or more deaths vs N. Nevertheless, the author did not find a PL behavior for flood data. Becerra et al. [2] use the same data set as Jonkman [15], but consider all disasters combined, both globally and disaggregated by continent. They obtained straight-line log-log plots for all disasters combined. The slopes of the casualties PL distributions were smaller than those for modern wars and terrorism. The explanation for this remained an open question. Another unsolved issue was the existence of PL behavior in combined disasters and not in individual disasters, such as floods. Here it is worth mentioning that casualties in earthquakes verified a PL distribution [2, 12, 15].

Johnson et al. [14] suggested a microscopic theory to explain similarity in patterns of violence, such as war and global terrorism. The similarity was observed regardless of underlying ideologies, motivations and the terrain in which events occurred. The authors introduced a model where the insurgent force behaved as a self-organizing system, which evolved dynamically through the continual coalescence and fragmentation of its constituent groups. They analyzed casualties' patterns arising within a given war, unlike previous studies that focused on the total casualty figure for one particular war [6, 18, 21, 22]. A PL behavior fitted well the data not only from Iraq, Colombia and non-G7 terrorism, but also with data obtained from the war in Afghanistan. The PL parameter for Iraq, Colombia and Afghanistan, was (close to) $\tilde{\alpha} = 2.5$. This value of the coefficient equalized the coefficient value characterizing non-G7 terrorism. In the literature, the PL parameter value was $\tilde{\alpha} = 2.51$ for non-G7 countries [7] and $\tilde{\alpha} = 1.713$ for G7 countries. This result suggested that PL patterns would emerge within any modern asymmetric war, fought by loosely-organized insurgent groups.

In 2006, Bogen and Jones [3] treated the severity of terrorist attacks, in terms of deaths and injured. They applied a PL distribution to victim/event rates and used the PL to predict mortality due to terrorism, through 2080. Authors claimed that these PL models could be used to improve strategies "to assess, prevent and manage terror-related risks and consequences".

Clauset et al. [8] studied the frequency and the number of casualties (deaths and injuries) of terrorist attacks, since 1968. They observed a scale-invariance behavior, with the frequency being an inverse power of the casualties. This behavior was independent of the type of weapon, economic development, and distinct time scales. The authors presented a new model to fit the frequency of severe terrorist attacks, since previous models in the literature failed to produce the heavy tail in the PL distribution. Their model assumed that the severity of an occurrence was a function of the execution plan, and that selection tools were better suited to model competition between states and non-state actors. Finally, researchers claimed that periodicity was a common feature in global terrorism, with period close to roughly 13 years.

Bohorquez et al. [4] studied the quantitative relation between human insurgency, global terrorism and ecology. They introduced a new model to explain the size distribution of casualties or the timing of within-conflict events. They considered insurgent populations as self-organized groups that dynamically evolved through decision-making processes. The main assumptions of the model were (i) being consistent with work on human group dynamics in everyday environments, (ii) a new perception of modern insurgencies, as fragmented, transient and evolving, (iii) the decision-making process about when to attack was based on competition for media attention. Authors applied a PL distribution to Iraq and Colombia wars, with parameter value close to $\tilde{\alpha} = 2.5$. A coefficient value of $\tilde{\alpha} = 2.5$ was in concordance with the coefficient value of $\tilde{\alpha} = 2.48 \pm 0.07$ obtained by Clauset et al. [8] on global terrorism. A PL fit to Spanish and American Civil Wars revealed a PL parameter value smaller (around $\tilde{\alpha} = 1.7$). Authors claimed that their model suggested a remarkable link between violent and non-violent human actions, due to its similarity to financial market models.

3 PL Behavior in Real Data

A double PL behavior is observed in various natural and man-made systems. In such cases, two different PLs, characterized by distinct PL parameters, fit to the real data. An example is given in Fig. 1, which represents the complementary cumulative distribution of the severity of worldwide terrorist attacks for the year 2005. The adopted measure to quantify the severity of an attack is the total number of fatalities. The depicted graph corresponds to a rank/frequency log-log plot. To construct the graph, we first sort the data (i.e., the terrorist attacks) in decreasing order according

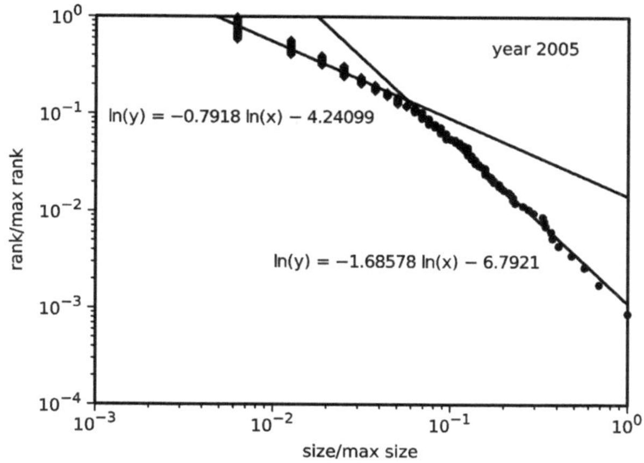

Fig. 1 Rank/frequency log-log plot corresponding to the distribution of the severity of terrorist attacks over the year 2005 (max size = 115; max rank = 2053)

to their severity, and number them, consecutively, starting from one [17]. Then a normalization of the values is carried out, meaning that, the number of fatalities (x-axis) is divided by the corresponding highest value, and the rank (y-axis) is divided by the rank of the smallest event. Finally, a PL is adjusted to the data using a least squares algorithm. All the log-log plots presented in this paper were made by following a similar procedure. The data used in this study was collected at the Global Terrorism Database (GTD) website (http://www.start.umd.edu/gtd). The GDT is an open-source database that includes information about more than 98,000 worldwide terrorist attacks, from 1970 through 2010 [18].

As can be seen in Fig. 1, a double PL distribution with parameters $(\tilde{C}_1, \tilde{\alpha}_1) = (0.0143, 0.7918)$ and $(\tilde{C}_2, \tilde{\alpha}_2) = (0.0011, 1.6858)$ fits well the data. The change in the behavior occurs at the relative value of $x = 0.0625$, approximately. Analyzing the period from 1980 up to 2010 (except 1993 because there is no data available), a similar behavior is found, meaning that in every year a double PL is observed. In Figs. 2, 3 the time evolution of the parameters $(\tilde{C}_i, \tilde{\alpha}_i)$, $i = 1, 2$ of the two PLs is shown. Regarding the parameters $\tilde{C}_1$ and $\tilde{C}_2$, it can be seen that they vary in a similar way, although $\tilde{C}_2$ is more *random* than $\tilde{C}_1$ and has values lower than it. With respect to $\tilde{\alpha}_1$ and $\tilde{\alpha}_2$, we have a similar evolution but, in this case, parameter $\tilde{\alpha}_2$ is always greater than $\tilde{\alpha}_1$. Generally speaking, a greater value of the exponent of the PL means that the events are more similar between each other than they are for lower values of $\tilde{\alpha}$. Hence, we can conclude that severe terrorist attacks are more evenly distributed, because they are characterized by a PL that has a higher exponent.

To complement the analysis with respect to the date of the occurrences, the parameters $(\tilde{C}_i, \tilde{\alpha}_i)$, $i = 1, 2$ of the PLs, corresponding to all events occurred in each year, were plotted (Fig. 4). It can be seen an interesting pattern emerging from the graph. The lines that connect each pair of points defining the double PLs are roughly aligned to each other. This graphical pattern reflects intrinsic properties of the recursive relationship behind each phenomenon. A comprehensive analysis for each type of application needs still further work to be clearly understood.

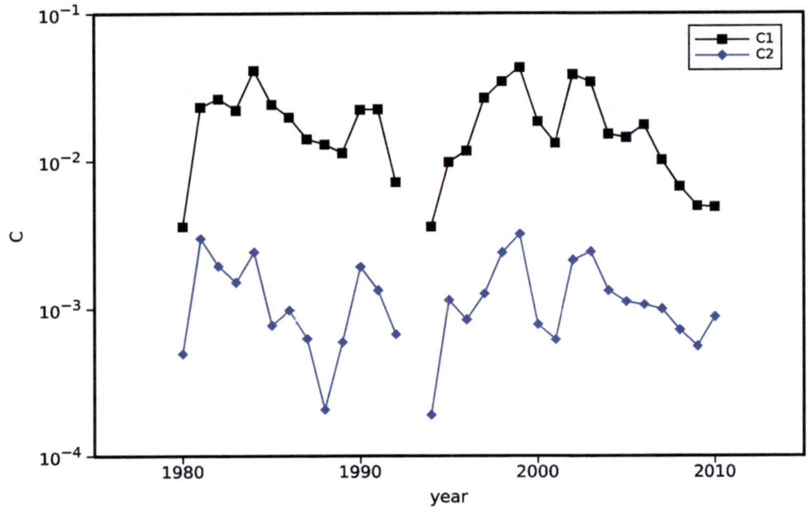

Fig. 2 Time evolution of parameters $\tilde{C}_1$ and $\tilde{C}_2$ over the period 1980–2010 (except 1993 because there is no data available)

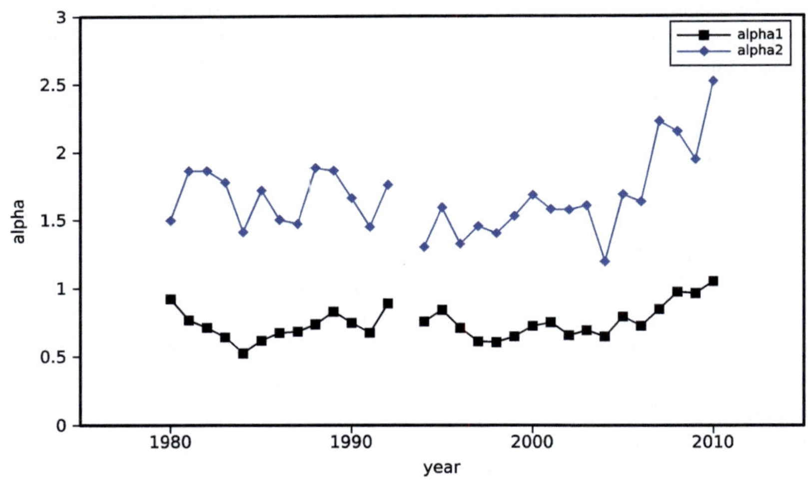

Fig. 3 Time evolution of parameters $\tilde{\alpha}_1$ and $\tilde{\alpha}_2$ over the period 1980–2010 (except 1993 because there is no data available)

4 Conclusion

In this paper we reviewed interesting and important results on PL distributions and their applications to the modeling of the number of victims in catastrophic events. We found double PL behavior in real data of catastrophic occurrences and, in particular, in terrorism. We have plotted the two PLs parameters, $(\tilde{C}_i, \tilde{\alpha}_i)$, $i = 1, 2$, corresponding

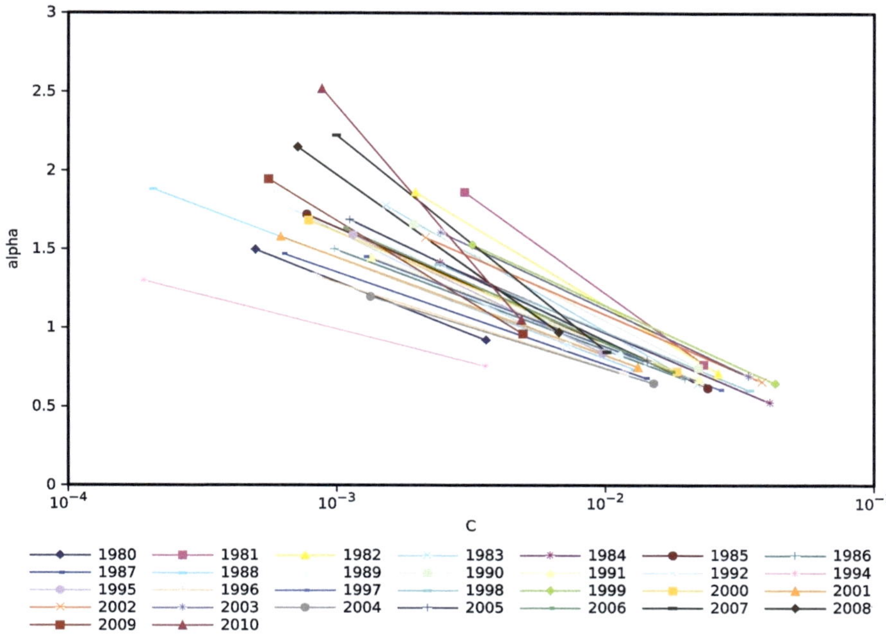

Fig. 4 Locus of the parameters $(\tilde{C}_i, \tilde{\alpha}_i)$, $i = 1, 2$, that characterize the distributions of terrorist attacks over the period 1980–2010 (except 1993 because there is no data available)

to all events occurred in each year, from 1980 to 2010. We observe an interesting pattern in the chart, where the lines, that connect each pair of points defining the double PLs that characterize every year, are roughly aligned to each other. More work is need in order to interpret these results.

References

1. Auerbach F (1913) Das Gesetz der Belvolkerungskoncentration. Petermanns Geogr Mitt 59: 74–76
2. Becerra O, Johnson N, Meier P, Restrepo J, Spagat M (2006) Casualties and power laws: a comparative analysis with armed conflict. Proceedings of the annual meeting of the American Political Science Association, Marriott, Loews Philadelphia, and the Pennsylvania Convention Center, Philadelphia. http://www.allacademic.com/meta/p151714_index.html
3. Bogen KT, Jones ED (2006) Risks of mortality and morbidity from worldwide terrorism: 1968–2004. Risk Anal 26:45–59
4. Bohorquez JC, Gourley S, Dixon AR, Spagat M, Johnson NF (2009) Common ecology quantifies human insurgency. Nature 462(7275):911–914
5. Carson M, Langer JS (1989) Mechanical model of an earthquake fault. Phys Rev A 40: 6470–6484
6. Cederman LE (2003) Modeling the size of wars: from billiard balls to sandpiles. Am Polit Sci Rev 97:135–150

7. Clauset A, Young M (2008) Scale invariance in global terrorism. *e-print* http://arxiv.org/abs/physics/0502014
8. Clauset A, Young M, Gleditsch KS (2007) On the frequency of severe terrorist events. J Confl Resolut 51(1):58–87
9. Davis DR, Weinstein DE (2002) Bones, bombs, and break points : the geography of economic activity. Am Econ Rev 92(5):1269–1289.
10. Estoup JB (1916) Gammes Stenographiques. Institut de France, Paris
11. Geller DS, Singer JD (1998) Nations at war: a scientific study of international conflict. Cambridge University Press, Cambridge
12. Gutenberg B, Richter CF (1944) Frequency of earthquakes in California. Bull Seismol Soc Am 34:185–188
13. Guzzetti F (2000) Landslide fatalities and the evaluation of landslide risk in Italy. Eng Geol 58:89–107
14. Johnson NF, Spagat M, Restrepo JA, Becerra O, Bohorquez JC, Suarez N, Restrepo EM, Zarama R (2006) Universal patterns underlying ongoing wars and terrorism. http://arxiv.org/abs/physics/0605035
15. Jonkman SN (2005) Global perspectives on loss of human life caused by floods. Nat Hazards 34:151–175
16. Li W (2003) References on Zipf's law. http://ccl.pku.edu.cn/doubtfire/NLP/Statistical_Approach/Zip_law/references%20on%20zipf%27s%20law.htm
17. National Consortium for the Study of Terrorism and Responses to Terrorism (START) (2011) Global Terrorism Database [Data file]. http://www.start.umd.edu/gtd/search/Results.aspx?start_month=0&end_month=12&start_year=2011&end_year=2011&start_day=0&end_day=31
18. Newman MEJ (2005) Power laws, Pareto distributions and Zipf's law. Contemp Phys 46: 323–351
19. Pareto V (1896) Cours d'Economie Politique. Droz, Geneva
20. Pinto CMA, Lopes AM, Tenreiro Machado JA (2012) A review of power laws in real life phenomena. Commun Nonlinear Sci Numer Simul 17(9):3558–3578
21. Richardson LF (1948) Variation of the frequency of fatal quarrels with magnitude. J Am Stat Assoc 43:523–546
22. Richardson LF (1960) Statistics of deadly quarrels. Quadrangle, Chicago
23. Roberts DC, Turcotte DL (1998) Fractality and selforganized criticality of wars. Fractals 6: 351–357
24. Sornette D (2003) Critical phenomena in natural sciences, 2nd edn. Springer, Heidelberg (Chap. 14)
25. Zipf G (1932) Selective studies and the principle of relative frequency in language. Harvard University Press, Cambridge
26. Zipf G (1949) Human behavior and the principle of least effort. Addison-Wesley, Cambridge

Optimization of Quadruped Robot Locomotion Gaits Through a Genetic Algorithm

Manuel F. Silva

Abstract During the last years research and development on legged robots has grown steadily. Legged systems present major advantages when compared with "traditional" vehicles, allowing locomotion in terrain inaccessible to vehicles with wheels and tracks. However, its energy consumption still lag being these vehicles, existing several aspects that need to be improved and optimized. One of them regards the parameters values that these machines should adopt to minimize the energy consumption. Due to the large number of parameters involved in this optimization process, one way to achieve meaningful results is using evolutionary strategies. Genetic Algorithms are a way to "imitate nature" replicating the process that nature designed for the generation and evolution of species. The objective of this paper is to present a genetic algorithm, running over a simulation application of legged robots, which allows the optimization of several parameters of a quadruped robot model, for distinct locomotion gaits.

Keywords Legged robots · Locomotion · Gait · Optimization · Genetic algorithms

1 Introduction

Several walking robots have been developed up to date [1]. Compared with traditional vehicles with wheels and tracks, their major advantage is the fact of allowing locomotion in terrain inaccessible to other type of vehicles, because they do not need a continuous support surface.

Since legged locomotion robots are inspired in animals observed in nature, a frequent approach to their design is to make a mechatronic mimic of the animal that is intended to replicate, either in terms of its physical dimensions, or in terms of characteristics such as the gait and the actuation of the limbs. Several examples of robots that have been developed based on this approximation are discussed by Silva and Machado [1].

M. F. Silva (✉)
INESC TEC — INESC Technology and Science (formerly INESC Porto)
and ISEP/IPP—School of Engineering, Polytechnic Institute of Porto, Porto, Portugal
e-mail: mss@isep.ipp.pt

N. M. Fonseca Ferreira, J. A. Tenreiro Machado (eds.), *Mathematical Methods in Engineering,* DOI 10.1007/978-94-007-7183-3_17,

However, in the present state of development, there are several aspects that need to be improved and optimized in these machines. With this idea in mind, different optimization strategies have been proposed and applied to these systems, either during its design and construction phases, or during its operation [2].

One possibility makes use of genetic algorithms (GAs) as the engine to generate robot structures. GAs are an alternative way of imitating nature. Animals characteristics are not directly copied but, instead, is replicated the process that nature conceives for its generation and evolution.

In some cases it is performed a GA modular approach to the robot design [3–5]. There are also works on which evolutionary strategies are used to optimize the structure of a specific robot [6, 7]. Other authors proposed the use of GAs for the simultaneous generation of the mechanical structure and the robot controller, for distinct types of robots [8–12].

Bearing these ideas in mind, the objective of this paper is to present a GA, running over a simulation application of legged robots, which allows the optimization of a quadruped robot model parameters, for distinct locomotion gaits often used by animals moving at different velocities.

The remainder of this paper is organized as follows. Section 2 presents the robot model and its control architecture. Sections 3 and 4 present the implemented GA, and some simulation results, respectively. Finally, Sect. 5 outlines the main conclusions of this study.

2 Robot Model and Control Architecture

2.1 Kinematics and Trajectory Planning

We consider the model of a legged robot (Fig. 1, left) with $n = 4$ legs, equally distributed along both sides of the robot body, having each one two rotational joints (*i.e.*, $j = \{1, 2\} \equiv \{\text{hip, knee}\}$) [13].

Motion is described by means of a world coordinate system. The kinematic model comprises: the cycle time T, the duty factor β, the transference time $t_T = (1-\beta)T$, the support time $t_S = \beta T$, the step length L_S, the stroke pitch S_P, the body height H_B, the maximum foot clearance F_C, the ith leg lengths L_{i1} and L_{i2} (being the total length of each robot leg equal to 1 m) and the foot trajectory offset O_i ($i = 1, \ldots, n$). Moreover, a periodic trajectory for each foot is considered, with body velocity $V_F = L_S/T$.

Gaits describe sequences of leg movements, alternating between transfer and support phases. In this work are considered three walking gaits (Walk, Chelonian Walk and Amble), two symmetrical running gaits (Trot and Pace) and five asymmetrical running gaits (Canter, Transverse Gallop, Rotary Gallop, Half-Bound and Bound). These are the gaits usually adopted by animals moving at low, moderate and high speed, respectively, being their main characteristics presented in Table 1.

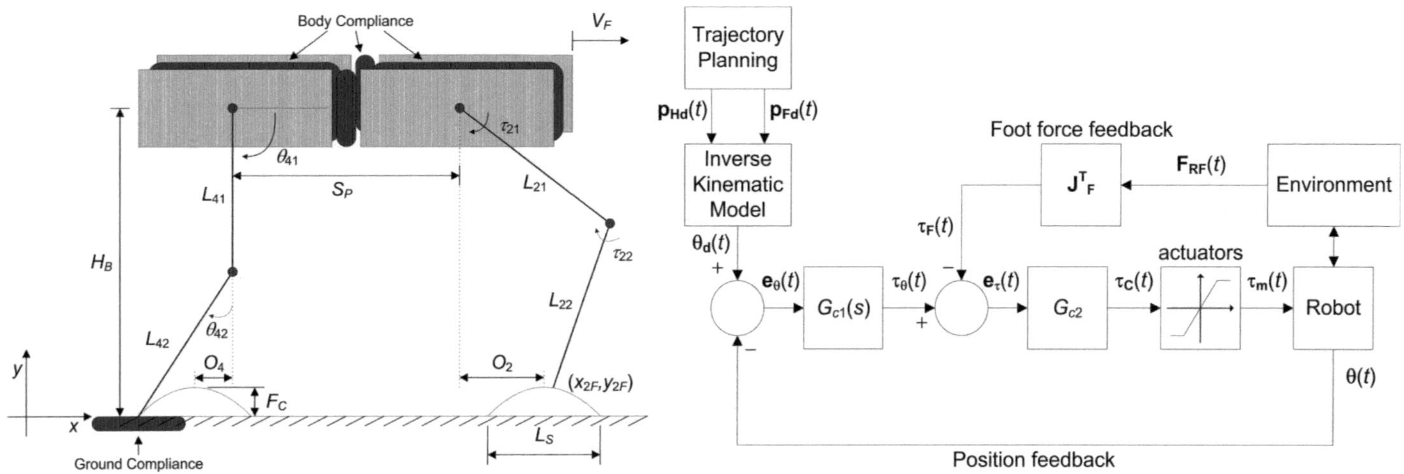

Fig. 1 Kinematic and dynamic quadruped robot model (*left*) and its control architecture (*right*)

Table 1 Quadruped gait parameters

Gait	ϕ_1	ϕ_2	ϕ_3	ϕ_4	β
Walk	0	0.5	0.75	0.25	0.65
Chelonian Walk	0	0.5	0.5	0	0.8
Amble	0	0.5	0.75	0.25	0.45
Trot	0	0.5	0.5	0	0.4
Pace	0	0.5	0	0.5	0.4
Canter	0	0.3	0.7	0	0.4
Transverse gallop	0	0.2	0.6	0.8	0.3
Rotary gallop	0	0.1	0.6	0.5	0.3
Half-bound	0.7	0.6	0	0	0.2
Bound	0	0	0.5	0.5	0.2

Given the particular gait and the duty factor β, it is possible to calculate, for leg i, the corresponding phase ϕ_i, the time instant where each leg leaves and returns to contact with the ground and the Cartesian trajectories of the tip of the feet (that must be completed during t_T) [14]. Based on this data, the trajectory generator is responsible for producing a motion that synchronises and coordinates the legs.

The robot body, and by consequence the legs hips, is assumed to have a desired horizontal movement with a constant forward speed V_F, being the Cartesian coordinates of the hip of the legs, for leg i, given by $\mathbf{p_{Hd}}(t) = [x_{iHd}(t), y_{iHd}(t)]^{\mathbf{T}}$ [13].

Regarding the feet trajectories, for each cycle, the desired trajectory of the foot of the swing leg is computed through a cycloid function and described by (for leg i) $\mathbf{p_{Fd}}(t) = [x_{iFd}(t), y_{iFd}(t)]^{\mathbf{T}}$ [13].

The algorithm for the forward motion planning accepts, as inputs, the desired Cartesian trajectories of the leg hips $\mathbf{p_{Hd}}(t)$ and feet $\mathbf{p_{Fd}}(t)$ and, by means of an inverse kinematics algorithm ψ^{-1}, generates as outputs the joint trajectories $\mathbf{\Theta_d}(t) = [\mathbf{\Theta}_{i1d}(t), \mathbf{\Theta}_{i2d}(t)]^{\mathbf{T}}$ [13], that constitute the reference for the robot control system. In this study it is adopted the mammal leg configuration, namely selecting in ψ^{-1} the solution corresponding to a forward knee.

In order to avoid the impact and friction effects, at the planning phase null velocities of the feet are considered in the instants of landing and taking off, assuring also the velocity continuity.

2.2 Robot Dynamic Model

2.2.1 Inverse Dynamics Computation

The model for the robot inverse dynamics is formulated as [13]:

$$\mathbf{\Gamma} = \mathbf{H}(\mathbf{\Theta})\ddot{\mathbf{\Theta}} + \mathbf{c}\left(\mathbf{\Theta}, \dot{\mathbf{\Theta}}\right) + \mathbf{g}(\mathbf{\Theta}) - \mathbf{F_{RH}} - \mathbf{J}^{\mathbf{T}}(\mathbf{\Theta})\mathbf{F_{RF}} \quad (1)$$

where $\mathbf{\Gamma}$ is the vector of forces/torques, $\mathbf{\Theta}$ is the vector of position coordinates, $\mathbf{H(\Theta)}$ is the inertia matrix and $\mathbf{c}\left(\mathbf{\Theta}, \dot{\mathbf{\Theta}}\right)$ and $\mathbf{g(\Theta)}$ are the vectors of centrifugal/Coriolis and gravitational forces/torques, respectively. The matrix $\mathbf{J^T(\Theta)}$ is the transpose of the robot Jacobian matrix, $\mathbf{F_{RH}}$ is the vector of the body inter-segment forces and $\mathbf{F_{RF}}$ is the vector of the reaction forces that the ground exerts on the robot feet, being null during the foot transfer phase.

Moreover, the joint actuators are not considered ideal, exhibiting a saturation, being the maximum torque that each actuator can supply τ_{ijMax}.

2.2.2 Robot Body Model

The dynamic model for the hexapod body and foot-ground interaction (Fig. 1) considers a compliant robot body, divided in n identical segments (each with mass $M_b n^{-1}$, while making the total mass of the robot equal to 100 kg) and a linear spring-damper system is adopted to implement the intra-body [13]. The parameters of this spring-damper system, $K_{\eta H}$ and $B_{\eta H}$ ($\eta = \{x, y\}$ in the {horizontal, vertical} directions, respectively), are defined so that the body behavior is similar to the one expected to occur on an animal [13].

2.2.3 Foot-Ground Interaction Model

The contact of the ith robot foot with the ground is modelled through a non-linear system (Fig. 1) with linear stiffness $K_{\eta F}$ and non-linear damping $B_{\eta F}$ ($\eta = \{x, y\}$ in the {horizontal, vertical} directions, respectively) [15]. The values for the parameters $K_{\eta F}$ and $B_{\eta F}$ are based on the studies of soil mechanics [15].

2.3 *Control Architecture*

The general control architecture of the multi-legged locomotion system is depicted in Fig. 1 (right), being $G_{c1}(s)$ a PD controller and G_{c2} a simple P controller [15]. The trajectory planning is held in the Cartesian space, but the control is performed in the joint space, which requires the integration of the inverse kinematic model in the forward path. The control algorithm considers an external position and velocity feedback and an internal feedback loop with information of foot-ground interaction force.

3 Developed Genetic Algorithm

GAs are adaptive methods which may be used to solve search and optimization problems [16]. By mimicking the principles of natural selection, GAs are able to evolve solutions towards an optimal one. Although the optimal is not guaranteed, the GA is a stochastic search procedure that, usually, generates good results. The

GA maintains a population of candidate solutions (the individuals). Individuals are evaluated and fitness values are assigned based on their relative performance. They are then given a chance to reproduce, i.e., replicating several of their characteristics. The offspring produced are modified by means of mutation and/or recombination operators before they are evaluated and reinserted in the population. This is repeated until some condition is satisfied.

3.1 Structure of the Used Chromosome

The chromosome used in the developed GA presents 34 genes (i.e., 34 robot parameters). The genes are organized as presented in Table 2: the first gene (L_s) contains information regarding the step length and the last gene (Kd_{22}) contains the derivative gain of joint 2 of the robot rear legs. These values are coded directly into real numbers (value encoding).

3.2 Measure for the Fitness Evaluation

For the fitness function is used the mean absolute density of energy per travelled distance E_{av} [17]. This index is computed assuming that energy regeneration is not available by actuators doing negative work (by taking the absolute value of the power). At a given joint j (each leg has $m = 2$ joints) and leg i ($n = 4$ legs since a quadruped is being considered), the mechanical power is the product of the motor torque and angular velocity. The global index E_{av} is obtained by averaging the mechanical absolute energy delivered over the travelled distance d:

$$E_{av} = \frac{1}{d} \sum_{i=1}^{n} \sum_{j=1}^{m} \int_{0}^{T} \left| \tau_{ij}(t)\, \dot{\Theta}_{ij}(t) \right| dt \quad \left[\mathrm{Jm}^{-1}\right] \tag{2}$$

The performance optimization is achieved through the minimization of the index E_{av}.

3.3 Base Structure of the Developed GA

The outline of the specific GA is as follows:

1. **Start:** Generate a random population of $v = 50$ chromosomes. The values for the genes that constitute the chromosome are uniformly distributed in the ranges mentioned in Table 2.
2. **Simulation:** Simulate the robot locomotion for all chromosomes in the population.
3. **Fitness:** Evaluate the fitness function for each chromosome by computing E_{av}.

Table 2 Interval of variation of the 34 genes used in the chromosome

Minimum Value	Variable	Maximum Value
$0 <$	L_s	≤ 10 m
$0 <$	H_B	≤ 1 m
$0 <$	F_C	≤ 1 m
$0 <$	L_{11}	≤ 1 m
$0 <$	L_{12}	≤ 1 m
$0 <$	L_{21}	≤ 1 m
$0 <$	L_{22}	≤ 1 m
$0 <$	O_1	≤ 10 m
$0 <$	O_2	≤ 10 m
$0 <$	M_b	≤ 100 kg
$0 <$	M_{11}	≤ 10 kg
$0 <$	M_{12}	≤ 10 kg
$0 <$	M_{21}	≤ 10 kg
$0 <$	M_{22}	≤ 10 kg
$0 <$	K_{xh}	≤ 10000 Nm
$0 <$	K_{yh}	≤ 10000 Nm
$0 <$	B_{xh}	≤ 10000 Nms^{-1}
$0 <$	B_{yh}	≤ 10000 Nms^{-1}
$-400 <$	$\tau_{11\ min}$	≤ 0 Nm
$0 <$	τ_{11Max}	≤ 400 Nm
$-400 <$	$\tau_{12\ min}$	≤ 0 Nm
$0 <$	τ_{12Max}	≤ 400 Nm
$-400 <$	$\tau_{21\ min}$	≤ 0 Nm
$0 <$	τ_{21Max}	≤ 400 Nm
$-400 <$	$\tau_{22\ min}$	≤ 0 Nm
$0 <$	τ_{22Max}	≤ 400 Nm
$0 <$	Kp_{11}	≤ 10000
$0 <$	Kd_{11}	≤ 1000
$0 <$	Kp_{12}	≤ 10000
$0 <$	Kd_{12}	≤ 1000
$0 <$	Kp_{21}	≤ 10000
$0 <$	Kd_{21}	≤ 1000
$0 <$	Kp_{22}	≤ 10000
$0 <$	Kd_{22}	≤ 1000

4. **New population:** Create a new population by repeating the following steps:
   - Selection—Select the four best parent chromosomes according to their fitness. These solutions are copied without changes to the new population (elitism).
   - Crossover—Select 90 % of the individuals to be replaced by the crossover of the parents: two random parents are chosen and an arithmetic mean operation is performed to produce one new offspring.
   - Mutation—Select 1 % of the individuals to be replaced by mutation of the parents: one random parent is chosen and a small number is added to selected values, to make a new offspring.
   - Spontaneous generation—The remaining individuals are replaced by new randomly generated ones (such as in step 1).
5. **Loop:** If this iteration is the 500th or the GA has converged (the value of the fitness function for the chromosome with the best fitness function is equal to the one that is in the position corresponding to 90 % of the population), stop the algorithm, else, go to step 2.

Table 3 Optimum values for the hexapod parameters while walking with the Walk Gait, being $V_F = 1\ ms^{-1}$, $E_{av} = 500.002$ J/m and the travelled distance $d = 1.281$ m

Parameter	Optimum Value
L_s	= 1.056 m
H_B	= 0.724 m
F_C	= 0.076 m
L_{11}	= 0.484 m
L_{12}	= 0.516 m
L_{21}	= 0.425 m
L_{22}	= 0.575 m
O_1	= − 0.383 m
O_2	= − 0.035 m
M_b	= 83.134 kg
M_{11}	= 4.976 kg
M_{12}	= 2.923 kg
M_{21}	= 6.485 kg
M_{22}	= 2.482 kg
K_{xh}	= 79991.055 Nm
K_{yh}	= 9084.575 Nm
B_{xh}	= 991.235 Nms^{-1}
B_{yh}	= 92.299 Nms^{-1}
τ_{11min}	= − 296.987 Nm
τ_{11max}	= 105.782 Nm
τ_{12min}	= − 136.718 Nm
τ_{12max}	= 145.311 Nm
τ_{21min}	= − 287.426 Nm
τ_{21max}	= 115.196 Nm
τ_{22min}	= − 283.489 Nm
τ_{22max}	= 342.611 Nm
Kp_{11}	= 3012.207
Kd_{11}	= 789.264
Kp_{12}	= 4395.400
Kd_{12}	= 165.975
Kp_{21}	= 3202.196
Kd_{21}	= 543.265
Kp_{22}	= 5429.295
Kd_{22}	= 156.955

4 Simulation Results

The main objective of this study was to find the optimal values for the robot model and controller parameters, considering that the robot was moving with $V_F = 1\ ms^{-1}$, while adopting the gaits Walk (1), Chelonian Walk (2), Amble (3), Trot (4), Pace (5), Canter (6), Transverse Gallop (7), Rotary Gallop (8), Half-Bound (9) and Bound (10).

This study started by determining the optimal values for the robot model and controller parameters, considering a robot moving at $V_F = 1\ ms^{-1}$, with the Walk Gait. Running the GA, with the parameters described in Sect. 3.3, the algorithm converged to the results given in Table 3.

Analyzing the results presented in Table 3 it should be referred that the length of the upper segment of the leg should be smaller than the corresponding length of the

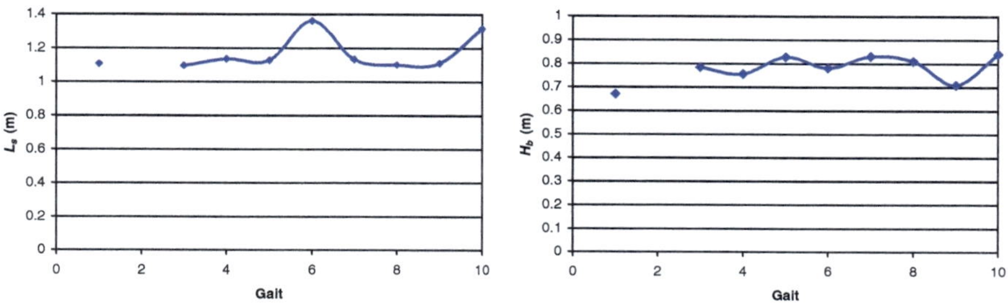

Fig. 2 Optimum values of the Step Length L_S (*left*) and Body Height H_B (*right*), for the gaits under study

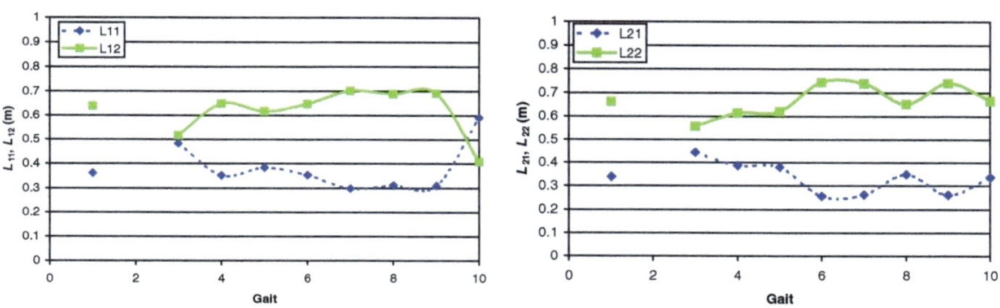

Fig. 3 Optimum values of the front legs links lengths L_{11} and L_{12} (*left*) and of the rear legs links lengths L_{21} and L_{22} (*right*), for the gaits under study

lower segment. In the same way, the upper segment of the leg should be heavier than the lower segment. Finally, the trajectory of the legs must be displaced to the rear of the moving direction, as indicated by the values of the parameters O_i.

Following, the GA was executed with the same parameters, for the distinct gaits under analysis. The algorithm converged to the results that are described in the sequel. There was one exception: for the Chelonian Walk gait the GA did not converge, although several attempts (distinct runs of the GA) were made.

Figure 2 depicts two charts with the optimum values of the Step Length (L_S) (left) and Body Height (H_B) (right), for the distinct gaits under consideration, determined by the GA. It is seen that the robot should move with a value of $L_S \approx 1.1$ m, except for the Canter and Bound gaits, for which case it should be $L_S \approx 1.3$ m. Concerning the value for the Body Height, analyzing the chart presented in the left part of this figure, one can conclude that the robot should adopt a value of $H_B \approx 0.8$ m.

Figure 3 depicts two charts with the optimum values of the front legs links lengths L_{11} and L_{12} (left) and of the rear legs links lengths L_{21} and L_{22} (right), for the distinct gaits under study, determined by the GA. Analyzing the results presented in these figures it should be referred that the length of the upper segment of the leg should be smaller than the corresponding length of the lower segment. The relation between the lengths of both segments is $L_{i1}/L_{i2} \approx 0.3/0.7$.

There are only two exceptions to these general results; for the case of the front legs, when the robots adopts the Amble gait both segments should be of similar lengths and

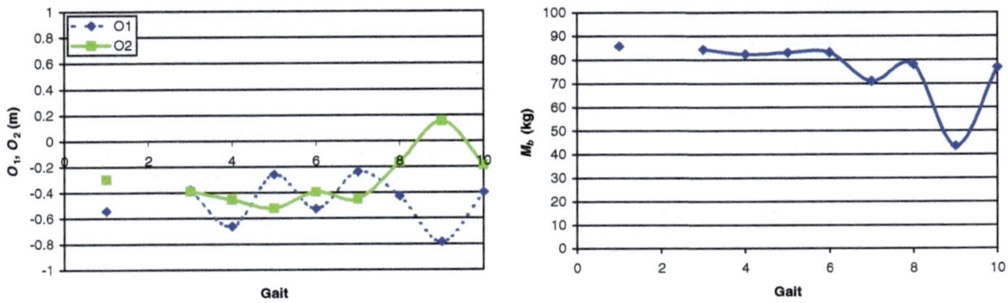

Fig. 4 Optimum values of the foot trajectory offset O_1 and O_2 (*left*) and of the Body Mass M_B (*right*), for the gaits under study

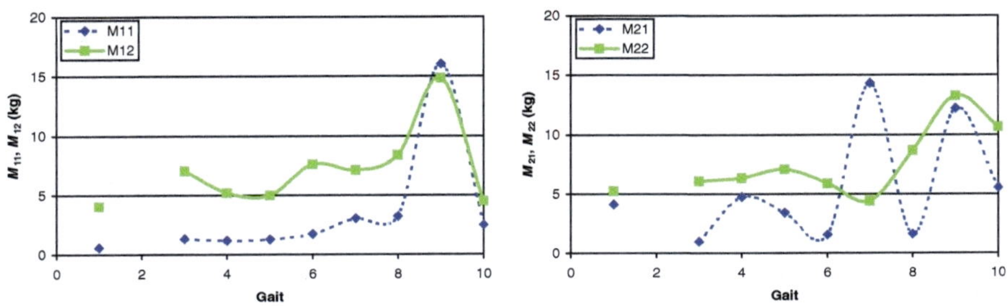

Fig. 5 Optimum values of the front legs link masses M_{11} and M_{12} (*left*) and of the rear legs link masses M_{21} and M_{22} (*right*), for the gaits under study

when the quadruped adopts the Bound gait the results obtained by the GA indicate that the lower segment of the leg should be smaller than the corresponding upper segment.

In Fig. 4 are presented two charts with the optimum values of the foot trajectory offset O_1 and O_2 (left) and of the Body Mass (M_B) (right), determined by the GA, for the gaits under study. Concerning the foot trajectory offset, the results presented in this chart indicate that the robot should move with the feet trajectory displaced to the rear of the hip trajectory (in the moving direction), as indicated by the values of the parameters O_i. Regarding the robot mass distribution, the body should concentrate most of its value (it is assumed that the total mass of the robot is equal to 100.0 kg) being $M_B > 70$ kg for all gaits under study, except for the Half-Bound.

Finally, Fig. 5 shows two charts with the optimum values of the front legs link masses M_{11} and M_{12} (left) and of the rear legs link masses M_{21} and M_{22} (right), determined by the GA, for the gaits under study. The left chart indicates that the lower segment of the front legs should be heavier than the upper segment ($M_{12} > M_{11}$), except for the Half-Bound gait.

In a similar manner, the right chart indicates that the lower segment of the rear legs should be heavier than the upper segment ($M_{22} > M_{21}$), except for the Transverse Gallop gait. These results seem to agree with the ones presented in Fig. 3, since longer legs links segments are heavier.

5 Conclusions

This paper presented a GA developed for the optimization of quadruped robot parameters. This GA runs over a simulation application of legged robots (developed in the C programming language), which allows the optimization of several parameters of the robot model and of its gaits for different locomotion speeds.

Based on this GA, were determined the optimum locomotion parameters for the quadruped robot and its controller, while the robot is moving at $V_F = 1 ms^{-1}$ with distinct gaits.

As ideas for future work, the author plans to develop several simulation experiments to find the parameters that optimize the robot locomotion, from the viewpoint of the index E_{av}, for different values of V_F in the range $0.1 \leq V_F \leq 10.0\, ms^{-1}$.

Acknowledgments To Sérgio Carvalho, for implementing the basic structure of the GA used in this work.

References

1. Silva MF, Machado JAT (2007) A historical perspective of legged robots. J Vib Control 13 (9–10):1447–1486
2. Silva MF, Machado JAT A literature review on the optimization of legged robots. J Vib Control, accepted for publication
3. Farritor S, Dubowsky S, Rutman N, Cole J (1996) A systems-level modular design approach to field robotics. Proceedings of the IEEE international conference on robotics and automation, pp 2890–2895
4. Leger C (2000) DARWIN2K: An evolutionary approach to automated design for robotics. Kluwer Academic, Hingham
5. Nolfi S, Floreano D (2000) Evolutionary robotics - the Biology, intelligence, and technology of self-organizing machines. The MIT Press, Cambridge
6. Juárez-Guerrero J, Muñoz-Gutiérrez S, Cuevas WWM (1998) Design of a walking machine structure using evolutionary strategies. Proceedings of the IEEE international conference on systems, man and cybernetics, pp 1427–1432
7. Ishiguro A, Kawasumi K, Fujii A (2002) Increasing evolvability of a locomotion controller using a passive-dynamic-walking embodiment. Proceedings of the IEEE/RSJ international conference on intelligent robots and systems, pp 2581–2586
8. Lipson H, Pollack JB (2000) Towards continuously reconfigurable self-designing robots. Proceedings of the IEEE international conference on robotic and automation, pp 1761–1766
9. Endo K, Yamasaki F, Maeno T, Kitano H (2002) A method for co-evolving morphology and walking pattern of biped humanoid robot. Proceedings of the IEEE international conference on robotic and automation, pp 2775–2780
10. Endo K, Maeno T, Kitano H (2002) Co-evolution of morphology and walking pattern of biped humanoid robot using evolutionary computation. Consideration of characteristic of the servomotors. Proceedings of the IEEE/RSJ international conference on intelligent robots and systems, pp 2678–2683
11. Marbach D, Ijspeert AJ (March 10–13, 2004) Co-evolution of configuration and control for homogenous modular robots. Proceedings of IAS'2004-8th conference on intelligent autonomous systems, Amsterdam, pp 712–719

12. Hollinger GA, Briscoe JM (April 2005) Genetic optimization and simulation of a piezoelectric pipe-crawling inspection robot. Proceedings of the IEEE international conference on robotic and automation, Barcelona, pp 486–491
13. Silva MF, Machado JAT, Lopes AM (2005) Modelling and simulation of artificial locomotion systems. ROBOTICA 23(5):595–606
14. Song S-M, Waldron KJ (1989) Machines that walk: the adaptive suspension vehicle. MIT Press, Cambridge
15. Silva MF, Machado JAT, Lopes AM (2003) Position/force control of a walking robot. Mach Intelligence Robotic Control 5(2):33–44
16. Fleming PJ, Purshouse RC (2002) Genetic algorithms in control systems engineering. IFAC Professional Brief
17. Silva MF, Tenreiro Machado JA (January 2008) Kinematic and dynamic performance analysis of artificial legged systems. ROBOTICA 26(1):19–39

Analysis of an Incomplete Information System Using the Rough Set Theory

C. I. Faustino Agreira, M. M. Travassos Valdez, C. M. Machado Ferreira and F. P. Maciel Barbosa

Abstract In this paper it is applied a Rough Set approach that takes into account an incomplete information system to study the steady-state security of an electric power system. The Rough Set Theory has been conceived as a tool to conceptualize, organize and analyze various types of data, in particular, to deal with inexact, uncertain or vague knowledge. The knowledge acquisition process is a complex task, since the experts have difficulty to explain how to solve a specified problem. So, an incomplete set of relevant information may arise. The study presents a systematic approach to transform examples in a reduced set of rules. These rules can be used successfully to avoid security problems and provides a deeper insight into the influence of parameters on the steady-state system performance.

Keywords Incomplete information systems · Rough set theory

1 Introduction

Recently, the Rough Sets theory (RST) has been used successfully to handle efficiently problems where large amounts of data are produced [1]. RST constitutes a framework for inducing minimal decision rules. These rules can be used in turn to perform a classification task. Important concepts include the elimination of redundant criteria to give more compact rules. The strength of a rule can be quantified using rough membership. The main goal of the rough set analysis is to search large databases for meaningful decision rules and, finally, acquire new knowledge. This approach is based in four main topics: indiscernibility, approximation, reducts and decision rules [1]. A reduct is a minimal set of attributes, from the whole attributes set, that preserves the partitioning of the finite set of objects and, therefore, the original classes. It means that the redundant attributes are eliminated. When the

C. I. Faustino Agreira (✉) · M. M. Travassos Valdez · C. M. Machado Ferreira
Instituto Superior de Engenharia de Coimbra, Instituto Politécnico de Coimbra, Coimbra, Portugal
e-mail: cif@isec.pt

F. P. Maciel Barbosa
Inesc Tec and Faculdade de Engenharia, University of Porto, Porto, Portugal
e-mail: fmb@fe.up.pt

N. M. Fonseca Ferreira, J. A. Tenreiro Machado (eds.), *Mathematical Methods in Engineering,* DOI 10.1007/978-94-007-7183-3_18,

reducts are evaluated, the task of creating definite rules for the value of the decision attribute of the information system is practically performed. Decision rules are generated combining the attributes of the reducts with the values. Decision rules extracte knowledge, that can be used when classifying new objects not in the original information system.

The RST has been conceived as a tool to conceptualize, organize and analyze various types of data, in particular, to deal with inexact, uncertain or vague knowledge. In the Rough Sets analysis the concept of an information system is used to construct the approximation space. It enables representation of data in a useful form of a table. The information system is, in fact, a finite data table where columns are labelled by attributes and rows are labelled by objects [2]. Attributes are generally classified into conditions and decisions. Usually, a number of condition attributes and a single decision attribute are presented. In an incomplete information system the attribute values for objects may be unknown (missing or null) [3].

The knowledge acquisition process is a complex task, since the experts have difficulty to explain how to solve a specified problem. So, an incomplete set of relevant information may arise. This study presents a systematic approach to transform examples in a reduced set of rules [4]. These rules can be used successfully to avoid security problems and provides a deeper insight into the influence of parameters on the system performance.

This paper is organised as follows. Section 1 presents an introduction to the problem. Section 2 is devoted to the Rough Set Theory considering an incomplete information system. In Sect. 3 is presented the test power network that was analysed and shows the results obtained using the proposed approach [5]. Finally, in Sect. 4, some conclusions that provide a valuable contribution to the understanding the RST applied to the security analysis of the electric power system are presented.

2 Rough Set Theory

Rough Set Theory can be considered as an extension of the Classical Set Theory, for use when representing incomplete knowledge. Rough sets can be considered sets with fuzzy boundaries—sets that cannot be precisely characterized using the available set of attributes. Many different problems can be addressed by RST. During the last few years this formalism has been approached as a tool used in connection with many different areas of research. It has also been used for, among many others, knowledge representation, data mining, dealing with imperfect data, reducing knowledge representation and for analysing attribute dependencies.

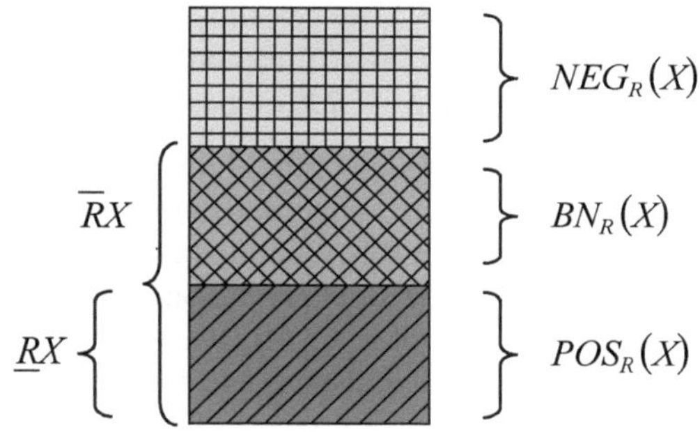

Fig. 1 Definition of *R*-approximation sets and *R*-regions

2.1 Information System

Information System (IS) can be defined as a K = (U, R, V, ρ), where U is a finite set of objects, R is a finite set of attributes, V is the domain of each attribute of R, and ρ is a total function that defines the following application: ρ. U × R→V, i.e, the examples. Va is called the value set of a.

2.2 Approximations Sets

The Rough Set Theory (RST) is a new mathematical tool presented to dispose incomplete and uncertainty problem [1]. It works with lower and upper approximation of a set as it is shown in Fig. 1. The discernibility relation is used for two basic operations in rough set theory i.e. upper $\overline{R}X$ and lower $\underline{R}X$ approximations, which defines crisp and vague manner in the sets. If any concept of the universe can be formed as a union of some elementary sets, it is referred as crisp (precise). On the contrary, if the concept cannot be presented in such a way, it is referred as vague (imprecise, rough). $\underline{R}X$ is defined as the collection of cases whose equivalence classes are fully contained in the set of cases to approximate. $\overline{R}X$ is defined as the collection of cases whose equivalence classes are at least partially contained in (i.e. overlap with) the set of cases to approximate [6].

There are five regions of interesting: $\overline{R}X$ and $\underline{R}X$, and $POS_R(X)$, $BN_R(X)$ and $NEG_R(X)$. These sets are defined as shown below.

Let a set $X \subseteq U$, R be an equivalence relation and knowledge. Two subsets base can be associated:

i) R—Lower: $\underline{R}X = U\{Y \in U/R : Y \subseteq X\}$
ii) R—Upper: $\overline{R}X = U\{Y \in U/R : Y \cap X \neq \varphi\}$

It means that the elements belonging to $\underline{R}X$ set can be with certainty classified as elements of X; while the elements belong to $\overline{R}X$ set can be possibly classified as elements of X. In the same way, $POS_R(X)$, $BN_R(X)$ and $NEG_R(X)$ are defined below [1].

iii) $POS_R(X) = \underline{R}X \Rightarrow$ certainly member of X
iv) $NEG_R(X) = U - \overline{R}X \Rightarrow$ certainly non member of X
v) $BN_R(X) = \overline{R}X - \underline{R}X \Rightarrow$ possibly member of X

Before the presentation of the algorithm, it is necessary to define two major concepts in Rough Set Theory, reduct and core. These concepts are important in the knowledge base reduction. Let R be a family of equivalence relations. The reduct of R, RED(R), is defined as a reduced set of relations that conserves the same inductive classification of set R. The core of R, CORE(R), is the set of relations that appears in all reduct of R, i.e., the set of all indispensable relations to characterize the relation R. As the core is the intersection of all reducts, it is included in every reduct, i.e., each element of the core belongs to some reduct. Therefore, in a sense, the core is the most important subset of attributes, since none of its elements can be removed without affecting the classification strength of attributes.

The approximation of classification is a simple extension of the definition of approximation of sets. Namely if $F = \{X_1, X_2, \ldots, X_N\}$ is a family of non empty sets, then $\underline{R}F = \{\underline{R}X_1, \underline{R}X_2, \ldots, \underline{R}X_n\}$ and $\bar{R}F = \{\bar{R}X_1, \bar{R}X_2, \ldots, \bar{R}X_n\}$, are called the RF—lower and the $\bar{R}F$—upper approximation of the family F [3].

Two measures can be defined to describe inexactness of approximate classification. The first one is the extension of the measure defined to describe accuracy of approximation sets.

The accuracy of approximation of F by R is defined as [1]:

$$\alpha R(F) = \frac{\sum \text{card}\underline{R}X_i}{\sum \text{card}\overline{R}X_i} \tag{1}$$

where card(X) denotes the cardinality of $X = \phi$.

The accuracy of approximation can be used to measure the quality of approximation of decision classes on the universe U. It is possible to use another measure of accuracy defined by $1 - \alpha R(X)$. Some other measures of approximation accuracy are also used based on entropy or some more specific properties of boundary regions. The choice of a relevant accuracy of approximation depends on a particular data set. The accuracy of approximation of X can be tuned by R.

The second measure, called the quality of approximation of F by R, is the following [1]:

$$\gamma R(F) = \frac{\sum \text{card}\underline{R}X_i}{\text{card } U} \tag{2}$$

The accuracy of classification expresses the percentage of possible correct decision, when classifying objects, employing the knowledge R. The quality of classification expresses the percentage of objects that can be correctly classified as belonging to classe F employing knowledge R. By selecting a proper balance between the accuracy of classification and the description size it is expected to define the classifier with the high quality of classification also on unseen objects.

One of the most important applications of RST is the generation of decision rules for a given information system for the prediction of classes for new objects which are

Fig. 2 Reduction algorithm

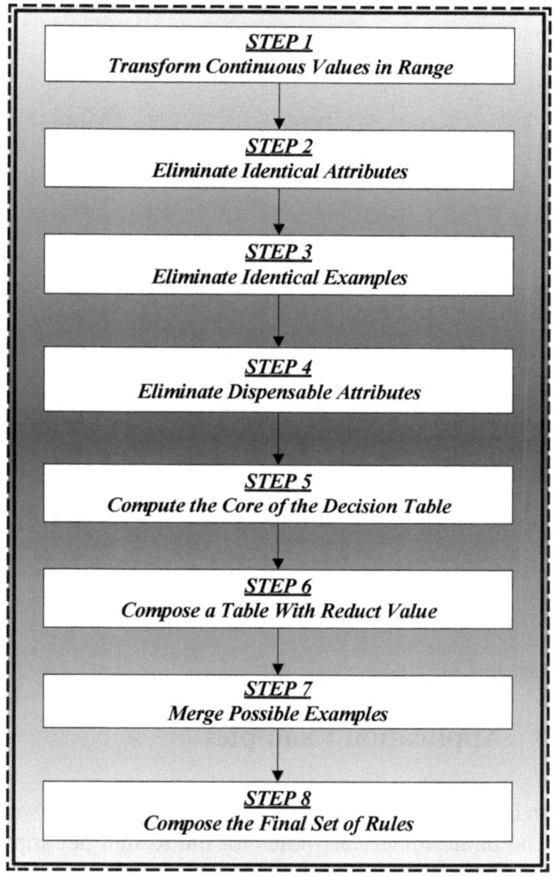

beyond observation. The rules are presented in an "If condition(s) then decision(s)" format.

2.3 Incomplete Information System

It may happen that some attribute values for an object are missing. To indicate such a situation a distinguished value, so-called null value, is usually assigned to those attributes [2]. If *Va* contains null value for at least one attribute $a \in U$ then K is called an incomplete information system, otherwise it is complete. Further on, we will denote null value by * [2]. The algorithm of the reduction of a decision table is shown in Fig. 2 [7].

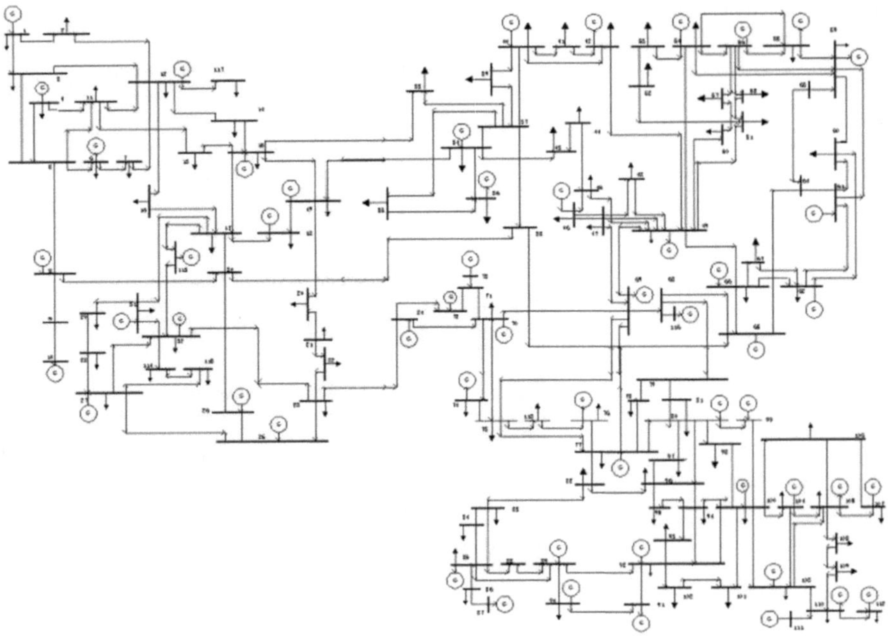

Fig. 3 IEEE 118 test power network

3 Application Examples

In Fig. 3 it is shown the 118IEEE Test Power Network that was used in this study [8]. The input numerical values for the Rough Set approach considering an incomplete information system were obtained using the software package *SecurMining 1.0*, developed by the authors [5]. The ROSE software package was used to perform the RST analysis [9].

In this section it is presented the final results using the Rough Set Theory. A first order contingency study was carried out and it was obtained a list of 231 contingencies that allows the construction of a contingency control database. The specified attributes are as follows:

A Overloads in the transmission lines
B Number of overloaded transmission lines
C Voltage levels
D Number of busbars with voltage violation
E Severity indices related to the power and the voltage
F Severity indices related to the power losses

Table 1 presents a set of information related to a contingency control database. Table 2 shows the chosen range for the coded qualitative attributes. The condition attributes are coded into three qualitative terms: Low, Medium and High. The decision attribute

Table 1 The attributes represented by the set

Cont No	Attributes						
	A	B	C	D	E	F	S
1	2	1	1	1	3	3	A
2	0	1	1	1	3	3	A
3	3	1	1	1	3	3	E_2
4	2	1	1	1	3	3	E_1
5	2	1	1	1	3	3	E_1
6	2	1	1	1	3	3	A
7	3	3	0	1	3	3	E_2
8	2	1	1	1	3	3	E_1
9	2	1	1	1	3	3	E_1
10	2	1	1	1	3	3	E_1
11	2	1	2	1	3	3	A
12	2	1	1	1	3	3	A
13	2	0	1	1	3	3	A
14	2	1	2	1	3	3	A
15	2	1	1	1	3	3	A
16	2	1	1	1	3	3	A
17	2	1	1	1	3	3	A
18	2	1	1	1	3	3	A
19	3	1	1	1	3	3	E_2
20	2	1	1	1	3	0	A
...	...	...	...	...	...	...	...
...	...	...	...	...	...	...	...
222	3	1	1	1	3	1	E_2
223	2	1	1	1	3	1	A
224	2	1	1	1	3	1	A
225	2	1	1	1	3	1	A
226	2	1	1	1	3	1	A
227	2	1	1	1	3	1	A
228	2	1	1	1	3	1	A
229	2	1	1	1	3	0	A
230	2	1	1	1	3	1	A
231	2	1	1	1	3	1	A

is coded into four qualitative terms: Normal (N), Alert (A), Emergency 1 (E1) and Emergency 2 (E2).

Step 1 The first step of the algorithm is to redefine the value of each attribute according to a certain metric that was described above. Using these redefinitions for each contingency of Table 1, Table 3 arises.

Step 2 and 3 The next step of the algorithm is to verify if any attribute can be eliminated by repetition. It can be verified that the attributes are different for all examples. Some examples are identical (for instance, contingencies 5–6, 9 and 10). The similar examples are also merged.

Step 4 The next step is to verify if the decision table contains only indispensable attributes. This task can be accomplished eliminating each attribute step–by–step

Table 2 Definition of range attributes coding

Attributs	Codes			
	0	1	2	3
A		90 % <	90 % ≤ a ≤ 110 %	> 110 %
B		2 ≤	3 ≤ b ≤ 4	> 4
C		0.85 <	0.85 ≤ c ≤ 1.05	> 1.05
D		2 ≤	3 ≤ d ≤ 5	> 5
E		0.800 <	800 ≤ e ≤ 0.900	> 0.900
F		0.800 <	0.800 ≤ e ≤ 0.900	> 0.900
S	N	A	E_1	E_2

and verifying if the table gives the correct classification. For example, if the attribute E is eliminated, the table continues to give a correct classification. So, it can be said that E is a dispensable attribute for this decision table. However, when the attribute C is eliminated it can be verify that the contingencies 1 and 4 have the same set of attributes but they give different classification. In this case, we say that the attribute C is indispensable for all attributes, so we can realize that the attributes A, B, C, D and F are indispensable, and E is dispensable for this decision table.

Step 5 and 6 Using the last information, can be computed the core of the set of contingencies. This computation can be done eliminating each attribute, step-by-step, and verifying if the decision table continues consistent. Using the compute package [9] it can be verified that the attributes A, B, C, D and F are the Core and the Reduct of the Problem.

Step 7 and 8 According to the step 5 and 6, and using logical arithmetic, we can compose the set of rules. Incorporating the range values the final set of rules and approximate rules that contains the knowledge of Table 1, can be expressed the quality of classification for all conditions and the attributes in the core is 0.1385.

The Table 4 shows the approximation of the objects in the Decision levels.

According to the algorithm described, and using logical arithmetic, it is possible to compose the set of rules. Also, incorporating the range values the final set of rules and approximate rules that contains the knowledge of a initial database range values were obtained with the software package *SecurMining 1.0* and the ROSE computer programme [5, 9].

Exact Rules:

1. *If (A is M and D is M) then* $S = A$.
2. *If (C is M and E is L) then* $S = A$.
3. *If (A is H) then* $S = E_2$.

Approximate Rules:

4. *If (A is M and C is L) then* $S = A$ *or* $S = E_1$.
5. *If (A is M and D is L) then* $S = A$ *or* $S = E_2$.

Table 3 Database with range values

Cont No	Attributes						
	A	B	C	D	E	F	S
1	M	L	L	L	H	H	A
2	*	L	L	L	H	H	A
3	H	L	L	L	H	H	E_2
4	M	L	L	L	H	H	E_1
5	M	L	L	L	H	H	E_1
6	M	L	L	L	H	H	E_1
7	H	H	*	L	H	H	E_2
8	M	L	L	L	H	H	E_1
9	M	L	L	L	H	H	E_1
10	M	L	L	L	H	H	E_1
11	M	L	M	L	H	H	A
12	M	L	L	L	H	H	A
13	M	*	L	L	H	H	A
14	M	L	M	L	H	H	A
15	M	L	L	L	H	H	A
16	M	L	L	L	H	H	A
17	M	L	L	L	H	H	A
18	M	L	L	L	H	H	A
19	H	L	L	L	H	H	E_2
20	M	L	L	L	H	*	A
...	...	...	...	...	...	...	...
...	...	...	...	...	...	...	...
223	M	L	L	L	H	L	A
224	M	L	L	L	H	L	A
225	M	L	L	L	H	L	A
226	M	L	L	L	H	L	A
227	M	L	L	L	H	L	A
228	M	L	L	L	H	L	A
229	M	L	L	L	H	*	A
230	M	L	L	L	H	L	A
231	M	L	L	L	H	L	A

Table 4 Approximation of the objects

Decision level	Number of objects	Approximation upper	Approximation lower	Precision the approximation. of classification
1—Alert	167	2	201	0.0100
2—Emerg. I	34	0	199	0.0000
3—Emerg. II	30	30	30	1.0000

The above rules can be written in a more clear way:

Exact Rules:

1. *If the overloads in the transmission lines present a medium value and the number of busbars with voltage violation are medium then the Power System is in* ***Alert state****.*

2. *If the voltage levels assume a medium values and the severity indices related to the power and the voltage are lower values then the Power System is also in* ***Alert state***.
3. *If the overloads in the transmission lines present a high values then the Power System is in* ***Emergency state II***.

Approximate Rules:

4. *If the voltage levels assume a medium values and the voltage levels present a Lower values then the Power System is in* ***Alert*** *or in* ***Emergency state I***.
5. *If the voltage levels present a medium values and the number of busbars with voltage violation assume lower values then the Power System is in* ***Alert*** *or in* ***Emergency state II***.

4 Conclusions

In this paper it was presented the RST applied to an Electric Power System considering an incomplete information system. The Knowledge acquisition process is a complex task, since the experts have difficulty to explain how to solve a specified problem. The proper definitions of reducts allow to define knowledge reduction that does not diminish the original system's abilities to classify objects or to make decisions. Both reduction of dispensable knowledge and finding of optimal decision rules are transformable to the problem of computing prime implicates discernibility functions. It was also shown that discernibility functions for incomplete information systems may be constructed in conjunctive normal form. Consequently, an incomplete set of relevant information may arise. In order to overcome this problem it is proposed a new methodology to study and analyse the steady-state contingency classification using the RST. The study presents a systematic approach to transform examples in a reduced set of rules.

Acknowledgments The first author would like to thank Fundação para a Ciência e Tecnologia, FCT, that partially funded this research work through the PhD grant no SFRH/BD/38152/2007.

References

1. Pawlak Z (1991) Rough sets—theoretical aspects of reasoning about data. Kluwer, Dordrecht
2. Kryszkiewicz M (1998) Rought set approach to incomplete information systems. Inform Sciences 112:39–49
3. Kryszkiewicz M (1999) Rules in incomplete information systems. Inform Sciences 113:271–292
4. Coutinho MP, Lambert-Torres G, da Silva LEB, Fonseca EF, Lazarek H (2007) A methodology to extract rules to identify attacks in power system critical infrastructure: New results. In: Proceedings. IEEE Power Engineering Society general meeting, IEEE PES GM, Tampa
5. Faustino Agreira CI (2010) Data mining techniques for security study and analysis to the electrical power systems. Ph.D dissertation, University of Porto

6. Ching-Lai, Crossley P, MIEEE, Dunand F (2002) Knowledge extraction within distribution substation using rough set approach. In: Power Engineering Society winter meeting, vol 1, pp 654–659
7. Lambert-Torres G, da Silva APA et al (1996) Classification of power system operating point using rough set techniques. In: IEEE international conference on systems, man and cybernetics
8. Power systems test case archive 118 bus power flow test case. Department of Electrical Engineering, University of Washington. http://www.ee.washington.edu/research/pstca/
9. ROSE2—Rough sets data explorer Laboratory of intelligent decision support systems of the Institute of Computing Science, Poznan. http://www.idss.cs.put.poznan.pl/software/rose/

Chain Drives Modelling Using Kinematic Constraints and Revolute Clearance Joints Formulations

Cândida Pereira and Jorge Ambrósio

Abstract Based on Multibody Dynamics two different formulations for modelling chain drive mechanisms are presented in this work: (i) one in which the revolute joints are considered as ideal joints, modelled as kinematic constraints; (ii) and another in which the kinematic constraints are removed and replaced by a pair of forces representing the contact between the connected bodies, i.e., modelled using the revolute clearance joint formulation. When the chain drive components' connections are modelled as kinematic joints, the integration of the equations of motion lead to constraint violations that grow to a point at which the chain seems to start vibrating with a very high frequency and ends up disintegrating, even when the Baumgarte stabilization method is used. This problem is, however, eliminated when the interaction between the chain drive components is modelled using the revolute clearance joint formulation, since any constraint violation is exhibited as the number of kinematic constraints used in the multibody model is kept to a minimum.

Keywords Multibody dynamics · Revolute clearance joints · Kinematic constraints · Constraint violations · Chain drives

1 Introduction

The dynamics of chain drives may be efficiently analyzed by employing multibody dynamics tools, since these mechanisms can be modelled as a constrained dynamic system composed by a large number of bodies, here taken as rigid bodies, interconnected by ideal or clearance revolute joints [1]. The links and rollers that compose the chain are connected to each other by revolute joints that constrain the relative motion of the links in different directions, as depicted in Fig. 1. The same definition can be extended to the chain engagement on the sprockets. When the roller is seated

C. Pereira (✉)
Coimbra Institute of Engineering, Polytechnic Institute of Coimbra, Coimbra, Portugal
e-mail: candida@isec.pt

J. Ambrósio
Instituto Superior Técnico, Technical University of Lisbon, Lisbon, Portugal
e-mail: jorge@dem.ist.utl.pt

N. M. Fonseca Ferreira, J. A. Tenreiro Machado (eds.), *Mathematical Methods in Engineering*, DOI 10.1007/978-94-007-7183-3_19,

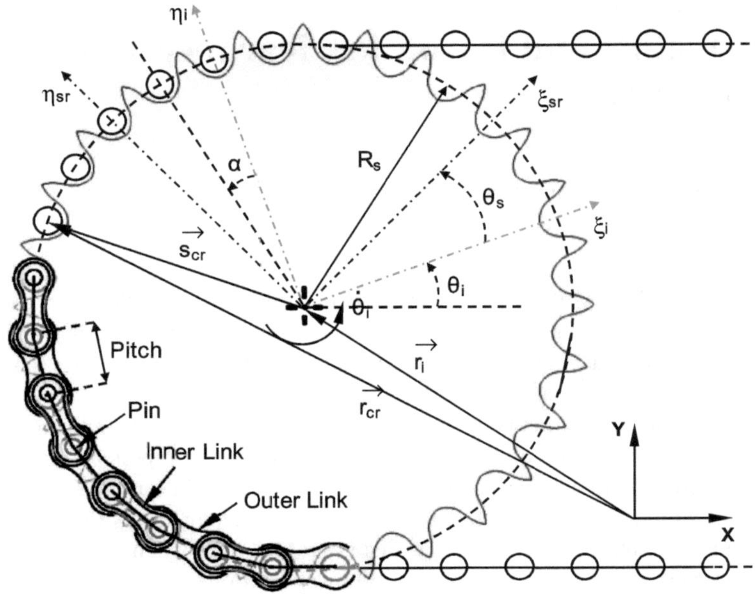

Fig. 1 Multibody representation of a chain drive

on the sprocket tooth, its relative motion is restricted and the connectivity of these two bodies modelled by a set of kinematic constraints or by a force relation.

The functionality of a kinematic joint relies on the relative motion allowed between the connected components, which in practice implies the existence of clearance between the mating parts. Accurate clearance modelling is thus required to correctly predict the behaviour of the impact process to fully describe the dynamic of systems, which is not possible to account for using kinematic constraints to describe joints [2–4]. In fact, no matter how small the clearance is, its presence in mechanical systems can lead to wear, vibration and fatigue phenomena. As a result, lack of precision or even random overall behaviour can be expected [2–10]. Despite these undesirable effects, clearances are necessary in a roller chain drive to allow relative motion between links and sprockets and to permit the link assemblage. It is important to quantify the effects of clearances on the global system response in order to define the minimum level of suitable tolerances that will allow the roller chain to achieve the required performance. However, regardless of clearance values, the overall dynamics of the chain drive may be insensitive to their presence but also the identification of the nominal behaviour of such drives is done by assuming perfect kinematic joints.

In this work and based on the multibody dynamics tools, two different formulations for modelling chain drive systems are presented: (i) one in which the revolute joints are considered as perfect joints, modelled as kinematic constraints; (ii) and another in which the contact between the connected bodies is modelled using the revolute clearance joint formulation. A joint with clearance is included in a multibody system much like a revolute joint. However, while a perfect or ideal joint imposes

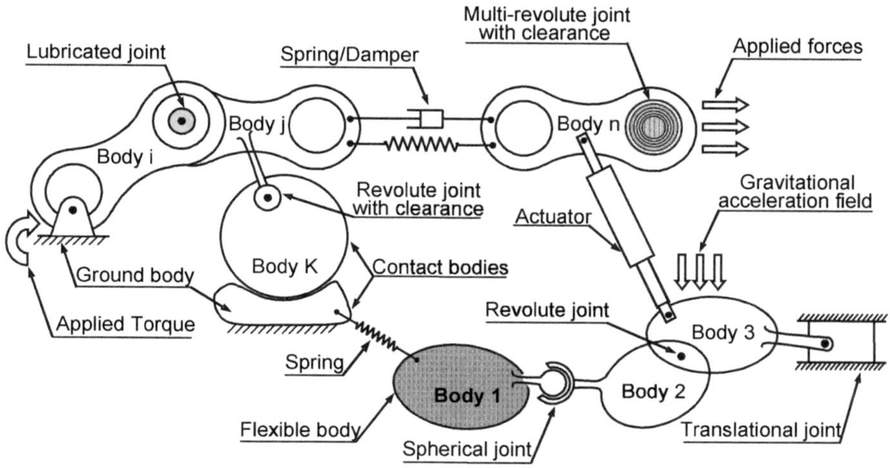

Fig. 2 Schematic representation of a generalized multibody system

two kinematic constraints, limiting the motion between bodies, the presence of clearance in a revolute joint implies that those kinematic constraints are removed and replaced by a pair of forces representing the interaction between the connected bodies [5–10]. Therefore, the clearance in a revolute joint implies that the system to which it is applied has two extra degrees of freedom relative to an ideal system, which are modelled using an appropriate contact force model [11]. The relative merits and drawbacks of the proposed formulations with clearance joints with respect to models with ideal joints are discussed in this context.

2 Multibody Systems Formulation

A multibody system can be represented schematically as a collection of rigid and/or flexible bodies interconnected by kinematic joints and by some force elements, as shown in Fig. 2. Thus, any mechanical system can be understood as a multibody system, where the kinematic joints control the relative motion between the bodies, while the force elements represent the internal forces that develop between bodies due to their relative motion. The forces applied to the system components may be the result of springs, dampers, and position and/or force actuators, or externally applied forces describing e.g. gravitational forces, friction forces and contact/impact forces. A wide variety of mechanical systems can be modelled in this way such as robots, heavy machinery, automotive suspensions and steering systems, machinery tools, satellites, sports vehicles or railway rolling stock, among others [1, 12–14] as well as the human body [15].

In order to analyze the dynamic response of a constrained multibody system, it is first necessary to formulate the equations of motion that govern its behaviour.

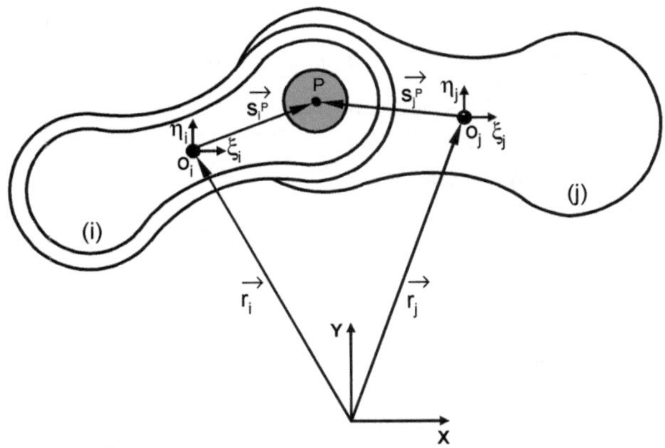

Fig. 3 Planar revolute joint connecting bodies *i* and *j*

In a broad sense, the best known methods to derive the equations of motion are: Newton-Euler's method [1], Lagrange's method [16], and Kane's method [17]. Two kinds of coordinates are frequently employed to formulate the equations of motion of multibody systems. The first one uses a minimum number of relative coordinates, corresponding to the degrees of freedom of the system. The second approach, a system of dependent coordinates, e.g. Cartesian coordinates, is used to describe the system configuration. The formulation of multibody system dynamics adopted in this work closely follows the Cartesian coordinates approach proposed by Nikravesh [1]. In this formulation, the Lagrange multipliers technique with 2D generalized Cartesian coordinates and the Newton-Euler equations of motion of rigid bodies are employed to govern the dynamics of roller chain drives.

2.1 Constraint Equations

A kinematic joint imposes kinematic conditions on the relative motion between adjacent bodies of the system. When these conditions are expressed in analytical form they are called constraint equations. Kinematic constraint types include revolute joints, translational joints, spherical joints and cylindrical joints. The kinematic constraints considered in this work are assumed to be holonomic, arising from geometric relations on the generalized coordinates [1, 16, 18]. In order to illustrate the methodology, and because this is the only type of joint used to model roller chain drive mechanisms, the formulation for the planar revolute joint is reviewed. Details on the formulation of other types of kinematic joints can be found e.g. in Nikravesh [1]. The revolute joint is a pin and bush type of joint that constrains the relative translation between the two bodies *i* and *j*, allowing only relative rotations, as illustrated in Fig. 3. The centers of mass of bodies *i* and *j* are O_i and O_j, respectively. Body-fixed coordinate systems $\zeta\eta$ are attached at their centers of mass, while the *XY* coordinate frame represents the global coordinate system.

The kinematic conditions for the revolute joint require that two distinct points, each belonging to a different body, share the same position in space all the time.

This means that the global position of a point P in body i is coincident with the global position of a point P in body j. That condition is expressed by two algebraic equations as

$$\mathbf{r}_i + \mathbf{s}_i^P - \mathbf{r}_j - \mathbf{s}_j^P = 0 \quad (1)$$

which is rewritten as

$$\mathbf{\Phi}^{(r,2)} \equiv \mathbf{r}_i + \mathbf{A}_i \mathbf{s}_i'^P - \mathbf{r}_j - \mathbf{A}_j \mathbf{s}_j'^P = 0 \quad (2)$$

where $\mathbf{\Phi}^{(r,2)}$ denotes the planar revolute (r) joint constraint, which contains two (2) independent equations, and $\mathbf{A}_i$ and $\mathbf{A}_j$ represent the transformation matrixes of body i and j, respectively.

2.2 *Kinematic Analysis*

In the study of multibody systems motion there are two different types of analysis that can be performed, namely, kinematic and dynamic analysis. Kinematics is a first step in the complete analysis of a mechanical system, dealing only with space and time and neglecting forces and their effects [19]. Kinematic analysis is thus the study of the system's motion regardless of the forces that produce it. Since the interaction between the forces and the system motion is not considered, the motion of the system needs to be specified by driving elements that govern it during the analysis. The position, velocity and acceleration are obtained using the kinematic constraint equations that describe the topology of the system.

When the configuration of a multibody system is described by nc Cartesian coordinates, a set of m algebraic kinematic independent holonomic constraints $\mathbf{\Phi}$ can be written in a compact form as [1],

$$\mathbf{\Phi}(\mathbf{q},t) = \mathbf{0} \quad (3)$$

where $\mathbf{q}$ is the vector of generalized coordinates and t is the time variable generally associated with the driving elements. The velocities and accelerations of the system elements are evaluated using the velocity and acceleration constraint equations. Thus, the first time derivative of (3) provides the velocity constraint equations,

$$\mathbf{\Phi}_q \dot{\mathbf{q}} = -\mathbf{\Phi}_t \equiv \boldsymbol{\upsilon} \quad (4)$$

where $\mathbf{\Phi}_q$ is the Jacobian matrix of the constraint equations, i.e. the matrix of the partial derivates $\partial\, \mathbf{\Phi}/\partial\, \mathbf{q}$, $\dot{\mathbf{q}}$ is the vector of generalized velocities and $\boldsymbol{\upsilon}$ is the right hand side of velocity equations, which contains the partial derivates of $\mathbf{\Phi}$ with respect to time, $\partial\, \mathbf{\Phi}/\partial t$. The second-order derivative of (3) with respect to time leads to the acceleration constraint equations, expressed as

$$\mathbf{\Phi}_q \ddot{\mathbf{q}} = -(\mathbf{\Phi}_q \dot{\mathbf{q}})_q \dot{\mathbf{q}} - 2\mathbf{\Phi}_{qt} \dot{\mathbf{q}} - \mathbf{\Phi}_{tt} \equiv \boldsymbol{\gamma} \quad (5)$$

where $\ddot{\mathbf{q}}$ is the acceleration vector and $\boldsymbol{\gamma}$ is the right hand side of acceleration equations, i.e. the vector of quadratic velocity terms, which contains the terms that are exclusively functions of velocity, position and time. In the case of holonomic scleronomic constraints, i.e. when $\boldsymbol{\Phi}$ is not explicitly dependent on time [1, 16, 18], the term $\boldsymbol{\Phi}_{\mathrm{t}}$ in (4) and the terms $\boldsymbol{\Phi}_{\mathrm{qt}}$ and $\boldsymbol{\Phi}_{\mathrm{tt}}$ in (5) vanish.

2.3 Equations of Motion for a Constrained Multibody System

Dynamic analysis of multibody systems, on the other hand, aims to understand the relation between the motion of the system components and the causes that produce it, including external applied forces and moments. The motion of the system is not usually prescribed and its calculation is one of the principal objectives of the analysis. Dynamic analysis also provides a way to estimate external forces that depend on the relative position between the system components such as the forces exerted by springs, dampers and actuators. Furthermore, in the process of the dynamic analysis the external forces that are developed through the interaction between the system components and the surrounding environment, such as contact-impact forces and friction forces, are evaluated. The internal reaction forces and moments generated at the kinematic joints are also obtained in the dynamic analysis.

The equations of motion for a constrained multibody system of rigid bodies are written as [1]

$$\mathbf{M}\ddot{\mathbf{q}} = \mathbf{g} + \mathbf{g}^{(\mathbf{c})} \tag{6}$$

where $\mathbf{M}$ is the system mass matrix, $\ddot{\mathbf{q}}$ represents the vector that contains the system accelerations, $\mathbf{g}$ is the generalized force vector, which contains all internal and external forces and moments, and $\mathbf{g}^{(\mathbf{c})}$ is the vector of constraint reaction equations. The joint reaction forces can be expressed in terms of the Jacobian matrix of the constraint equations and the vector of Lagrange multipliers as [1]

$$\mathbf{g}^{(\mathbf{c})} = -\boldsymbol{\Phi}_{\mathbf{q}}^{\mathrm{T}}\boldsymbol{\lambda} \tag{7}$$

where $\boldsymbol{\lambda}$ is the vector that contains m unknown Lagrange multipliers associated with m holonomic constraints. The Lagrange multipliers are physically related to the reaction forces and moments generated between the bodies interconnected by kinematic joints. Thus, substituting (7) in (6) yields

$$\mathbf{M}\ddot{\mathbf{q}} + \boldsymbol{\Phi}_{\mathbf{q}}^{\mathrm{T}}\boldsymbol{\lambda} = \mathbf{g} \tag{8}$$

In dynamic analysis, a unique solution is obtained when the constraint equations and the differential equations of motion are considered simultaneously for a proper set of initial conditions. With this objective, (5) is appended to (8), yielding a system of differential algebraic equations. Thus, the mathematical simulation of a constrained multibody system requires the solution of a set of nc differential equations coupled with a set of m algebraic equations, written as

$$\begin{bmatrix} \mathbf{M} & \mathbf{\Phi}_{\mathbf{q}}^{T} \\ \mathbf{\Phi}_{\mathbf{q}} & \mathbf{0} \end{bmatrix} \begin{Bmatrix} \ddot{\mathbf{q}} \\ \boldsymbol{\lambda} \end{Bmatrix} = \begin{Bmatrix} \mathbf{g} \\ \boldsymbol{\gamma} \end{Bmatrix} \tag{9}$$

From the initial values for positions and velocities, (9) is solved for accelerations, $\ddot{\mathbf{q}}$, and Lagrange multipliers, $\boldsymbol{\lambda}$, using the L-U factorization in conjunction with forward and backward substitution. The positions and velocities in the next time step are then obtained by integration of the velocity and acceleration vectors, $\dot{\mathbf{q}}$ and $\ddot{\mathbf{q}}$. This procedure is repeated until the final time is reached. The integration process is performed here using a predictor-corrector algorithm with both variable step and order [20, 21].

3 Dynamics of Chain Drives Using Kinematics Constraints

The system of equations of motion, described by (9), does not use explicitly the position and velocity constraint equations, i.e. equations (3) and (4). As a result, for moderate or long simulations the position and velocity constraint equations start to be violated due to problems such as: inaccurate initial conditions for positions and velocities [22], constraints that are treated introducing forces of constraint within the equations of motion [23, 24], numerical integration [25], large number of bodies involved and/or the stiffness of the system [26]. Since the cause is unknown, the constraints violation problem happens when the pin-bushings, the bushing-rollers and the rollers-sprockets connections are modelled as kinematic joints. Special procedures must be followed to avoid or minimize this drift, including the Constraint Violation Stabilization Method, proposed by Baumgarte [23]. This constraint stabilization method has been extensively applied to the dynamic analysis of mechanical systems in order to suppress the growth of error and achieve a stable response [1]. Therefore, it is also implemented here.

When (9) is solved, (5) and (6) are only solved implicitly. In fact, only the second derivatives of constraint equations are satisfied in each integration step. It is known that (5) represents an unstable system [27, 28]. The Baumgarte Stabilization Method includes in the differential equation (5) the feedback terms from the position and velocity constraint violations as [23]

$$\ddot{\mathbf{\Phi}} + 2\alpha\dot{\mathbf{\Phi}} + \beta^2\mathbf{\Phi} = \mathbf{0} \tag{10}$$

Equation (10) is the differential equation for a closed-loop system in terms of kinematic constraint equations, where the terms $2\alpha\dot{\mathbf{\Phi}}$ and $\beta^2\mathbf{\Phi}$ play the role of control terms. The equations of motion for a dynamic system subjected to holonomic constraints is rewritten as

$$\begin{bmatrix} \mathbf{M} & \mathbf{\Phi}_{\mathbf{q}}^{T} \\ \mathbf{\Phi}_{\mathbf{q}} & \mathbf{0} \end{bmatrix} \begin{Bmatrix} \ddot{\mathbf{q}} \\ \boldsymbol{\lambda} \end{Bmatrix} = \begin{Bmatrix} \mathbf{g} \\ \boldsymbol{\gamma} - 2\alpha\dot{\mathbf{\Phi}} - \beta^2\mathbf{\Phi} \end{Bmatrix} \tag{11}$$

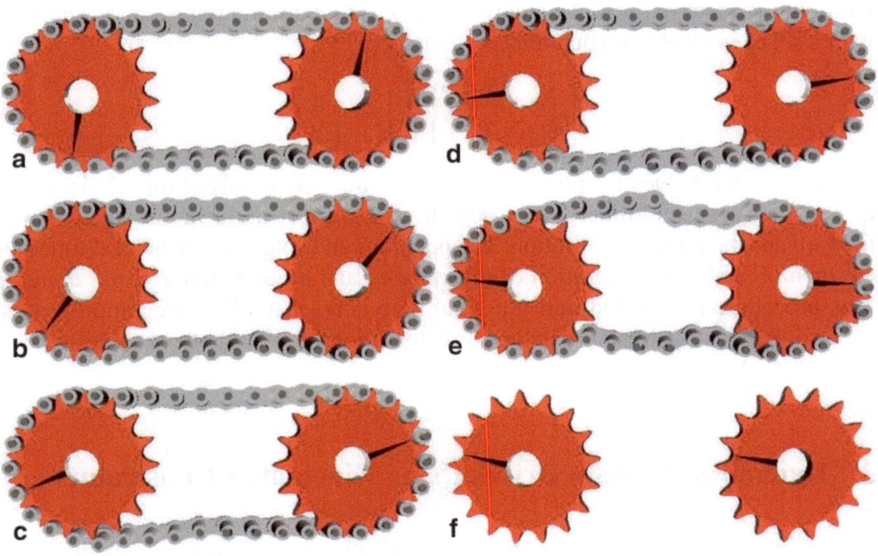

Fig. 4 Sequence of images of a complex multibody system configuration from **a** the correct initial positions, **b–f** accumulation of error until the total disintegration of the system takes place

For nonzero values of the feedback coefficients, i.e. if α and β are chosen as positive constants, the solution vector oscillates around the exact solution, and the stability of the general solution of (11) is usually guaranteed. Moreover, when α is equal to β, critical damping is achieved, which generally stabilizes the system response more rapidly [1]. Generally, for rigid multibody dynamics the values of $\alpha = \beta = 5$ are used and also here taken [8–10].

3.1 Elimination of Initial Constraint Violations

Regardless of the original cause and for the chain drive multibody system, as time progresses the constraint violation errors increase and even the use of the Baumgarte Stabilization Method is unable to control the problem. For instance, for the chain drive shown in Fig. 4, in which the pin/bushing hinges are represented by ideal revolute joints, the integration of the equations of motion leads to constraint violations that grow to a point at which the chain seems to start vibrating with a very high frequency, as depicted by Fig. 4e) and ends up disintegrating, as depicted by Fig. 4f).

In order to attempt to minimize the problem of the constraint violations, a procedure to ensure the kinematic consistency of the initial conditions of the problem proposed by Nikravesh [25] has been used. In the initial condition correction method the correct positions and velocities are expressed as

$$\mathbf{q}^0 = \mathbf{q}^e + \Delta\mathbf{q} \tag{12a}$$

$$\dot{\mathbf{q}}^0 = \dot{\mathbf{q}}^e + \Delta\dot{\mathbf{q}} \tag{12b}$$

where $\mathbf{q}^e$ and $\dot{\mathbf{q}}^e$ are the initial guesses for the positions and velocities, respectively, and $\Delta\mathbf{q}$ and $\Delta\dot{\mathbf{q}}$ are the corrections of the positions and velocities that need to be evaluated. Based on the minimization of the sum-of-squares of the corrections, Nikravesh [25] shows that the iterative process to correct the positions is done as follows:

For $\mathbf{q} = \mathbf{q}^i$:
Evaluate $\boldsymbol{\sigma}^i = \boldsymbol{\Phi}(\mathbf{q}^i)$;
Compute $\Delta\mathbf{q}^i = -\boldsymbol{\Phi}^T_{\mathbf{q}^i}\left(\boldsymbol{\Phi}_{\mathbf{q}^i}\boldsymbol{\Phi}^T_{\mathbf{q}^i}\right)^{-1}\sigma^i$;
Correct $\mathbf{q}^{i+1} = \mathbf{q}^i + \Delta\mathbf{q}$;

Repeat if necessary

The correction of the velocities is done by first evaluating the velocity constraint violations

$$\boldsymbol{\varepsilon} = \boldsymbol{\Phi}_{\mathbf{q}}\dot{\mathbf{q}} \tag{13}$$

Then, the correction of the velocities required for (12b) is obtained as

$$\Delta\dot{\mathbf{q}} = -\boldsymbol{\Phi}^T_{\mathbf{q}^i}\left(\boldsymbol{\Phi}_{\mathbf{q}^i}\boldsymbol{\Phi}^T_{\mathbf{q}^i}\right)^{-1}\boldsymbol{\varepsilon} \tag{14}$$

Although the initial conditions guarantee the non-violation of constraint equations on position and velocity level, (3) and (4) are only satisfied at the initial instants of time. Unfortunately, even for the case of initial consistent kinematic conditions, the constraint violations that accumulate during the integration process cannot be handled by the constraint stabilization method for this type of mechanism. Therefore, the simulation of chain drives requires that a constraint correction method is used throughout the analysis and not only a constraint stabilization method.

4 Dynamics of Chain Drives Using the Revolute Clearance Joint Formulation

When the chain drive mechanism is described as a dynamic system composed of a large number of rigid bodies, links and rollers, connected to each other by revolute clearance joints, the kinematic joints used in the multibody model with perfect joints are eliminated from the system and contact forces are applied in each contact pair. In fact, a joint with clearance is included in a multibody system much like a revolute joint. However, while for a perfect or ideal revolute joint it is assumed that the

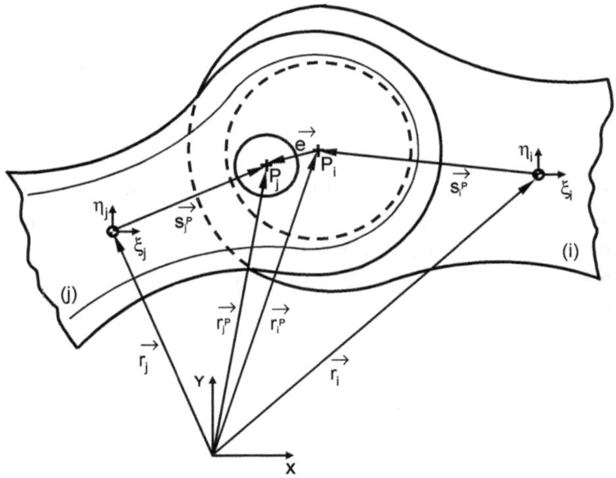

Fig. 5 Generic revolute joint with clearance in a multibody system

connecting points of the two bodies are coincident, as shown in Fig. 3, the inclusion of a clearance in a joint separates the connecting points and, as a result, two extra degrees of freedom are added to the mechanical system. It is, therefore, necessary to establish the geometric condition that defines whether the joint elements are in contact. Figure 5 illustrates two bodies, *i* and *j*, connected by a generic revolute joint with clearance, in which part of body *i* can represent e.g. the bushing link and part of body *j* represents the pin link. The centers of mass of bodies *i* and *j* are O_i and O_j, respectively, and body-fixed coordinate systems $\xi\eta$ are attached at their center of mass. Point P_i indicates the center of the revolute body *i*, while the center of the joint in body *j* is denoted by point P_j.

When the centers of the bushing and pin, given by points P_i and P_j respectively, separate, there is the possibility for contact to take place if their distance exceeds the existing radial clearance of the joint, denoted by *c*. A penetration between bodies *i* and *j* exist if

$$\delta = \sqrt{\mathbf{e}^T\mathbf{e}} - c > 0, \tag{15}$$

where the vector eccentricity **e**, that connects the centers of the bodies, is

$$\mathbf{e} = \mathbf{r}_\mathrm{j}^\mathrm{P} - \mathbf{r}_\mathrm{i}^\mathrm{P} \tag{16}$$

in which the positions of points P_i and P_j, $\mathbf{r}_\mathrm{j}^\mathrm{P}$ and $\mathbf{r}_\mathrm{i}^\mathrm{P}$, are described in the global coordinates reference frame as [1, 5–10]

$$\mathbf{r}_\mathrm{k}^\mathrm{P} = \mathbf{r}_\mathrm{k} + \mathbf{A}_\mathrm{k}\mathbf{s}_\mathrm{k}^{\prime\mathrm{P}}, \ (\mathrm{k} = \mathrm{i}, \mathrm{j}) \tag{17}$$

where $\mathbf{A}_\mathrm{k}$ is the transformation matrix from the local to global coordinates. In (15) $\mathbf{e}^\mathrm{T}$ is the transpose of vector **e**.

Based on the continuous contact force method, the dynamics of revolute clearance joints is controlled by the contact forces developed on each contacting rigid body, which are evaluated using an appropriate contact force model [11]. Thus, when (15) is fulfilled, the resultant contact force in each body is evaluated by summing the contributions of both the normal f_n, that results from deformation of the impacting bodies, and tangential forces f_t, associated to friction phenomenon, as expressed by

$$\mathbf{f}_\mathrm{i} = f_n\mathbf{n} + f_t\mathbf{t} \tag{18}$$

$$\mathbf{f}_\mathrm{j} = -\mathbf{f}_\mathrm{i} \tag{19}$$

for bodies i and j, respectively. The contributions of contact forces are incorporated into the equation of motion of a multibody system, given by (9), through the generalized vector of forces and moments $\mathbf{g}$. The mathematical models that represent the contact conditions between the chain and the sprocket teeth have been recently presented [29].

The dynamics of revolute clearance joints is then controlled by the contact-impact forces developed on each contacting rigid body and included into the equations of motion during the contact-impact period [5–10]. The chain drive modelled in this form does not exhibit any constraint violation as the number of kinematic constraints used in the multibody model is kept to a minimum. Dynamic results show that this formulation is robust and describe thoroughly all the features of chain drives, including the polygonal effect, the pretension effect, the vibration patterns and the impact on the chain due to chain engagement on the sprockets, which can be found in Ref [30].

5 Conclusions

Based on the multibody dynamics formulations two different approaches for modelling chain drive systems: (i) one in which the revolute joints are considered as perfect joints; (ii) and another in which the contact between the connected bodies is modelled as a revolute clearance joint are presented in this work.

When modelling the pin-bushings, the bushing-rollers and the rollers-sprockets connections as kinematic joints, these are included in the equations of motion that govern the behaviour of the chain drive constrained multibody system using an Augmented Lagrangian formulation. However, because such form of the equations of motion are solved together with the acceleration constraints, the explicit form of the position and velocity constraint equations are not present. The integration of the equations of motion lead to constraint violations that grow to a point at which the chain seems to start vibrating with a very high frequency and ends up disintegrating even when the Baumgarte Stabilization Method is used. A procedure to ensure the kinematic consistency of the initial conditions of the problem, proposed by Nikravesh, is implemented to prevent initial violation constraints. Unfortunately, even for the case of initial consistent kinematic conditions, the constraint violations that accumulate during the integration process cannot be handled by the constraint stabilization method for this type of mechanism. The simulation of chain drives requires, therefore, that a constraint correction method is used throughout the analysis and not only a constraint stabilization method. This means that much more work must be pursued to solve the constraints violation problem. However, this problem is eliminated when the chain drive component connections are modeled as revolute clearance joints, since the kinematics constraints are replaced by contact forces, which are evaluated by penalty contact force models. The dynamics of revolute clearance joints is then controlled by the contact-impact forces developed on each contacting rigid body and included into the equations of motion during the contact-impact period. The chain drive modelled in this form does not exhibit any constraint violation as the number of kinematic constraints used in the multibody model is kept to a minimum.

References

1. Nikravesh P (1988) Computer-aided analysis of mechanical systems. Prentice-Hall, Englewood-Cliffs
2. Earles S, Kilicay O (1980) A design criterion for maintaining contact at plain bearings. Proc Inst Mech Eng 194:249–258
3. Earles S, Seneviratne L (1990) Design guidelines for predicting contact loss in revolute joints of planar mechanisms. Proc Inst Mech Eng 204:9–18
4. Zhu S, Zwiebel S, Bernhardt G (1999) A theoretical formula for calculating damping in the impact of two bodies in a multibody system. Proc Inst Mech Eng 213:211–216
5. Ravn P (1998) A continuous analysis method for planar multibody systems with joint clearance. Multibody Syst Dyn 2:1–24
6. Ravn P, Shivaswamy S, Alshaer B, Lankarani H (2000) Joint clearances with lubricated long bearings in multibody mechanical systems. J Mech Design 122:484–488
7. Schwab A, Meijaard J, Meijers P (2002) A comparison of revolute joint clearance models in the dynamic analysis of rigid and elastic mechanical systems. Mech Mach Theory 37:895–913
8. Flores P, Ambrósio J, Claro JCP (2004) Dynamic analysis for planar multibody mechanical systems with lubricated joints. Multibody Syst Dyn 12:47–74
9. Flores P, Ambrósio J (2004) Revolute joints with clearance in multibody systems. Comput Struct 82:1359–1369
10. Flores P, Ambrósio J, Claro JP, Lankarani H (2008) Kinematics and dynamics of multibody systems with imperfect joints: models and case studies. Springer, Dordrecht
11. Lankarani H, Nikravesh P (1994) Continuous contact force models for impact analysis in multibody systems. Nonlinear Dyn 5:193–207
12. Wittenburg J (2008) Dynamics of multibody systems. Springer Berlin Heidelberg, New York
13. Lankarani HM, Olivares G, Nagarajan H (2003) A virtual multibody and finite element method analysis environment in the field of aerospace crashworthiness. In: Schiehlen W, Valasek M (eds) Virtual nonlinear multibody systems, NATO science series, II. Mathematics, Physics and Chemistry. Kluwer Academic Publishers, Prague, p 103, pp 187–212
14. Ambrósio JAC, Gonçalves JPC (2001) Vehicle crashworthiness design and analysis by means of nonlinear flexible multibody dynamics. Int J Vehicle Des 26(4):309–330
15. Silva MPT, Ambrósio JAC, Pereira MS (1997) Biomechanical model with joint resistance for impact simulation. Mulibody Syst Dyn 1:65–84
16. Shabana AA (1989) Dynamics of multibody systems. Wiley, New York
17. Kane TR, Levinson DA (1985) Dynamics theory and applications. McGraw Hill, New York
18. Rahnejat H (2000) Multi-body dynamics: historical evolution and application. J Mech Eng Sci 214:149–173

19. Drazetic P, Level P, Canaple B, Mongenie P (1996) Impact on planar kinematic chain of rigid bodies: application to movements of anthropomorphic dummy in a crash. Int J Impact Eng 18(5):505–516
20. Gear CW (1981) Numerical solution of differential-algebraic equations. IEEE Trans Circuit Theory 18(1):89–95
21. Shampine L, Gordon M (1975) Computer solution of ordinary differential equations: the initial value problem. Freeman, San Francisco
22. Nikravesh PE (1984) Some methods for dynamic analysis of constrained mechanical systems: a survey. In: Haug EJ (ed) Computer aided analysis and optimization of mechanical system dynamics. Springer, Berlin, pp 351–368
23. Baumgarte J (1972) Stabilization of constraints and integrals of motion in dynamical systems. Comput Method Appl M 1:1–16
24. Baumgarte JW (1983) A new method of stabilization for holonomic constraints. J Appl Mech 50:869–870
25. Nikravesh P (2007) Initial condition correction in multibody dynamics. Multibody Syst Dyn 18:107–115
26. Garcia de Jálon J, Bayo E (1994) Kinematic and dynamic simulations of multibody systems. Springer, New York
27. Cochin I, Cadwallender W (1997) Analysis and design of dynamic systems, 3rd edn. Addison Wesley, New Jersey
28. Franklin GF, Powel JD, Enami Naeini A (2002) Feedback control of dynamic systems, 4th edn. Prentice Hall, Englewood Cliffs
29. Pereira C, Ambrósio J, Ramalho A (2010) Contact mechanics in a roller chain drive using a multibody approach. In proceedings of the 11th Pan-American Congress of applied mechanics
30. Pereira C, Ambrósio J, Ramalho A (2009) A methodology for the generation of models for multibody chain drives. Multibody Syst Dyn 24:303–324

Jacobi Polynomials and Some Related Functions

Mariana Marčoková and Vladimír Guldan

Abstract The classical Jacobi orthogonal polynomials (especially their special case—the Legendre polynomials) appear as the solutions of some problems of mathematical physics. In the contribution we deal with some relations connecting generalized Legendre polynomials of a certain type and the classical Jacobi polynomials orthogonal with respect to two different special weight functions. We also point out relations between the classical Legendre polynomials, the associated Legendre functions of the first kind, the Legendre functions of the first kind and the generalized g-Legendre functions obtained by Mirevski et al. using fractional calculus.

Keywords Jacobi polynomial · Legendre polynomial · Legendre function

1 Basic Properties of Orthogonal Polynomials

In this section we recall the definitions concerning orthogonal polynomials and some theorems on their basic properties.

Definition 1 Let $(a, b) \subset R$ be a finite or infinite interval. A function $w(x)$ is called the weight function if at this interval it fulfills the following conditions:

(i) $w(x)$ is nonnegative at (a, b), i.e.

$$w(x) \geq 0,$$

(ii) $w(x)$ is integrable at (a, b) and

$$0 < \int_a^b w(x)dx < \infty,$$

M. Marčoková(✉)
Department of Mathematics, University of Žilina, Žilina, Slovak Republic
e-mail: mariana.marcokova@fpv.uniza.sk

V. Guldan
Department of Applied Mathematics, University of Žilina, Žilina, Slovak Republic
e-mail: vladimir.guldan@fstroj.uniza.sk

N. M. Fonseca Ferreira, J. A. Tenreiro Machado (eds.), *Mathematical Methods in Engineering,* DOI 10.1007/978-94-007-7183-3_20,

(iii) if $(a,\ b)$ is an infinite interval, then for every $n = 0, 1, 2, \ldots$

$$0 < \int_a^b |x|^n w(x) dx < \infty$$

is necessary condition for the weight function.

Definition 2 Let $\{P_n(x)\}_{n=0}^{\infty}$ be a system of polynomials, where every polynomial $P_n(x)$ has the degree n. If for all polynomials of this system

$$\int_a^b P_n(x) P_m(x) w(x) dx = 0 \ n \neq m$$

then the polynomials $\{P_n(x)\}_{n=0}^{\infty}$ are called orthogonal in $(a,\ b)$ with respect to the weight function $w(x)$. If moreover

$$||P_n(x)||_{w(x)} = \left[\int_a^b P_n^2(x) w(x) dx \right]^{\frac{1}{2}} = 1$$

for every $n = 0, 1, 2, \ldots$, then the polynomials are called orthonormal in $(a,\ b)$ with respect to $w(x)$.

Remark 1 The condition of the orthonormality of the system $\{P_n(x)\}_{n=0}^{\infty}$ has the form

$$\int_a^b P_n(x) P_m(x) w(x) dx = \delta_{nm}$$

where $\delta_{nm} = 1$ for $n = m$ and $\delta_{nm} = 0$ for $n \neq m$.

Theorem 1 For every weight function $w(x)$ one and only one system of polynomials $\{P_n(x)\}_{n=0}^{\infty}$ orthonormal in $(a,\ b)$ exists, where

$$P_n(x) = \sum_{k=0}^{n} a_k^{(n)} x^{n-k} \quad a_0^{(n)} > 0.$$

Proof E.g. in [1] or [2].

Theorem 2 A polynomial $P_n(x)$ is orthogonal in $(a,\ b)$ with respect to the weight function $w(x)$, if and only if for arbitrary polynomial $S_m(x)$ of the degree $m < n$ the following condition is fulfilled

$$\int_a^b P_n(x) S_m(x) w(x) dx = 0.$$

Proof E.g. in [1] or [2].

Theorem 3 If the interval of orthogonality is symmetric according to the origin of coordinate system and weight function $w(x)$ is even function, then every orthogonal polynomial $P_n(x)$ is even function and odd function, respectively, depending on evenness and oddness of its degree n, respectively, i.e.

$$P_n(-x) = (-1)^n P_n(x).$$

Proof E.g. in [1] or [2].

2 Jacobi Polynomials, Legendre Polynomials, Legendre Associated Functions

It is well-known that the classical Jacobi polynomials $\{P_n(x;\alpha,\beta)\}_{n=0}^{\infty}$ are orthogonal in the interval $I = (-1,\ 1)$ with respect to the weight function

$$J(x) = (1-x)^{\alpha}(1+x)^{\beta},\ x \in (-1,\ 1), \tag{1}$$

where $\alpha > -1, \beta > -1$. Very important special case of the Jacobi polynomials are the classical Legendre polynomials $\{P_n(x;0,0\}_{n=0}^{\infty}$, for which $\alpha = \beta = 0$ in the weight function $J(x)$. In the next we denote them by $\{P_n(x\}_{n=0}^{\infty}$.

Classical orthogonal polynomials are solutions of the second order linear homogeneous differential equations of the form (cf. e.g. [1])

$$a(x)y_n''(x) + b(x)y_n'(x) + \lambda_n y_n(x) = 0,$$

where $a(x)$ is a polynomial of the degree at most 2, $b(x)$ is a polynomial of the degree 1 and λ_n does not depend of x. For the classical Jacobi polynomials this equation has the form

$$(1-x^2)y_n''(x) + [\beta - \alpha - (\alpha+\beta+2)x]\, y_n'(x) + n(n+\alpha+\beta+1)y_n(x) = 0,$$

which in the case of the classical Legendre polynomials is reduced to the equation

$$(1-x^2)y_n''(x) - 2xy_n'(x) + n(n+1)y_n(x) = 0. \tag{2}$$

Associated Legendre equation occurs e.g. in applications described by Laplace or Helmholtz equation in spherical coordinates. For $m = 0, 1, 2, \ldots, n$ it has the form

$$(1-x^2)y_n''(x) - 2xy_n'(x) + \left[n(n+1) - \frac{m^2}{1-x^2}\right] y_n(x) = 0. \tag{3}$$

Observe that for $m = 0$ it reduces to the Legendre Eq. (2). Its solutions are called the associated Legendre functions of the first and second kind, respectively. They are defined by (cf. [3])

$$P_n^m(x) = (1-x^2)^{\frac{m}{2}} \frac{d^m}{dx^m} P_n(x) \tag{4}$$

and

$$R_n^m(x) = (1 - x^2)^{\frac{m}{2}} \frac{d^m}{dx^m} R_n(x),$$

respectively. Here $P_n(x)$ and $R_n(x)$ are solutions of the Legendre Eq. (2), the first of them are the classical Legendre polynomials.

The classical Jacobi polynomials are often defined by the Rodrigues' formula (cf. [1])

$$P_n(x;\alpha,\beta) = \frac{(-1)^n}{2^n\, n!\, J(x)} \frac{d^n}{dx^n}[J^n(x)] \tag{5}$$

where $J(x)$ is given by (1). In the case of the classical Legendre polynomials it reduces to

$$P_n(x;\alpha,\beta) = \frac{(-1)^n}{2^n\, n!} \frac{d^n}{dx^n}[(1-x^2)^n]. \tag{6}$$

3 Generalized Legendre Polynomials of a Certain Type and Classical Jacobi Polynomials with Different Weight Functions

As it is seen from preliminaries, the Legendre classical polynomials $\{P_n(x\}_{n=0}^{\infty}$ are orthogonal in $I = (-1,\ 1)$ with respect to the weight function $L(x) = 1$.

Now we introduce the system of polynomials $\{Q_n(x\}_{n=0}^{\infty}$ which will be the polynomials orthonormal in I with respect to the weight function

$$Q(x) = (x^2)^{\gamma},$$

where $\gamma > 0$ and $Q_n(+\infty) > 0$. It is clear that these polynomials are generalization of the classical Legendre polynomials, which can be obtained by substituting $\gamma = 0$ in the weight function $Q(x)$.

Further, we introduce two classes of orthonormal polynomials:

(a) polynomials $\{P_n(x;0,\gamma-\frac{1}{2})\}_{n=0}^{\infty}$ orthonormal in I with respect to the weight function

$$J_1(x) = (1+x)^{\gamma-\frac{1}{2}}$$

and

(b) polynomials $\{P_n(x;0,\gamma)\}_{n=0}^{\infty}$ orthonormal in I with respect to the weight function

$$J_2(x) = (1+x)^{\gamma}.$$

In both these cases we have classical Jacobi polynomials orthogonal with the weight function (1) for $\alpha = 0,\ \beta = \gamma - \frac{1}{2}$ and $\alpha = 0,\ \beta = \gamma$, respectively. In the next we establish relations between them and the polynomials $\{Q_n(x\}_{n=0}^{\infty}$.

Theorem 4 In the notations introduced in the previous sections

$$Q_{2n}(x) = 2^{\frac{\gamma}{2}-\frac{1}{4}} P_n\left(2x^2 - 1; 0, \gamma - \frac{1}{2}\right) \tag{7}$$

and

$$Q_{2n+1}(x) = 2^{\frac{\gamma}{2}} x P_n(2x^2 - 1; 0, \gamma). \tag{8}$$

Proof According to the Theorem 3, the function $Q_{2n}(x)$ is even function. Putting $t = x^2$ we denote $W_n(t) = Q_{2n}(x)$. The orthogonality of the polynomials $\{Q_n(x\}_{n=0}^{\infty}$ for $r = 0, 1, \dots, n-1$ and $n > 0$ yields

$$0 = \int_0^1 x^{2r} Q_{2n}(x) x^{2\gamma} dx = \frac{1}{2} \int_0^1 t^r W_n(t) t^{\gamma - \frac{1}{2}} dt$$

$$= \frac{1}{2^2} \int_{-1}^1 \left(\frac{\tau+1}{2}\right)^r W_n\left(\frac{\tau+1}{2}\right)\left(\frac{\tau+1}{2}\right)^{\gamma-\frac{1}{2}} d\tau$$

$$= \frac{1}{2^{\gamma+\frac{3}{2}}} \int_{-1}^1 \left(\frac{\tau+1}{2}\right)^r W_n\left(\frac{\tau+1}{2}\right)(\tau+1)^{\gamma-\frac{1}{2}} d\tau .$$

From that it is clear that the polynomials $W_n\left(\frac{x+1}{2}\right)$ are orthogonal in I with respect to the weight function $J_1(x)$. According to the Theorem taking into account the uniqueness of these polynomials, we have

$$W_n\left(\frac{x+1}{2}\right) = k P_n\left(x; 0, \gamma - \frac{1}{2}\right),$$

where $k > 0$ in consequence of the fact that $P_n\left(\infty; 0, \gamma - \frac{1}{2}\right) > 0$ and $W_n(+\infty) > 0$.

From the orthonormality of the polynomials $W_n(t)$ we derive

$$\frac{1}{2} = \int_0^1 W_n^2(t) t^{\gamma-\frac{1}{2}} dt = k^2 \int_{-1}^1 P_n^2\left(\tau; 0, \gamma - \frac{1}{2}\right)\left(\frac{\tau+1}{2}\right)^{\gamma-\frac{1}{2}} \frac{1}{2} d\tau$$

$$= \frac{1}{2^{\gamma+\frac{1}{2}}} k^2 \int_{-1}^1 P_n^2\left(\tau; 0, \gamma - \frac{1}{2}\right)(\tau+1)^{\gamma-\frac{1}{2}} d\tau$$

from where we have $k = 2^{\frac{\gamma}{2}-\frac{1}{4}}$ and the relation (7), i.e.

$$Q_{2n}(x) = 2^{\frac{\gamma}{2}-\frac{1}{4}} P_n\left(2t - 1; 0, \gamma - \frac{1}{2}\right), \quad t = x^2.$$

Now we prove the relation (8). Putting $t = x^2$ we have

$$\overline{W}_n(t) = x^{-1} Q_{2n+1}(x),$$

where $\overline{W}_n(t)$ is the polynomial of the degree n and $Q_{2n+1}(x)$ is odd function. For $r=0,1,\dots,n-1$ and $n>0$ the orthogonality of the polynomials $\{Q_n(x)\}_{n=0}^{\infty}$ yields

$$0=\int_0^1 x^{2r+1}Q_{2n+1}(x)\,x^{2\gamma}\,dx=\frac{1}{2}\int_0^1 t^r\,\overline{W}_n(t)\,t^{\gamma+\frac{1}{2}}\,dt$$

$$=\frac{1}{2^2}\int_{-1}^1\left(\frac{\tau+1}{2}\right)^r\overline{W}_n\left(\frac{\tau+1}{2}\right)\left(\frac{\tau+1}{2}\right)^{\gamma+\frac{1}{2}}d\tau$$

$$=\frac{1}{2^{\gamma+\frac{5}{2}}}\int_{-1}^1\left(\frac{\tau+1}{2}\right)^r\left(\frac{\tau+1}{2}\right)^{\frac{1}{2}}\overline{W}_n\left(\frac{\tau+1}{2}\right)(\tau+1)^{\gamma}\,d\tau\,.$$

From there

$$\left(\frac{x+1}{2}\right)^{\frac{1}{2}}\overline{W}_n\left(\frac{x+1}{2}\right)=\overline{k}\,P_n(x;0,\gamma)$$

where $\overline{k}>0$ and from the orthonormality of the polynomials $t^{\frac{1}{2}}\,\overline{W}_n(t)$ we derive

$$\frac{1}{2}=\int_0^1 x^{-2}\,Q^2_{2n+1}(x)\,x^{2\gamma}\,dx=\int_0^1 t\,\overline{W}^2_n(t)\,t^{\gamma}\,dt$$

$$=\frac{1}{2}\int_{-1}^1\left(\frac{\tau+1}{2}\right)\overline{W}^2_n\left(\frac{\tau+1}{2}\right)\left(\frac{\tau+1}{2}\right)^{\gamma}d\tau$$

$$=\frac{1}{2^{\gamma+1}}\overline{k}^2\int_{-1}^1 P^2_n(\tau;0,\gamma)(\tau+1)^{\gamma}\,d\tau.$$

Finally we get $\overline{k}=2^{\frac{\gamma}{2}}$ and the relation (8) of the theorem.

4 Jacobi and Legendre Polynomials Related to Fractional Calculus

In [4], the authors have defined so-called g-Jacobi functions by the formula

$$P_\nu(x;\alpha,\beta)=\frac{(-1)^\nu}{2^\nu\,\Gamma(\nu+1)\,(1-x)^\alpha(1+x)^\beta}D^\nu[(1-x)^{\nu+\alpha}(1+x)^{\nu+\beta}],\tag{9}$$

where $\nu>0,\ \alpha>-1,\ \beta>-1$ and D^ν is the Riemann–Liouville fractional differentiation operator defined for $x>0$, m natural and $m-1\le\mu<m$ as

$$D^\mu f(x)=D^m[I^{m-\mu}f(x)]$$

with the Riemann–Liouville fractional integral $I^{m-\mu} f(x)$ of the function $f(x)$ of order $m-\mu > 0$

$$I^{m-\mu} f(x) = \frac{1}{\Gamma(m-\mu)} \int_0^x (x-t)^{m-\mu-1} f(t)dt.$$

In (9) the functions $P_\nu(x;\alpha,\beta)$ are generalization of the classical Jacobi polynomials defined by Rodrigues' formula (5) in which natural n is substituted by real $\nu > 0$ and the derivative $\frac{d^n}{dx^n}$ is substituted by D^ν.

In virtue of (6) for corresponding generalized g-Legendre functions we can write

$$P_\nu(x;0,0) = \frac{(-1)^\nu}{2^\nu\, \Gamma(\nu+1)} D^\nu[(1-x^2)^\nu].$$

According to [4, Theorem 12] the g-Legendre functions can be expressed by means of Gauss hypergeometric functions (cf. [3]) in the form

$$P_\nu(x;0,0) = {}_2F_1\left(-\nu, \nu+1; 1; \frac{1-x}{2}\right), \tag{10}$$

where

$${}_2F_1(a,b;c;x) = \sum_{n=0}^{\infty} \frac{(a)_n (b)_n}{(c)_n} \frac{x^n}{n!}\; |x| < 1.$$

As it is seen in [3] the expression (10) presents the definition of the Legendre function of the first kind of degree ν denoted in [3] by $P_\nu(x)$ reached by generalization of the relationship between the nth Legendre polynomial $P_n(x)$ and the Gauss hypergeometric function ${}_2F_1\left(-n, n+1; 1; \frac{1-x}{2}\right)$. So, we can write

$$P_\nu(x;0,0) = P_\nu(x). \tag{11}$$

In [4], the authors proved that the g-Jacobi functions $P_\nu(x;\alpha,\beta)$ satisfy the linear homogeneous differential equation

$$(1-x^2)y_\nu''(x) + [\beta - \alpha - (\alpha+\beta+2)x]\, y_\nu'(x) + \nu(\nu+\alpha+\beta+1)y_\nu(x) = 0. \tag{12}$$

From (11) and (12) we can conclude that the g-Legendre functions $P_\nu(x;0,0)$ as well as the Legendre functions $P_\nu(x)$ of the first kind of the degree ν satisfy the differential equations

$$(1-x^2)y_\nu''(x) - 2xy_\nu'(x) + \nu(\nu+1)y_\nu(x) = 0.$$

Finally, taking into account the proof of the fact that the associated Legendre functions (4) of the first kind satisfy the Eq. (3) we can conclude that the functions

$$P_\nu^m(x) = (1-x^2)^{\frac{m}{2}} \frac{d^m}{dx^m} P_\nu(x)$$

satisfy the equations

$$(1-x^2)y_\nu''(x) - 2xy_\nu'(x) + \left[\nu(\nu+1) - \frac{m^2}{1-x^2}\right] y_\nu(x) = 0, \tag{13}$$

where $\nu > 0$, $m = 0, 1, 2, \ldots, [\nu]$ and $[\nu]$ is integer part of ν. The proof is similar as it is done in [3] for the Eq. (3) and the function (4).

Because the associated Legendre functions of the first kind $P_n^m(x)$ and the Legendre functions of the first kind $P_\nu(x)$ have many properties in common with the Legendre polynomials $P_n(x)$, we can suppose that also the functions $P_\nu^m(x)$ have some similar properties like the above functions given in (4).

5 Applications

The classical Jacobi orthogonal polynomials and their special case—the Legendre polynomials and the associated Legendre functions as well as their generalizations occur in various applications of mathematics in engineering sciences.

In [5] we suggested to compute the length of a junction curve of a railroad between two sections of a rail with different curvatures by means of classical Legendre polynomials. Junction curve is a curve inserted between a straight line and a circle arc. Junction curve enables fluent transition between two sections of a rail with different curvatures. At Slovak railroads the cubic parabola is used as the junction curve. In [6] the length of such junction curve is computed by using binomial series, because the integral of the function

$$\sqrt{1 + [f'(x)]^2}, \; f(x) = cx^3,$$

(c is the specific constant) had to be computed. We suggested to use the Legendre polynomials, because the best approximation of a function by means of a polynomial is the approximation by means of an expansion of the function into a series of orthogonal polynomials followed by substituting the function by the partial sum of such series (cf. [1]). So we have used the following expansion of a function $\sqrt{1+x}$ (cf. [1])

$$\sqrt{1+x} = \frac{4}{3\sqrt{2}} P_0(x) - \frac{4}{\sqrt{2}} \sum_{n=1}^{\infty} \frac{(-1)^n P_n(x)}{(2n-1)(2n+3)}$$

where $\{P_n(x)\}_{n=0}^{\infty}$ are the classical Legendre polynomials.

Another application of the classical Legendre polynomials can be found in [7]. The authors used the Legendre polynomials for the approximation of cylindrical surfaces which was submitted as a strategy of measurement supported by experimental verification. The results were very good—the precision was high.

The Legendre associated functions were used in [8] for the approximation of the Earth shape. The author uses the functions $P_n^m(sin\phi)$ (ϕ is the geocentric coordinate of the geocentric radiusvector of a point of the Earth) defined by (4). The gravity potential of the Earth is expressed by the expansion into the series of spherical functions, where the functions $P_n^m(sin\phi)$ occur.

6 Conclusion

In this paper we proved relations (7) and (8) which may be used as expressions of the classical Jacobi polynomials of the argument $2x^2 - 1$ by means of generalized Legendre polynomials $\{Q_n(x)\}_{n=0}^{\infty}$ of the argument x. We have also pointed out some known relations of the classical Legendre polynomials and the associated Legendre functions related to certain their generalizations including those obtained by using fractional calculus. All the functions dealt with above are useful in applications especially those described by the Laplace equation.

Acknowledgment This research has been supported by the Slovak Grant Agency KEGA through the project No. 057ŽU-4/2012 and by the Slovak Grant Agency VEGA through the project No. 1/1069/12.

References

1. Sujetin PK (1979) Classical orthogonal polynomials. Nauka, Moskva [Russian]
2. Szegö G (1969) Orthogonal polynomials. Nauka, Moskva [Russian]
3. Andrews LC (1992) Special functions of mathematics for engineers. McGraw-Hill Inc., New York
4. Mirevski SP, Boyadjiev L, Scherer R (April 2007) On the Riemann–Liouville fractional calculus, g-Jacobi functions and F-Gauss functions. Appl Math Comput 187(1):315–325
5. Guldan V (June 2006) Infinite series, special functions and study programme Geodesy and cartography. In Proceedings of the 3rd Žilina's Didactic Conference DIDZA on CD, Žilina, Slovakia, [Slovak], 5 p
6. Bitterer L (1997) Geometry of rail. University of Žilina [Slovak]
7. Janecki D, Stepien K (December 2005) Legendre polynomials used for the approximation of cylindrical surfaces. Commun Sci Lett Univ Zilina 4:59–61
8. Tenzer R (December 2001) Geopotential model of Earth—approximation of Earth shape, geopotential model testing methods. Commun Sci Lett Univ Zilina 4:50–58

Descartes Rule of Signs and Linear Programming

Carla Fidalgo and Alexander Kovačec

Abstract Let $\Delta = \{(x, y) : x + y = 1, x, y \geq 0\}$ be the 1-simplex and for $m \geq 2$ consider the (binary) form

$$F(x, y) = u_n x^n + u_0 y^n - \sum_{\substack{i, j \geq 1 \\ i + j = n}} u_i x^i y^j.$$

Using linear programming and a little known refinement of Descartes' rule of signs due to Laguerre, it is shown that if all $u_i \geq 0$ and F is nonzero and nonnegative on Δ, then it assumes there exactly one global minimum. The investigation is motivated by a question concerning sum of squares representation.

Keywords Roots of polynomials · Linear programming · Positive semidefinitness · Sums of squares

1 Introduction

By a *diagonal minus tail* form we understand a real homogeneous polynomial

$$F(\underline{x}) = F(x_1, \ldots, x_k) = a_1 x_1^n + \cdots + a_n x_k^n - \sum_{\substack{i_1, \ldots, i_k \geq 0 \\ i_1 + \cdots + i_k = n}} a_{i_1 i_2 \cdots i_k} x_1^{i_1} \cdots x_k^{i_k},$$

with $n \geq 2$ and all occurring $a_i, a_{i_1 i_2 \cdots i_k} \geq 0$. The $(k-1)$-simplex is defined as $\Delta_{k-1} = \{\underline{x} \in \mathbb{R}^k_{\geq 0} : x_1 + \cdots + x_k = 1\}$. The authors conjecture that if F is positive semidefinite and of even degree then it is a sum of squares of polynomials. An

C. Fidalgo (✉)
Departamento de Física e Matemática, Instituto Superior de Engenharia de Coimbra, Coimbra, Portugal
e-mail: cfidalgo@isec.pt

A. Kovačec
Departamento de Matemática, Universidade Coimbra, Coimbra, Portugal
e-mail: kovacec@mat.uc.pt

N. M. Fonseca Ferreira, J. A. Tenreiro Machado (eds.), *Mathematical Methods in Engineering*, DOI 10.1007/978-94-007-7183-3_21,

algorithm for such a representation relying on finding the global minimum of F on the simplex has as yet performed successfully and has prompted the conjecture that in fact there is only one local, and hence global minimum for F on Δ.

For the binary case we write

$$\Delta = \Delta_1 = \{(x, y) : x + y = 1, x, y \geq 0\}, \qquad \underline{u} = (u_0, u_1, \ldots, u_{n-1}, u_n),$$

and the form as

$$F(x, y) = F(x, y, \underline{u}) = u_n x^n + u_0 y^n - \sum_{\substack{i,j \geq 1 \\ i+j=n}} u_i x^i y^j.$$

We will see that $F|\Delta \geq 0$ implies uniqueness of a critical point and hence in this case uniqueness of the global minimum.

The proof of this seems not as easy as may be expected. First the problem is reduced to a problem in one variable. We have to show that the nonnegativity of a certain polynomial function $\mathbb{R}_{>0} \ni t \mapsto f(t, \underline{u})$ obtained from dehomogenizing $F(x, y, \underline{u})$ implies existence and unicity of the positive root of a polynomial $t \mapsto p(t, \underline{u})$ related to the natural derivative of F in Δ. This latter claim is proved as a strengthening of its contrapositive. Supposing that $\underline{u}$ is such that $t \mapsto p(t, \underline{u})$ has more than one positive root, f would be negative in 1 or on a root where p (tendentially) changes from positive to negative; a '$\pm$-root of p', for short: see theorem 3.5.

In both of these polynomials the coordinates of $\underline{u}$ enter linearly. Fixing t_0, u_0, u_n, we consider the polyhedron $P = P(t_0, u_0, u_n)$ of all nonnegative $\underline{u}$ (of given u_0, u_n) that define functions $t \mapsto p(t, \underline{u})$ that have t_0 as a '$\pm$-root' and identify its vertices. Using that the affine function $P \ni u_{1:n-1} = (u_1, \ldots, u_{n-1}) \mapsto f(t_0, \underline{u})$ takes its maximum at some vertex of the polyhedron, and showing that the corresponding maximum value is negative will complete the proof in the cases where t_0 is not too large, in particular for $t_0 \leq 1$.

2 Descartes' Rule of Signs

The well known Descartes rule of signs states that the number of positive roots of a polynomial $p(x)$ with real coefficients does not exceed the number of sign changes of the nonzero coefficients of $p(x)$. So it gives an upper bound for the number of positive roots in terms of sign changes of the coefficients of a polynomial.

Despite its intuitive plausibility, Descartes' rule of signs was not directly proven until over a century after its original statement, in 1637. Since then many refinements and extensions where found by mathematicians like Laguerre, Pólya and Szegö among others and its interest remains nowadays to both mathematicians and computer scientists areas like isolation of the real roots of polynomial equations, polynomial real root-finding algorithms, etc.

We learned of refinements of Descartes' rule found by Laguerre, Polya and Szegö that we formulate here.

Theorem LPS Let $f(t) = \sum_{j=1}^{n} a_j t^j$ be a polynomial with real coefficients and $u > 0$. Then

a. the number of roots of $f_{|]0,u[}$ is by an even number less than the number of signchanges of the coefficient sequence of the power series $f(tu)/(1-t)^k$, where $k \in \mathbb{Z}_{\geq 0}$.
b. furthermore as $k \to \infty$ the number of signchanges 'converges' to the number of roots of f in $]0, u[$.

Let us see an example that shows that this theorem is stronger than Descartes' rule of signs.

Example 2.1 The polynomial $f(t) = 1+20t-90t^2+140t^3-70t^4$ has by Descartes' rule of signs at most three positive roots. But in fact the following table, where in row $k = 0, 1, \dots, 4$ we have the first few coefficients of the powerseries $f(t)/(1-t)^k$, shows that it is positive in the whole interval $]0, 1[$.

$\underline{a}$:	1	20	− 90	140	− 70	0	0	...
$\text{lps}^{(1)}(\underline{a})$:	1	21	− 69	71	1	1	1	...
$\text{lps}^{(2)}(\underline{a})$:	1	22	− 47	24	25	26	27	...
$\text{lps}^{(3)}(\underline{a})$:	1	23	− 24	0	25	51	78	...
$\text{lps}^{(4)}(\underline{a})$:	1	24	0	0	25	76	154	...

To study the number of roots in $]1, +\infty[$ one considers the reciprocal polynomial $x^4 f(1/x) = -70+140x-90x^2+20x^3+x^4$. It has roots in $]0, 1[$ that via inversion correspond to roots of f in $]1, +\infty[$. Looking at the first three rows of its associated table,

$\underline{a}$:	−70	140	−90	20	1	0	0	...
$\text{lps}^{(1)}(\underline{a})$:	−70	70	−20	0	1	1	1	...
$\text{lps}^{(2)}(\underline{a})$:	−70	0	−20	−20	−19	−18	−17	...

we find that the last row will eventually turn positive. So the original polynomial which evidently has at least one positive root, has in fact precisely one positive root somewhere in $]1, +\infty[$, and hence only one in $]0, +\infty[$. The slow convergence towards a positive value is consequence of the fact that the original polynomial and the reciprocal polynomial have as (only) (approximate) positive roots 1.04 and 0.96, respectively. If one chooses $u = 1.3$ (instead of 1) then the second table reads

$\underline{a}$:	−70	182	−152.1	43.94	2.86	0	0	...
$\text{lps}^{(1)}(\underline{a})$:	−70	112	−40.1	3.84	6.70	6.70	6.70	... ,
$\text{lps}^{(2)}(\underline{a})$:	−70	42	1.9	5.74	12.44	19.13	25.83	...

showing in the last row an earlier sign change. It estimates the roots of the reciprocal in $]0, u[$ hence of the original in $]1/u, +\infty[$.

3 Precise Statements

In connection with the representation of multivariate polynomials as sums of squares we came up with the following conjecture.

Let $\Delta = \{(x, y) : x+y = 1, x, y \geq 0\}$ be the 1-simplex, $\underline{u} = (u_0, u_1, \ldots, u_{n-1}, u_n) \in \mathbb{R}^{n+1}_{\geq 0}$ and consider the binary form

$$F(x, y) = F(x, y, \underline{u}) = u_n x^n + u_0 y^n - \sum_{\substack{i,j\geq 1 \\ i+j=n}} u_i x^i y^j.$$

Conjecture 3.1 If $F_{|\Delta} \geq 0$ then it has only one critical point (which is necessarily a unique global minimizer).

We present one of our solutions because we think it is quite surprising and possibly applicable in other contexts.

It is natural to approach this problem by trying to make use of the vast literature on (real) roots of univariate polynomials and hence to dehomogenize the original question.

To study a linear combination of monomials $x^i y^j$ under the condition $x+y = 1$ is the same as studying combinations of monomials $t^i (1-t)^j$ (which figure prominently in Bernstein polynomials) on the real line. If one does so introducing $t = x/y$ and defining

$$\mu_i(t) = -\frac{i}{n} t^{i-1} + \left(1 - \frac{i}{n}\right) t^i,$$

$$p(t, \underline{u}) = -u_0 + \sum_{i=1}^{n-1} u_i \mu_i(t) + u_n t^{n-1} \tag{1}$$

$$= \left(-u_0 - \frac{u_1}{n}\right) + \sum_{i=1}^{n-2} \left(\left(1 - \frac{i}{n}\right) u_i - \frac{i+1}{n} u_{i+1}\right) t^i + \left(\frac{u_{n-1}}{n} + u_n\right) t^{n-1},$$

$$f(t, \underline{u}) = u_0 - \sum_{i=1}^{n-1} u_i t^i + u_n t^n,$$

some elementary calculations lead to

Lemma 3.2 Supposing $y \neq 0$ and putting $t = \frac{x}{y}$ we have

i. $DF(x, y) = ny^{n-1} p(t, \underline{u})$;
ii. If $u_0, u_n > 0$ then $DF(1, 0) > 0$ and $DF(0, 1) < 0$, that is, the points $(1, 0)$ and $(0, 1)$ are not critical;
iii. $F(x, y) = y^n f(t, \underline{u})$.

One can reformulate conjecture 1 as follows

Conjecture 3.3 If $\mathbb{R}_{\geq 0} \ni t \mapsto f(t, \underline{u})$ is nonnegative then $\mathbb{R}_{\geq 0} \ni t \mapsto p(t, \underline{u})$ has exactly one positive root.

Note that nonnegativity of f implies $f(1, \underline{u}) \geq 0$ which means $\sum_{i=1}^{n-1} u_i \leq u_0 + u_n$. For some time we conjectured that this inequality is sufficient for unicity of roots.

As in Sect. 2 the LPS-theorem can be used for estimating the number of zeros in $]u', +\infty[$ of any polynomial. To this end define similarly as above the polynomial

$\widetilde{p}(x) = x^{\deg(p)} p(1/x)$ and note that $\widetilde{p}(x)$ has in $]0, 1/u'[$ as many zeros as p has in $]u', +\infty[$.

By means of the LPS-theorem we were able to show that under the hypothesis $\sum_{i=1}^{n-1} u_i < u_0 + u_n$ one of the sets $]0, 1[, \{1\},]1, \infty[$ must contain all the positive roots of the polynomial $p(t, \underline{u})$. Unfortunately we discovered that as we increased the degree $n-1$ of our polynomials the only way to show that $p(t, \underline{u})$ has only one root was to increase k in the calculation of the power series $p(t, \underline{u})/(1-t)^k$. Worse than this, after considerable theoretical insight we managed to find a counter example for the strengthened conjecture that

$$\sum_{i=1}^{n-1} u_i < u_0 + u_n \text{ implies unicity of roots of } p(t, \underline{u}).$$

If $u_0 = u_n = 1$ the first such example seems to occur for $n = 636$.

Example 3.4 Assume $n = 750$, $u_0 = u_n = 1$, $u_1 = 1.01412$, $u_{500} = 0.942$, and all other $u_i = 0$. Then

$$\begin{aligned} p(t, \underline{u}) &= -1 + u_1 \mu_1(t) + u_{500} \mu_{500}(t) + t^{749} \\ &= -1.00135216 + 1.01276784\, t - 0.628\, t^{499} + 0.314\, t^{500} + t^{749} \end{aligned}$$

and

$$f(t, \underline{u}) = 1 - u_1 t - u_{500} t^{500} + t^{750}.$$

Here $\sum_{i=1}^{n-1} u_i = 1.95612 < u_0 + u_n = 2$, but $p(t, \underline{u})$ has roots $(p) = \{0.991366, 0.992061, 0.992877\}$.

The following graphics show the behavior of p near its root $t_0 = 0.992061$ at two scales in the intervals $]0.98, 1[$ and $]0.991, 0.993[$. (The pairs of numbers in left upper and lower right corners delimited the figures.)

So we were back to square one.

Still not knowing how to accommodate the requirement that f be nonnegative, we formulated the conjecture as a contrapositive.

Let us say that a polynomial f has a *weak $\pm$-root* in t_0 if $f(t_0) = 0$ and $\dot{f}(t_0) \leq 0$.

We show now how we proved (surprisingly by means of theory of Linear Programming)

Theorem 3.5 Assume $p(t, \underline{u})$ has at least two roots in $\mathbb{R}_{>0}$. Then $f(1, \underline{u}) < 0$ or there is a weak $\pm$-root t_0 of $p_{|\mathbb{R}_{>0}}$ such that $f(t_0, \underline{u}) < 0$.

Note that it is clear that this theorem implies our conjecture, since it guarantees that if p has more than one root, then f is negative somewhere; since $p(0, \underline{u}) < 0$, it is also clear that the existence of two (distinct) roots for p, imply the existence of a weak $\pm$-root.

Given a polynomial $p(t, \underline{u})$ and a weak $\pm$-root t_0 one can define a polyhedron $P = P(t_0, u_0, u_n)$. Actually to say that t_0 is a weak $\pm$-root is to say that $p(t_0, \underline{u}) = 0$ and $\dot{p}(t_0, \underline{u}) := \frac{\partial p}{\partial t}(t_0, u) \leq 0$. By the formulae for p, in (1), this means that $\underline{u}$ satisfies the following system of linear inequalities; the matrix at the left being $(n+2) \times (n-1)$

$$\begin{bmatrix} -\mu_1(t_0) & -\mu_2(t_0) & \cdots & -\mu_{n-1}(t_0) \\ \mu_1(t_0) & \mu_2(t_0) & \cdots & \mu_{n-1}(t_0) \\ \dot{\mu}_1(t_0) & \dot{\mu}_2(t_0) & \cdots & \dot{\mu}_{n-1}(t_0) \\ -1 & & & \\ & -1 & & \\ & & \ddots & \\ & & & -1 \end{bmatrix} \begin{bmatrix} u_1 \\ u_2 \\ \vdots \\ u_{n-1} \end{bmatrix} \leq \begin{bmatrix} -u_0 + u_n t_0^{n-1} \\ u_0 - u_n t_0^{n-1} \\ -(n-1) u_n t_0^{n-2} \\ 0 \\ \vdots \\ 0 \end{bmatrix}.$$

The first two rows encoding $p(t_0, \underline{u}) = 0$ are evidently linearly dependent; the $n-1$ lower rows encode the nonnegativity of $\underline{u}_{1:n-1}$. The system defines a polyhedron $P = P(t_0, u_0, u_n)$ which may well be empty. Indeed if $n = 3$ for example, then it is easy to see that for any $u_0, u_n, t_0 > 0$, $P(t_0, u_0, u_n)$ is empty, but there are examples for large n where $P \neq \emptyset$.

Given this formalization one can show by means of the theory of linear programming that every vertex of P has exactly two nonzero coordinates. More $P \ni u_{1:n-1} \mapsto f(t_0, u_0 e_0 + u_{1:n-1} + u_n e_n)$ takes its maximum at some vertex of P, and then, by means of Cramer's rule that in each such vertex $f(t_0, \underline{u}) < 0$, which concludes the proof.

Multidimensional Markov Chains Usage in the Radio Resource Management Problem

Victor D. N. Santos, N. M. Fonseca Ferreira and F. Moita

Abstract This paper presents an analytical model that evaluates the performance of the Maximum Packing channel allocation technique on linear and planar cellular systems. The main innovations introduced by this model are the deterministic identification of the system space-state SMP and its application to a multi-dimensional Markov chain.

The model was thereafter applied to several cellular systems with different characteristics: number of cells, number of channels, interference constraints and offered traffic values. Simulation results have validated the model and shown that Maximum Packing technique provides the best performance among all the available algorithms.

Keywords Markov chains · Maximum packing · Models

1 Introduction

The exponential growth observed on the number of current mobile cellular systems users, has however, exposed several capacity and performance problems of those networks originated by an inefficient usage of the assigned spectrum. The inclusion of dynamic channel allocation technique on the mobile network protocols increases the utilization of the radio resources mitigating earlier mentioned problems and leading to more efficient solutions.

This paper presents a new analytical model to evaluate MP (Maximum Packing) assignment technique performance based on multidimensional Markov Chains. A Markov chain is a special case of a Markov process, which itself is a special case of a

V. D. N. Santos (✉) · N. M. Fonseca Ferreira · F. Moita
Dept. of Electrical Engineering, Institute of Engineering of Coimbra,
Polytechnic Institute of Coimbra, Quinta da Nora, Apartado 10057,
3031-601 Coimbra Codex, Portugal
e-mail: vsantos@isec.pt

N. M. F. Ferreira
e-mail: nunomig@isec.pt

F. Moita
e-mail: moita@isec.pt

N. M. Fonseca Ferreira, J. A. Tenreiro Machado (eds.), *Mathematical Methods in Engineering*, DOI 10.1007/978-94-007-7183-3_22,

random or stochastic process. A Markov chain is a discrete-state random process in which the evolution of the state of the process beginning at a time t (continuous-time) or n (discrete-time) depends only on the current state $X(t)$ or X_n, and not how the chain reached its current state or how long it has been in that state.

The analytical model is based on the system space-state S_{MP}, evaluation based only on the cellular geometry, interference constraints and amount of radio resources. The blocking probabilities at the system and cell level are determined by adding the occurrence probabilities, of all the states that correspond to specific cell and system blocking conditions.

2 MP Analytical Evaluation

2.1 Maximum Packing

The MP technique [1] is an idealized DCA algorithm that will only block a call if there are no possible reassignments in order to free a channel for the new call. This algorithm tries to find a possible solution to serve each call at each moment. The large amount of information required by this technique, in order to search for all possible reallocations, makes this algorithm impractical for implementation.

2.2 The Dimensionality of the MP Problem

In mobile cellular networks that adopt the FCA (Fixed Channel Allocation) technique as a resource management method, the existing radio resources are divided among the cells of the different clusters in a predefined and permanent way. Having in mind that in this channel allocation technique the cells are independent from each other, we conclude that the performance evaluation problem has only one dimension. In this case the blocking probability of each cell is given by the Erlang B formula where it is only necessary to know the number of available channels and the offered traffic load values per cell.

DCA techniques employ a different approach. In addition to the knowledge of the channel usage in a particular cell, the system must also be aware of the resource utilization in the neighboring cells. The acceptance of a new call depends on the global state of the system i.e. the number of calls accepted in each cell and the information with respect to the channels used in supporting ongoing calls.

The MP algorithm however precludes the necessity of that information because a call is admitted only if there are possible reassignments of the existing calls, in order to free a channel in that cell. Therefore, the dimension of the MP performance evaluation problem is equal to the number of cells on the considered mobile cellular system.

2.3 Mathematical Model Definition

The scenario presented in this paper points for a cellular system built up by a finite set of N non-overlapping cells that share a common pool of radio resources according to the MP channel allocation technique presented in the former section. It is also assumed that the offered traffic values of each cell are known, being independent from each other.

A non-negative integer vector $\vec{x}$ of length N was introduced to represents the radio resources utilization in the system at a given instant of time. Each of its component x_i represents the total number of active calls on a given cell [2].

After the scenario and model definitions the next step in our analysis is the identification of all the states that a given cellular system can take with respect to the radio resources utilization, which altogether form the space-state S_{MP}. The resulting space-state S_{MP} is a sub-space of Z_+^N restricted only by the number of available channels on the pool, the channel reuse constraints and the cellular system geometry i.e. the number of cells and its spatial distribution. These limitations on the system space-state S_{MP} can be expressed by (1)

$$S_{MP} = \left\{ \vec{x} \in Z_+^N : \bigcap_{i=1}^{N} \left(\sum_{j \in C_i} x_j \leq M \quad \forall C_i \right) \right\} \tag{1}$$

This equation declares that the maximum number of active calls inside any cluster containing a particular cell must be smaller or equal to M (the total number of channels on the system). It is necessary to take into account that this condition must be fulfilled simultaneously for all the system cells. In Eq. (1) the symbol C_i was introduced to represents an arbitrary cluster that contains the cell i.

The following step is the system stochastic property characterization. Under the usual accepted assumptions, a *Poisson* call arrival process and an exponential call-holding time distribution, the system stochastic process $\{x(t),\ t \geq 0\}$ can be considered as a *Markov* chain on the space-state S_{MP}, being the states occurrence probability in statistical balance $p(x)$ given [3]

$$p\left(\vec{x}\right) = \prod_{i=1}^{N} \frac{{A_i}^{x_i}}{x_i!} \cdot p\left(0, 0, \cdots, 0\right) \qquad \text{for all} \quad \vec{x} \in S_{MP} \tag{2}$$

As usual in this type of problems, that set of equations is not sufficient to evaluate the state occurrence probabilities, being necessary to introduce the statistical normalization equation.

$$\sum_{\vec{x} \in S_{MP}} p\left(\vec{x}\right) = 1 \tag{3}$$

The blocking probability will be now evaluated adding the occurrence probabilities of all the states that correspond to a given blocking situation. A call will be blocked

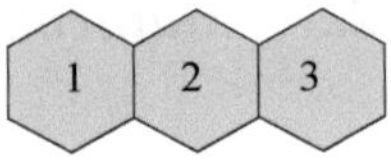

Fig. 1 Three cell system

if it finds the system in a state for which the maximum number of ongoing calls in at least one cluster, that contains this particular cell, is equal to the number of available channels.

The system blocking probability measured at a global basis is obtained by

$$GoS = \frac{1}{L} \cdot \sum_{i=1}^{N} A_i \cdot P_{bi} \quad \text{with} \quad L = \sum_{i=1}^{N} A_i \tag{4}$$

3 Highway Cellular Systems Analysis

This section is devoted to the evaluation of the MP channel allocation technique performance on linear cellular systems. The study of linear arrays of cells was separated from planar layouts of cells, due to certain particularities, which facilitate the evaluation of the system space-sate S_{MP} on this specific case leading to very simple equations. A cluster in a linear cellular system is made of K consecutive cells, being the cluster size equal to the system reuse factor. The total number of distinct clusters on a linear system is given by $N - K + 1$. The restrictions on the number of simultaneous calls in a cluster imposed by the channel reuse concept are now used to establish a very simple set of independent equations that will deterministically determine the system space-state.

Example 1: The 3-Cell Highway System This simplest possible scenario was selected among others because it provides a simple starting point for more complex system's studies. The system presented in Fig. 1 has the following parameters: $N = 3$; $M = 4$; $K = 2$; TYPE = LINEAR, where N represents the number of cell in the system, M the number of channel and K the system reuse factor.

The MP channel allocation technique works as follows: if a call arrives at one of the extremity cells and there is an available channel on the system, this channel is selected and locked in that cell and in the central cell in order to avoid interference. On the other hand, if a channel is assigned to the central cell, this channel must be locked in all the cells.

The space-state S_{MP} is a sub-space of Z_+^3, *i.e.* $S_{MP} \subset Z_+^3$ being its definition given by the following equation:

$$S_{MP} = \left\{ \vec{x} \to (x_1, x_2, x_3) \in Z_+^3 : x_1 + x_2 \leq M \wedge x_2 + x_3 \leq M \right\} \tag{5}$$

Figure 2 presents all the 55 states that belong to S_{MP}, the states that represent cell-blocking conditions in the *clusters* $C_{1,2}$ and $C_{2,3}$ and the allowed transitions among the system states.

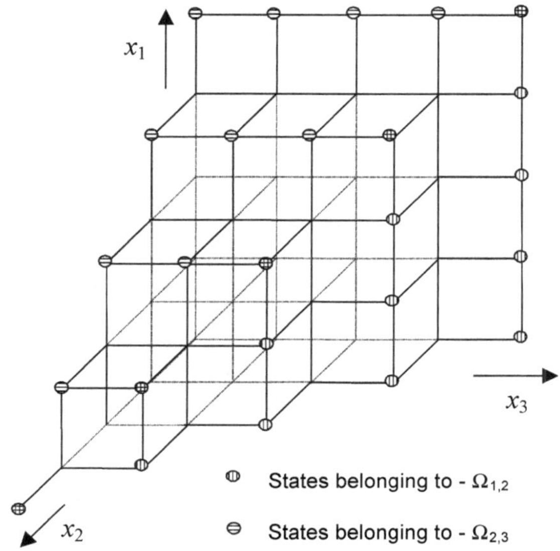

Fig. 2 Space S_{MP} and *Markov* chain for the three-cell system

The steady state probabilities of each one of the system states is now easily evaluated appealing only to the normalizing constant G assessment

$$G = p(0,0,0)^{-1} = \sum_{x_1=0}^{M} \sum_{x_2=0}^{M-x_1} \sum_{x_3=0}^{M-x_2} \prod_{i=1}^{3} \frac{A_i^{x_i}}{x_i!} \tag{6}$$

The occurrence probability values of each one of the 55 states belonging to S_{MP}, was presented in Fig. 3 considering a uniform offered traffic distribution per cell of 1.0 Erl. Those results were obtained analytically, appealing to Eqs. (2) and (6) and by simulation [4].

The next step is the blocking sub-spaces for each one of the system cells identification. Cell #1 is on a blocked state if the sum of active calls on the *cluster* $C_{1,2}$, is equal to the number of radio resources in the system. The set B_1 that represents the conjunction of all the states for which the cell #1 is blocked is equal to $\Omega_{1,2}$ and is given by the following expression

$$B_1 = \{\vec{x} \in S_{MP} :\ x_1 = M - x_2\} \tag{7}$$

The associated blocking probability value P_{b1}, is obtained adding the occurrence probability of all the elements of B_1, which leads to the following equation

$$P_{b1} = \frac{\sum\limits_{x_2=0}^{M} \sum\limits_{x_3=0}^{M-x_2} \left(\frac{A_1{}^{(M-x_2)}}{(M-x_2)!} \cdot \prod\limits_{i=2}^{3} \frac{A_i{}^{x_i}}{x_i!} \right)}{\sum\limits_{x_1=0}^{M} \sum\limits_{x_2=0}^{M-x_1} \sum\limits_{x_3=0}^{M-x_2} \prod\limits_{i=1}^{3} \frac{A_i{}^{x_i}}{x_i!}} \tag{8}$$

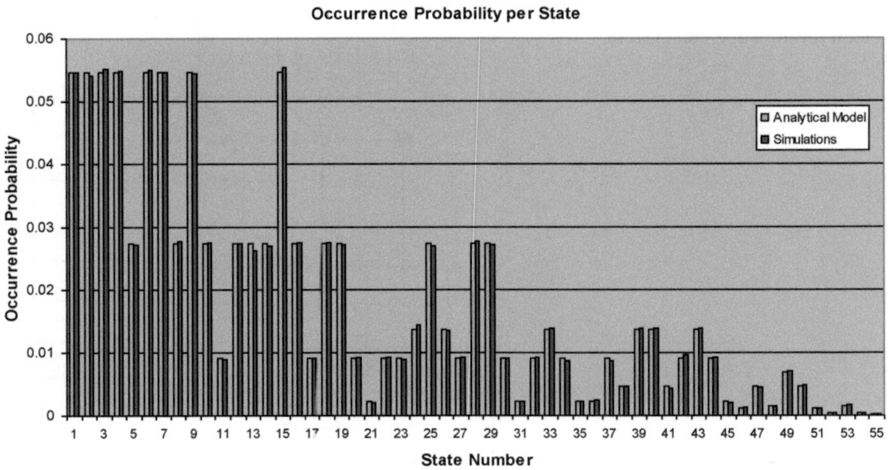

Fig. 3 State space S_{MP} occurrence probability values

On the other hand the cell #3 is in a blocked state if the sum of ongoing calls in the *cluster* $C_{2,3}$, is equal to the number of radio resources in the system.

$$B_3 = \{\vec{x} \in S_{MP} : \ x_3 = M - x_2\} \tag{9}$$

The blocking probability value P_{b3} for the cell 3 is given by

$$P_{b3} = \frac{\sum\limits_{x_1=0}^{M} \sum\limits_{x_2=0}^{M-x_1} \left(\prod\limits_{i=1}^{2} \frac{A_i^{\ x_i}}{x_i!} \cdot \frac{A_3^{\ (M-x_2)}}{(M-x_2)\,!} \right)}{\sum\limits_{x_1=0}^{M} \sum\limits_{x_2=0}^{M-x_1} \sum\limits_{x_3=0}^{M-x_2} \prod\limits_{i=1}^{3} \frac{A_i^{\ x_i}}{x_i!}} \tag{10}$$

The central cell blocking probability evaluation follows the same steps earlier presented. However, in opposition to the extremity cells, the central cell #2 is affected by the radio resources utilization in the distinct *clusters* $C_{1,2}$ and $C_{2,3}$. The blocking probability space-state for the central cell is given by the reunion of the states contained on the sets $\Omega_{1,2}$ and $\Omega_{2,3}$ resulting in the following expression.

$$B_2 = \{\vec{x} \in S_{MP} : x_1 + x_2 = M \vee \ x_2 + x_3 = M\} \tag{11}$$

In the evaluation of B_2 it is necessary to take into account that the elements of the sub-spaces $\Omega_{1,2}$ and $\Omega_{2,3}$ are not mutually exclusive. Thus, we should use the following equation retrieved from the set theory:

$$P\left(\Omega_{1,2} \cup \Omega_{2,3}\right) = P\left(\Omega_{1,2}\right) + P\left(\Omega_{2,3}\right) - P\left(\Omega_{1,2} \cap \Omega_{2,3}\right) \tag{12}$$

The sub-spaces $\Omega_{1,2}$ and $\Omega_{2,3}$ probabilities were already evaluated on the Eqs. (7) and (9) being necessary only to assess the sub-space $\Omega_{1,2} \cap \Omega_{2,3}$ and its corresponding probability. The set $\Omega_{1,2} \cap \Omega_{2,3}$ is given by

$$\Omega_{1,2} \cap \Omega_{2,3} = \{\vec{x} \rightarrow (x_1, x_2, x_3) \in S_{MP} : \ x_1 = x_3 = M - x_2\} \tag{13}$$

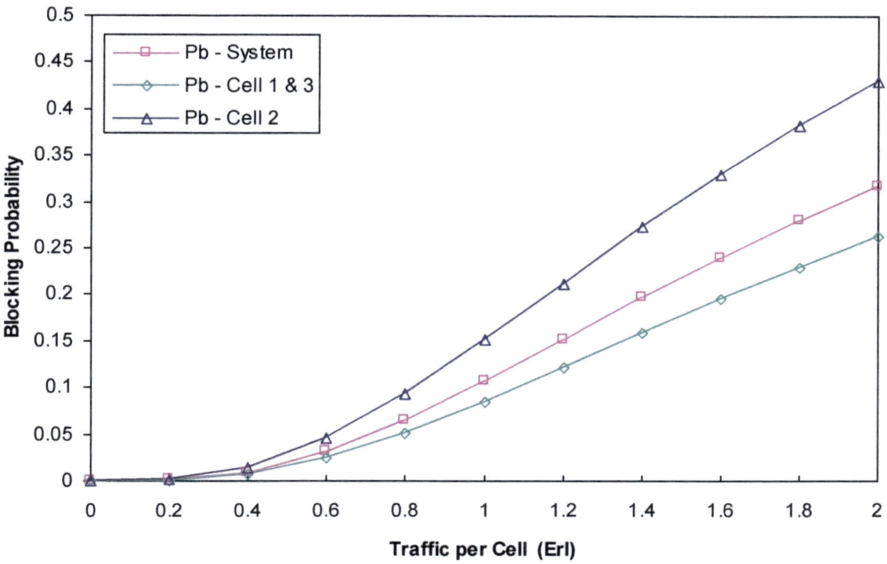

Fig. 4 Three cells highway system blocking probabilities

The blocking probability on the central cell #2 is obtained replacing (8), (10) and (13) on (12).

$$P_{b2} = P_{b1} + P_{b3} - \frac{\sum_{x_2=0}^{M} \frac{{A_1}^{(M-x_2)}}{(M-x_2)!} \cdot \frac{{A_2}^{x_2}}{x_2!} \cdot \frac{{A_3}^{(M-x_2)}}{(M-x_2)!}}{\sum_{x_1=0}^{M} \sum_{x_2=0}^{M-x_1} \sum_{x_3=0}^{M-x_2} \prod_{i=1}^{3} \frac{{A_i}^{x_i}}{x_i!}} \tag{14}$$

Cells blocking probabilities and the system GoS of the linear cellular system ($N=3$; $M=4$; $K=2$, TYPE = LINEAR) are depicted on Fig. 4.

Besides the blocking probabilities values evaluation it is also necessary to assess the algorithm complexity by counting the total number of states of each one of the earlier mentioned space-states S_{MP}, B_1, B_2 and B_3.

For the considered example (a system with three cells, and four channels), the sub-spaces B_1 and B_3 have 15 states less than the 25 states of central cell blocking sub-space B_2. The number of states belonging to S_{MP} is generically a function of the number of available channels in the system being given by

$$\#S_{MP} = \sum_{k=1}^{M+1} k^2 = \frac{(M+1) \cdot (M+2) \cdot (2 \cdot M + 3)}{6} \tag{15}$$

On the other hand, the number of states of the sub-spaces B_1 and B_3 is obtained from

$$\#B_1 = \#B_3 = \sum_{k=1}^{M+1} k = \frac{(M+1) \cdot (M+2)}{2} \tag{16}$$

Finally the number of states on the sub-space B_2 is given by

$$\#B_2 = (M+1)^2 \tag{17}$$

One can verify from the above equations, that the extremity cells present always a smaller number of blocking states than the central cell. This propriety is common to others cellular systems and arises from the fact that the boundary cells do not have the complete set of interference cells. The evaluation of the MP technique performance in more complex cellular systems, with a larger number of cells and other interference reuse constraints, will be considered now using the knowledge retrieved from this simple example.

Example 2: The n-Cell Highway System Lets consider now a highway cellular system with the same constraints as the above one but with an increased number of cells in its layout ($N=n$; $M=4$; $K=2$, TYPE = LINEAR). Under these assumptions the system space-state S_{MP} is defined by the following equation:

$$S_{MP} = \left\{ \vec{x} \in Z_+^N : \bigcap_{i=1}^{N-1} (x_i + x_{i+1} \leq M) \right\} \tag{18}$$

The occurrence probability of the system states is evaluated as before appealing to a new normalizing constant.

$$G = p\Bigg(\underbrace{0, \cdots, 0}_{nelements}\Bigg)^{-1} = \underbrace{\sum_{x_1=0}^{M} \sum_{x_2=0}^{M-x_1} \cdots \sum_{x_n=0}^{(M-x_{N-1})}}_{nsums} \prod_{i=1}^{n} \frac{{A_i}^{x_i}}{x_i!} \tag{19}$$

The peculiar structure of the last equation conducts to a very simple and efficient algorithm based on nested loops. The unique enigma on the proposed implementation is the loops index values selection. As it can be seen in (19) that the initial value of all the vector components is always zero. On the other hand the upper limit value of the vector components is given by an expression that contains the outer loops variable values only.

With respect to the blocking probabilities evaluation, for example the cell #1, is on a blocked state if the Eq. (7) is fulfilled, the single difference holds on the vector dimension that is now n instead of 3, being its blocking probability given by

$$P_{b1} = \frac{1}{G} \cdot \underbrace{\sum_{x_2=0}^{M} \sum_{x_3=0}^{M-x_2} \cdots \sum_{x_n=0}^{(M-x_{n-1})}}_{n-1\ sums} \left(\frac{{A_1}^{(M-x_2)}}{(M-x_2)!} \cdot \prod_{i=2}^{n} \frac{{A_i}^{x_i}}{x_i!} \right) \tag{20}$$

Figure 5 shows the cells blocking probabilities and the system GoS concerning to a particular cellular system represented by the parameters ($N=7$; $M=4$; $K=2$; TYPE = LINEAR).

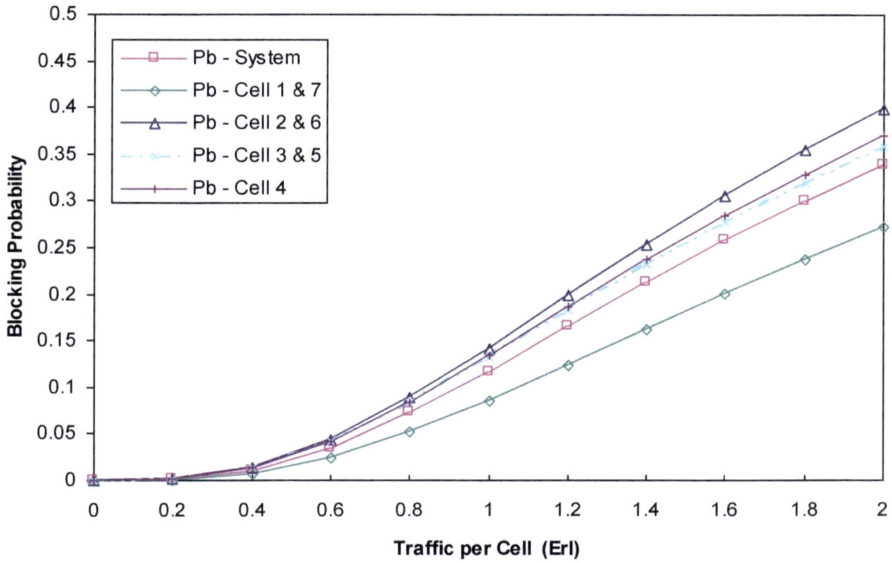

Fig. 5 Seven cells highway system blocking probabilities

Fig. 6 Seven cells planar network

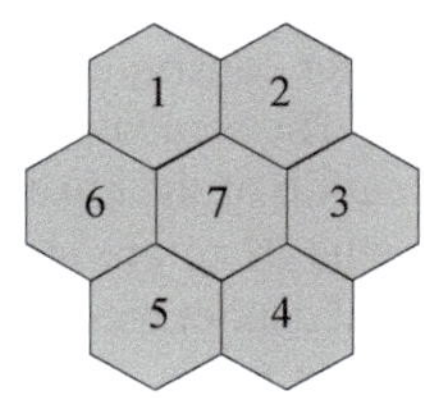

It was observed that the extremity cells present a lower blocking probability than the cells situated on inner positions. This trend was earlier explained based on the number of states of each cell. From the results retrieved from Figs. 4 and 5 we can also conclude that the blocking probabilities of the inner and outer cells present similar values irrespectively of the number of cells on the system.

4 Planar Cellular Systems Analysis

The deterministic evaluation of planar cellular systems space-state S_{MP} and blocking probabilities is not so simple as for linear cellular systems. It is only possible under certain restricted reuse factor values and system geometry.

Example 3: The Planar 7-Cell System Lets consider as a planar system example a 7-cell cellular system, represented in Fig. 6 with the following parameters ($N = 7$; $M = 10$; $K = 3$, TYPE = PLANAR).

Six cluster configurations were identified for this particular example: $C_{1,2,7}$; $C_{2,3,7}$; $C_{3,4,7}$; $C_{4,5,7}$; $C_{5,6,7}$; $C_{1,6,7}$ resultant from different arrangements of the system cells. The system's space-state, $S_{MP,}$ is now easily defined considering the constraints embedded on the cluster definition

$$S_{MP} = \{\vec{x} \to (x_1, x_2, \cdots, x_7) \in Z_+^7 : x_1 + x_6 + x_7 \leq M \wedge \\ \cdots \wedge x_i + x_{i+1} + x_7 \leq M : i = 1, 2, \cdots, 5\} \tag{21}$$

The normalization constant G for this particular system is given by the following formula

$$G = \sum_{x_7=0}^{M} \sum_{x_6=0}^{\chi} \sum_{x_5=0}^{\delta} \sum_{x_4=0}^{\varepsilon} \sum_{x_3=0}^{\varphi} \sum_{x_2=0}^{\phi} \sum_{x_1=0}^{\min(\delta;\gamma)} \prod_{i=1}^{7} \frac{A_i^{\,x_i}}{x_i!} \tag{22}$$

where

$$\chi = M - x_7; \quad \delta = M - (x_6 + x_7); \quad \varepsilon = M - (x_5 + x_7);$$
$$\varphi = M - (x_4 + x_7); \quad \phi = M - (x_3 + x_7); \quad \gamma = M - (x_2 + x_7);$$

The Eq. (22) presents however one particularity, the variable x_1 upper limit, that is given by the minimum of two numbers δ and γ. This operation arises from the planar systems intrinsic nature, for which, at end it remains one cell that is restricted by two other constraints already defined.

The evaluation of the cells blocking probability begins as usual with the identification of the clusters that contain that particular cell. The outer cells of the system, cells #1 to #6, are on a blocking condition if at least one cluster that contains that cell cannot receive an incoming call. For example the cell #1 blocking space-state is given by

$$B_1 = \{\vec{x} \in S_{MP} : x_1 + x_6 + x_7 = M \ \vee \ x_1 + x_2 + x_7 = M\} \tag{23}$$

For the central cell of the system, cell #7, all the clusters must be considered in the analysis. Cell #7 blocking space-state is given by the reunion of $\Omega_{1,6,7}$, $\Omega_{1,2,7}$, $\Omega_{2,3,7}$, $\Omega_{3,4,7}$, $\Omega_{4,5,7}$, and $\Omega_{5,6,7}$

$$B_7 = \{\vec{x} \in S_{MP} : x_1 + x_6 + x_7 = M \vee x_1 + x_2 + x_7 = M \\ \cdots \vee x_i + x_{i+1} + x_7 = M : i = 1, 2, \cdots, 5\} \tag{24}$$

Figure 7 shows the blocking probabilities of the cells and the system GoS as function of the offered traffic, concerning the planar cellular system presented on the example 3.

From the figure one can verify that under uniform traffic load conditions the outer cells, cell #1–#6, blocking probability presents the same value approximately 3 times less than the observed for central cell # 7.

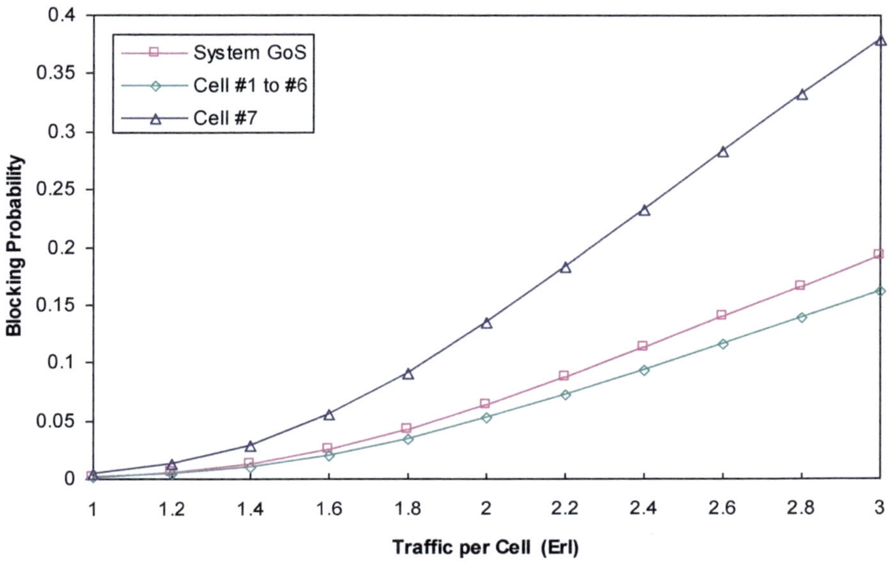

Fig. 7 Seven cells planar system performance

5 Conclusions

This paper proposes an analytical model that is expected to be so useful to evaluate the MP technique performance, as the classical Erlang B formula is to evaluate the FCA technique GoS. The model is the first known method that evaluates the MP technique performance and was validated by simulations performed for all the considered examples: highway and planar cellular systems. The main disadvantage is the enormous computational effort needed to evaluate its performance even on small systems with a reduced number of cells and channels.

References

1. David E, David M (October 1989) Performance analysis of cellular mobile communication system with dynamic channel assignment. IEEE J Select Areas Commun 7(8):1172–1179
2. Scott J, Varaiya PP (August 1991) Throughput in multiple service multiple resource communication networks. IEEE Trans Commun 39(8):1216–1222
3. Kulshreshtha A, Sivarajan KN (May 1999) Maximum packing channel assignment in cellular networks. IEEE Trans Veh Technol 48(3):858–872
4. Victor S et al (May 2000) A new dynamic channel allocation technique: simplified maximum packing. VTC'Spring, Tokyo, Japan, vol 2, pp 1390–1394, May 2000

A Manufacturing Scheduling Approach by Combining Simulation Technique with the Hodgson's Algorithm

Telmo Pinto and Leonilde Varela

Abstract The main objective of this paper consists on presenting an application of The Hodgson's manufacturing scheduling algorithm applied on a production system model, through simulation technique. The simulation, executed on Arena, is applied to a model of a production environment for producing three new products, by using five different components. The components are processed on several work centres, which include different machines, organized on a job shop environment. The Hodgson's algorithm is applied on a particular machine of the job shop, which consists on a bottleneck, and the main objective consists on minimizing the total number of tardy jobs on that bottleneck machine.

Keywords Manufacturing scheduling · Simulation · ARENA · Hodgson's algorithm

1 Introduction

Simulation techniques have been greatly applied on manufacturing scheduling over the last 30 years, by establishing itself as a high practical application on solving this kind of problems occurring on real world scenarios, which is highly due to its quality on enabling a powerful decision making process [1, 2]. In this work we used ARENA, because it is a user-friendly environment, which enables partners from different areas to easily work together, without having to know a programming language [3]. The ARENA uses SIMAN (SIMulation ANalysis) as a language for elaborating a model, by using a graphical interface, and enables powerful reports for different kind of results analysis.

The aim of this work consists on simulating a manufacturing environment where three different articles are to be produced by using five distinct components. Two production models are defined, one integrating machines, expressed by Mi, and another integrating machine centres, CMi. These models are going to be compared

T. Pinto (✉) · L. Varela
Department of Production and Systems, University of Minho, Braga, Portugal
e-mail: telmoppinto@gmail.com

L. Varela
e-mail: leonilde@dps.uminho.pt

N. M. Fonseca Ferreira, J. A. Tenreiro Machado (eds.), *Mathematical Methods in Engineering*, DOI 10.1007/978-94-007-7183-3_23,

through decision analysis, so that: if they execute equivalent operations the time/cost balancing has to be analysed in order to obtain acceptable time and cost values.

Another concern about this work has to due with the combination of the Hodgson's algorithm within the simulation process, in order to minimize the total number of tardy jobs on a bottleneck machine [4].

2 Problem Description

Suppose a company pretends to manufacture three new products, A1, A2 e A3, in a modular production from five components: C1, C2, C3, C4 e C5. The quantity of weekly search expected of the results is variable and normally distributed with the parameters (100; 5) for A1, (200; 10) for A2 and (400; 15) for A3. The processing times for each machine vary accordingly with the component that is being processed.

The assemblage of the three products is done in a assemblage flow shop, Pi (i = 1, ..., 3) while the components are produced in work posts (of a generic job shop) having machines designated by Mi (i = 1, ..., 6). It is also possible to use machine centres, CMi (i = 1, 2) alternatively to the machines, which are twice as fast. The sequence of manufacture depends according with the component. After each operation, the component can immediately follow the next machine. The process of add in pile is done at the launching of the component, that is, at the transportation to the assemblage flow shop. The components transportation from the warehouse of raw materials to the machines (or alternatively, to the machine centres) is done by an Automatic Guided Vehicle (AGV). Every transport has the capacity for two components (of any type).

2.1 Assumptions

The following assumptions were considered on the resolution of the proposed problem:

1. Each machine does only one specific type of operation.
2. A machine can only execute an operation at a certain moment in time and it implies a permanent employee.
3. An operation can't be interrupted after being initiated.
4. There aren't alternative sequences for each component.
5. Each AGV can transport a maximum of two pallets for trip.
6. Mean Time Between Failures (MTBF) for any machine, machine centre or AGV is 200 h.
7. Mean Time To Recovery (MTTR) of each equipment is 10 h.
8. The velocity of any vehicle or transportation of production material in the system is fixed during the simulation time and it is 1 m per second. The acceleration and slowdown are ignored.

9. The weekly activity time is 40 h.
10. The transportation of the components between the machines is done in wagons that follow unidirectional paths fixated in rails.

3 Methodology

The simulation model was developed using ARENA 10.0 software.

The problem described will be simulated 3 times and each simulation will last a total of 1000 h. The Warm Up time is 20 h.

The processing time varies according with the component and the machine/machine centre in which that is being processed.

There were two identical production models implemented: the first executes processes in machines, and the second uses machines centres.

At a first phase, every existing waiting queues on the model are ordered following the FIFO (First In First Out) technique, that is, the entity that is waiting for a longer period will be the first to be processed when the machine gets free. The goal of this paper is to change, for each specific processor, the priority of that waiting queue.

The amount of each machine type varies according with the exposed scenario at the results analysis. This variation will have an effect not only on the production costs but also on the quickness of the whole production process. Thus, the time/cost components should be heuristically analyzed in order to produce the best possible, and at an acceptable cost and duration.

3.1 EDD Rule and Hodgson's Algorithm

There are several priority measures. A priority rule is a rule that specifies the priority of how the entities present in the waiting queue of a processor are processed. The Earliest Due Date rule specifies that when a machine is free, it is selected the entity that has the earliest due date to be processed first.

The Hodgson algorithm determines the sequence of tasks whose number of delayed jobs is minimum [5–7].

Let E be the set that contains all the jobs that must be processed and L the empty set. Thus, the algorithm is constituted by the following sequence of steps:

Step 1: Sort the jobs that belong to the E set and sort by increasing deliver date (earliest due date rule).

Step 2: If none of the jobs is delayed the sequence is optimal. If the opposite happens, the delayed job and its k position (i.e. [k]) are indentified.

Step 3: Identify the entity of bigger duration in the set of the first k jobs. Remove it from the set E and put it on the L set. Establish the new times for the conclusion of the remained jobs on E and return to step 2.

For this work was implemented the Hodgson algorithm, by using JAVA 5 language.

Table 1 Problem data—processing time (*Ti*) and due date (*Ei*) for each component (*Ci*)

Ci	Ei	Ti
1	1	2
2	5	7
3	3	8
4	9	13
5	7	11

Table 2 Components of each article

A1	2c1	2c3	1c5	–
A2	3c1	2c2	1c3	1c4
A3	2c1	2c3	3c5	–

As described the Hodgson algorithm ensures the sequence with the least number of delays in a unique machine. The modulated system illustrates a job shop (Machines/Machine centre 1) and also an assemblage flow shop (Assemblages Posts).

The goal is to restrict the problem to a single machine and evaluate the impact in the priority of the waiting queue. This approach needs demands new considerations:

- Restrict the machine process to only five considered components;
- Ignore the existing sequences that were causing intervening periods and randomization of the components in the waiting queue;
- Consider the components of that machine, after processed, as finished products;
- Consider that the due dates are only related to the components we wish to evaluate.
- Consider that there are five components with known delivery dates that go through the machine.

For the implementation of the Hodgson Algorithm in Arena, besides the considerations cited above, it was necessary to execute the following procedures:

- Wait t time units, during which the processor is inactive.
- Apply the Hodgson Algorithm for the existing components in the waiting queue;
- Keep the values of the due dates for the components that belong to the E set;
- Increase at a large scale the due dates values of the L set;
- Change the choice rule for the machine waiting queue (until now FIFO) in order to allow priority to the lower values of due date (EDD rule);

The example considered the following data shown on Table 1:

4 Simulation Model in Arena

The input in the system is performed by the demand of three types of articles (A1, A2 and A3). If there are products in stock, it won't be necessary to produce more. The demand of an article implies the necessity to produce the respective components, as seen in Table 2.

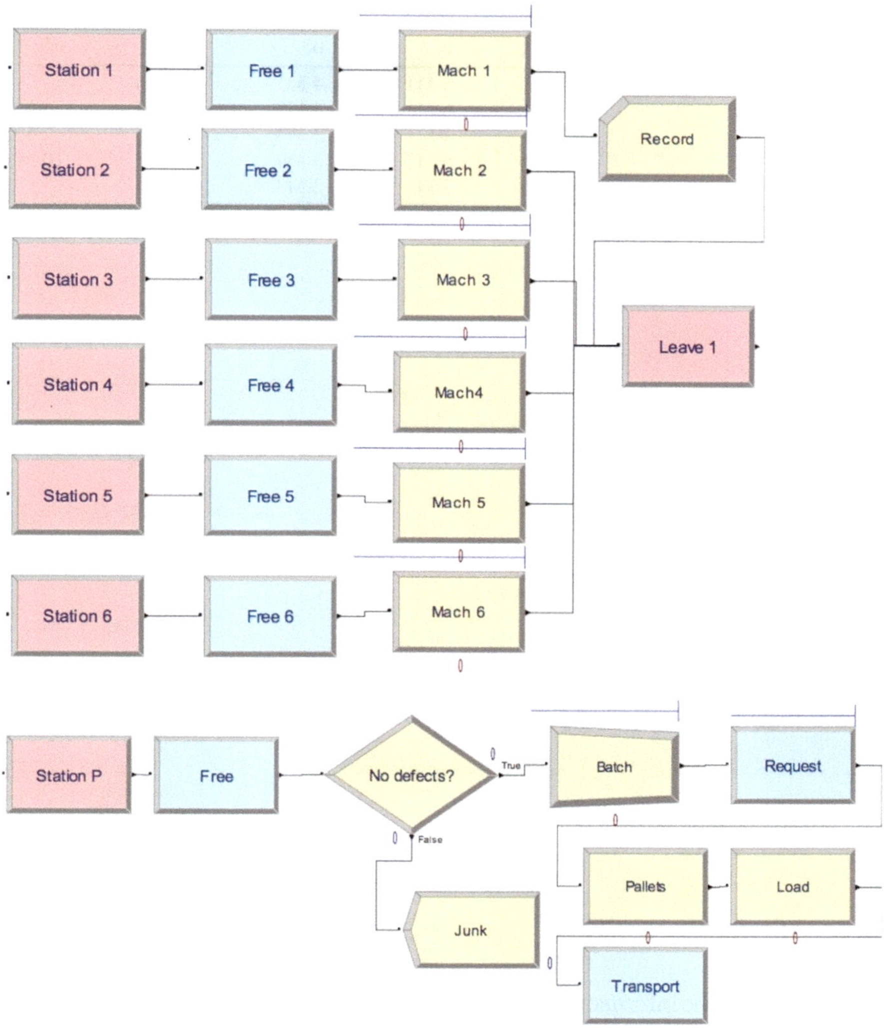

Fig. 1 Machines in the work centre

The components will be produced in machines or machine centres, proceeding afterwards to the flow shop (assemblage posts) in order to manufacture the final articles.

Table 3 Processing times in minutes of the components in each machine

cmp	Op	o1	o2	o3	o4	o5
c1		3 M2	5 M1	4 M5	10 M6	–
c2		12 M1	6 M3	4 M2	3 M4	6 M6
c3		5 M3	1 M1	2 M4	8 M5	7 M6
c4		5 M2	6 M3	3 M4	5 M5	10 M6
c5		5 M1	7 M2	2 M3	4 M5	–

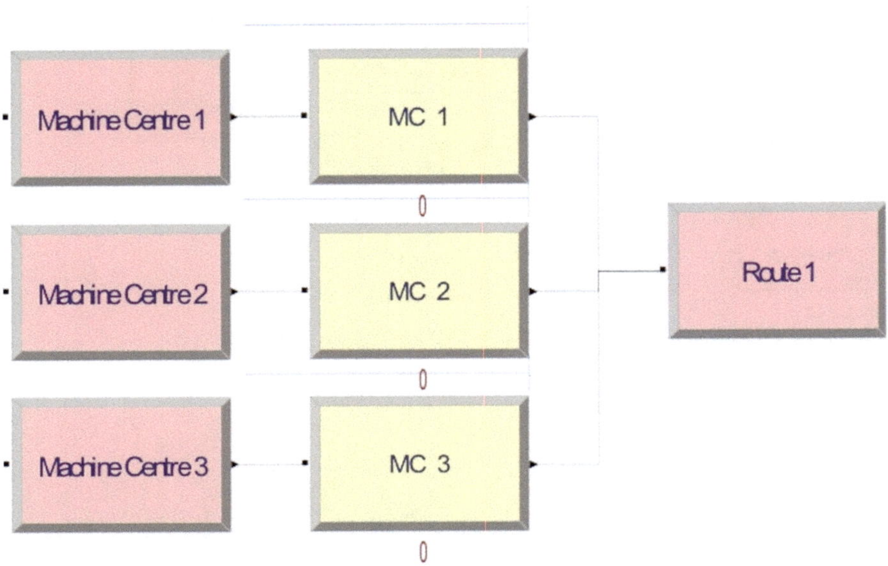

Fig. 2 Machine centres

4.1 Machines

After receiving the information about the article demand, the raw material is required from the warehouse. Then, it will be sent to the machines (Fig. 1), where it will be processed in order to obtain the necessary components to produce articles.

The production sequence of the five components and their processing times (in minutes) of each machine can be seen in Table 3.

4.2 Machine Centre

The machine centres work as machine substitutes. The raw material is sent to those machine centres and processed in order to obtain the necessary components to produce the articles (Fig. 2).

Table 4 Components processing times, in minutes, on each machine centre

Component	Operation	o1	o2	o3
c1		4 CM1	2 CM2	10 M6
c2		11 CM1	1.5 CM2	6 M6
c3		3 CM1	5 CM2	7 M6
c4		5.5 CM1	4 CM2	10 M6
c5		7 CM1	2 CM2	–

The machine centres are twice as fast to produce than machines. The machine centre 1 (CM1) will replace the machines M1, M2 e M3. The machine centre 2 (CM2) replaces M4 and M5. The machine 6 (M6) doesn't has substitute. The components processing times in minutes and their sequence in the machine centres are defined in Table 4.

4.3 *Assembling Centres*

After processing the components in the machines or in the machine centres, the components are sent to a flow shop with three posts (Fig. 3) where are produced the final articles.

5 Results Analysis

The results presented below highlight the existing processes in the machines and in the machines centres. The simulation results suggest parameters to evaluate the system state, like waiting queues length, mean waiting time and mean occupancy. So it is necessary to change the resource parameters and verify the impact in the system.

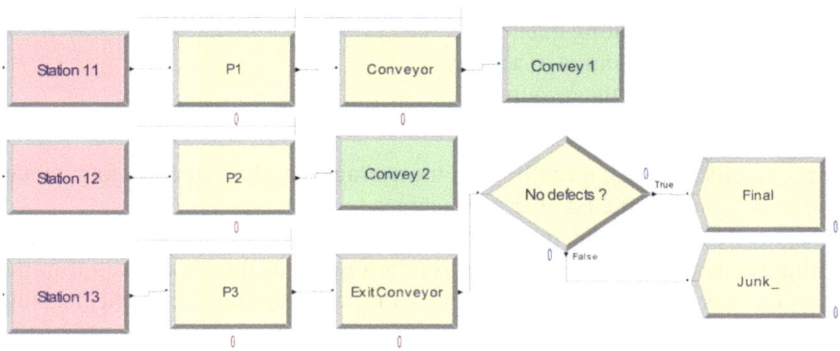

Fig. 3 Assembly centres

Table 5 Results for each machine (M1 = M2 = M3 = M4 = M5 = M6 = 1 unit)

(a)	M1	M2	M3	M4	M5	M6
(b)	1	1	1	1	1	1
(c)	6800	1071	4.59	0.20	470.51	977.85
(d)	22.75	22.75	16.46	8.07	22.75	22.76
(e)	0.96	0.96	0.79	0.34	0.96	0.96

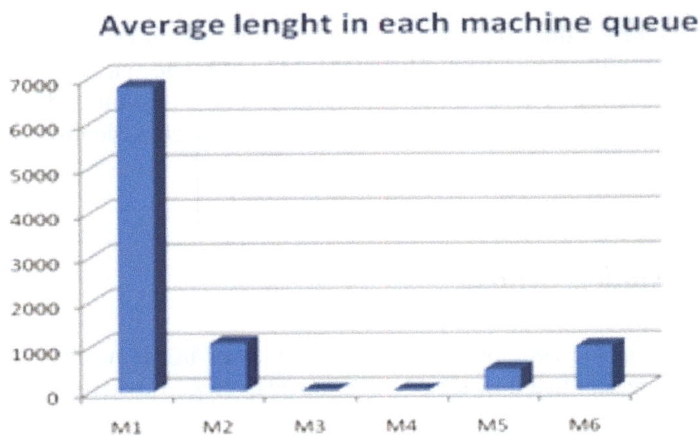

Fig. 4 Average length for each machine queue

At the following tables are presented the corresponding results of the various machines there are present in the model. The values presented are the arithmetic mean of the values from the three simulations. The number of machines is modified to check the variation produced in the various measures of the resource.

In this first Table (5), the machine number of each type is one unit, and:

a. Types of machines;
b. Number of resources;
c. Average size of the waiting queue;
d. Average cost of the resource in activity;
e. Average occupancy of the resource;
f. Types of machines centres

In the next graphic (Fig. 4), it can be seen the average length in each waiting queue.

In the next Table (6), the number of machines for each type is 2 units.

The next tables indicate the machine centre results. It is also adjusted the number of machine centres in the model to confirm the impact of this modification in the resource measures. In the next Table (7) and in the next graphic (Fig. 5), the machine centres number for each type is one unit.

In the next Table (8), the machines centre number for each type is two units.

As can be seen, the average length in the waiting queues for some machines and machine centre is very high.

Table 6 Results for each machine (M1 = M2 = M3 = M4 = M5 = M6 = 2 units)

(a)	M1	M2	M3	M4	M5	M6
(b)	2	2	2	2	2	2
(c)	12	1.47	0.23	0.18	59.88	106.49
(d)	47.50	38.515	22.9	15.34	45.51	45.51
(e)	1	0.81	0.48	0.32	0.96	0.96

Table 7 Results for each machine centres

(f)	CM1	CM2	M6
(b)	1	1	1
(c)	227.91	0.01	175.02
(d)	71.25	42.17	71.25
(e)	1.0	0.59	1.0

Fig. 5 Average length of queues on the machine centres

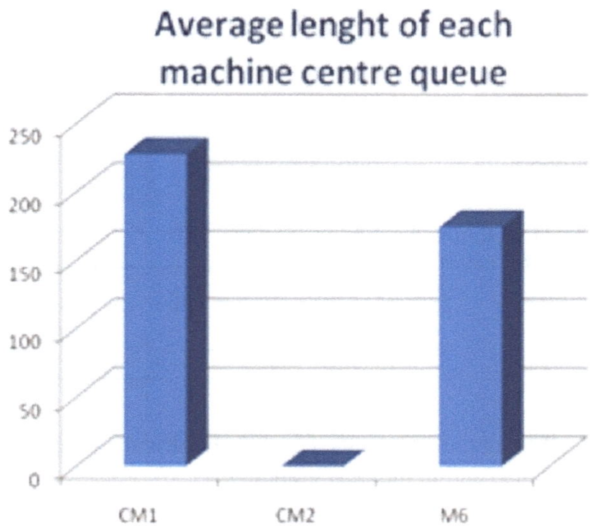

Table 8 Results for each machine centre (CM1 = CM2 = M6 = 2 units)

(f)	CM1	CM2	M6
(b)	2	2	2
(c)	5.71	0	2419
(d)	125.86	25.37	142.5
(e)	0.88	0.18	1.0

5.1 *Sensitivity Analysis*

In order to minimize the average time in the waiting queues for each resource, the machine number and machine centres are adjusted. Those resources number are changed to reasonable values using empirical methods.

Table 9 Results for each machine

(a)	M1	M2	M3	M4	M5	M6
(b)	3	3	2	1	6	6
(c)	0.57	0.41	0.38	0.59	0.32	0.36
(d)	47.51	37.82	22.39	14.47	47.96	58.47
(e)	0.67	0.53	0.47	0.61	0.34	0.41

Fig. 6 Average length for each waiting queues of each machine

Table 10 Results for each machine centres

(f)	CM1	CM2	M6
(b)	3	2	6
(c)	0.11	0.02	0
(d)	125.85	74.47	188.96
(e)	0.59	0.52	0.44

The obtained results for that empirical modification can be seen in the next Table (9) and in the next graphic (Fig. 6).

Another option is adjust the number of machine centres Table (10).

It can be seen in the graphics and tables above that the average length of waiting queues for each machine and for each machine centres is very low. Therefore, the resource average costs increase.

6 Conclusion

Applying the Hodgson's algorithm combined with the simulation technique enables to optimize the number of late jobs.

Moreover, combining the simulation technique with the Hodgson algorithm enables to obtain more attractive results for the analyzed manufacturing scheduling problem, namely related with the accomplishment of imposed orders due dates at the same time we can easily change any other problem parameter, regarding the jobs processing times as assigned machines for its execution.

We can also realize that, if the time between job arrivals is not very long, the results obtained are considerable good.

Considering the dispatching rule, based on the earliest due date, for the Hodgson's algorithm initialization also performed very well, enabling almost every job to be delivered on the corresponding deadline but this approach based on the combination of this method with the simulation technique also enables to easily change rules, for other kind of performance measure selection, namely related with the minimization of total completion time of jobs. In terms of future work we are also considering to implement additional computational rules for enriching alternative approaches, based on other kind of methods, namely based on neighborhood search techniques, to be used in a comparative basis to the work presented on this paper.

Acknowledgements The authors wish to acknowledge the support of: (1) The Foundation for Science and Technology—FCT, under the scope of the financed Project on "Ubiquitous oriented embedded systems for globally distributed factories of manufacturing enterprises"—PTDC/EME-GIN/102143/2008, and (2) EUREKA, under the Project E!4177-Pro-Factory UES.

References

1. Kelton WD, Randall RP, Sturrock DT (2004) Simulation with arena. McGraw-Hill, New York
2. Koh K, Souza R, Ho N (May 1995) Direct database-simulation of a job-shop. Int J Prod Econ 39(3):281–287
3. Almeida MS et al (2006) Utilização da simulação em ARENA 7.0 no auxílio do balanceamento da célula de montagem de uma fábrica de calçados. In: XXVI ENEGEP, Fortaleza, Brasil
4. Armentano VA, e Ronconi DP (2000) Minimização do tempo total de atraso no problema flowshop combuffer zero através de busca tabu. Gestão e Produção 7(3):352–363
5. Baker KR (1974) Introduction to sequencing and scheduling. Wiley, New York
6. Vinod V, e Sridharan R (2009) Simulation-based metamodels for scheduling a dynamic job shop with sequence-dependent setup times. Int J Prod Res 47(6):1425–1447
7. Wang J-B (2008) Single-Machine scheduling with past-sequence-dependent setup times and time dependent learning effect. Comput Ind Eng 55:584–591

Long Time Numerical Approximation of Coherent-Structure Solutions of the Cubic Schrödinger Equation

I. Alonso-Mallo, A. Durán and N. Reguera

Abstract The purpose of this work is to determine suitable numerical methods to simulate the evolution of coherent structures for the cubic nonlinear Schrödinger equation with Dirichlet boundary conditions on a finite one-dimensional interval. We consider different time integrators, some of them preserving one or two invariants of the problem. We show that the preservation of these invariants is essential for a good long time simulation.

Keywords Cubic nonlinear Schrödinger equation · Dirichlet boundary conditions · Ground state · Invariants · Finite element methods · Conservative integration

1 Introduction

In this work we are going to consider the nonlinear Schrödinger equation on an interval $[0, L]$ with homogeneous Dirichlet boundary conditions:

$$\left.\begin{array}{l} i\Psi_t + \Psi_{xx} + f(|\Psi|^2)\Psi = 0, \quad t > 0, \quad 0 \le x \le L, \\ \Psi(0,t) = \Psi(L,t) = 0, \quad t > 0, \\ \Psi(x,0) = \Phi(x), 0 \le x \le L. \end{array}\right\} \tag{1}$$

I. Alonso-Mallo(✉)
Departamento de Matemática Aplicada, Facultad de Ciencias,
Universidad de Valladolid, Valladolid, Spain
e-mail: isaias@mac.uva.es

A. Durán
Departamento de Matemática Aplicada, E.T.S.I, Telecomunicación,
Universidad de Valladolid, Valladolid, Spain
e-mail: angel@mac.uva.es

N. Reguera
Departamento de Matemáticas y Computación, Escuela Politécnica Superior,
Universidad de Burgos, Burgos, Spain
e-mail: nreguera@ubu.es

N. M. Fonseca Ferreira, J. A. Tenreiro Machado (eds.), *Mathematical Methods in Engineering,* DOI 10.1007/978-94-007-7183-3_24,

where $\Psi = \Psi(x,t)$ and f is the nonlinear term. It is well known the importance of these kind of equations in many applications such as waves in plasmas and propagation in optical fibers [1, 4, 6, 7, 8, 13, 14]. Although this work can be carried out for more general non linear terms (see [3]) we are going to consider $f(x) = x$, which corresponds to the cubic nonlinear Schrödinger equation.

Of special relevance for this work are the two invariant quantities given by the Hamiltonian energy

$$H(\Psi) = \frac{1}{2}\int_0^L |\Psi_x|^2 dx - \frac{1}{4}\int_0^L |\Psi|^4 dx, \tag{2}$$

and the particle number

$$N(\Psi) = \frac{1}{2}\int_0^L |\Psi|^2 dx. \tag{3}$$

Problem (1) admits ground state solutions of the form

$$\Psi(x,t) = \Phi(x)e^{-i\lambda t} \tag{4}$$

where Φ is the solution of the problem

$$\left.\begin{array}{l} \Phi_{xx} + f(|\Phi|^2)\Phi + \lambda\Phi = 0, \quad 0 \le x \le L \\ \Phi(0) = \Phi(L) = 0 \end{array}\right\} \tag{5}$$

Some works [9–11] suggest a relevant role of these structures in the dynamics of (1), in the sense that the solutions of these NLS models tend to form coherent structures that persist along with small turbulences. On the other hand (see [11]), the profiles $\Phi(x)$ are minimizers of the Hamiltonian subject to a fixed value of the particle number, that is, $\Phi(x)$ is the solution of the problem

$$H(\varphi) \to \min \quad \text{subject} \quad \text{to} \quad N(\varphi) = N_0. \tag{6}$$

2 Spatial Discretization

For the numerical integration of the problem (1) we are going to consider the method of lines. We first discretize in space with cubic finite elements [2]. For this, let us establish a spatially discretized version of the weak formulation of (1): finding $u_h(t) \in \mathcal{V}_h$, such that,

$$\left.\begin{array}{rcl} \langle \frac{du_h}{dt}, w_h\rangle & + & i\langle \partial_x u_h, \partial_x w_h\rangle = i\langle |u_h|^2 u_h, w_h\rangle,\ \forall w_h \in \mathcal{V}_h,\ t \ge 0 \\ u_h(0) & = & u_{0,h} \end{array}\right\} \tag{7}$$

where $\mathcal{V}_h$ (h denotes the parameter of the spatial discretization) is the space of Hermite piecewise cubic polynomial functions, which are continuous and with continuous derivative, and satisfying homogeneous Dirichlet boundary conditions.

Therefore, we are looking for approximations $u_h(x,t) \in \mathcal{V}_h$, to the solution $u(x,t)$ of (1). More precisely, if $h = L/J > 0$ for a natural J, $x_j = jh$ for $0 \le j \le J$ are the spatial nodes, and $\{\sigma_j\}_{j=1}^{J-1} \bigcup \{\widetilde{\sigma}_j\}_{j=0}^{J}$ are the shape functions, then the approximation $u_h(x,t)$ will be of the form:

$$u_h(x,t) = \sum_{j=1}^{J-1} u_j(t)\sigma_j(x) + \sum_{j=0}^{J} \widetilde{u}_j(t)\widetilde{\sigma}_j(x) \tag{8}$$

where $u_j(t)$, $1 \le j \le J-1$ and $\widetilde{u}_j(t)$, $0 \le j \le J$ are the approximations to $u(x_j,t)$, $1 \le j \le J-1$ and $\partial_x u(x_j,t)$, $0 \le j \le J$ respectively.

After applying this spatial discretization, we obtain a system of ordinary differential equations

$$R_h \frac{dU}{dt} = M_h U + \phi(U) \tag{9}$$

with

$$U = [\widetilde{u}_0, u_1, \widetilde{u}_1, u_2, \widetilde{u}_2, \ldots, u_{J-1}, \widetilde{u}_{J-1}, \widetilde{u}_J]^T,$$

and where R_h and M_h are symmetric matrices and $\phi(\cdot)$ is the nonlinear term.

3 Time Integration

In order to choose a convenient method for the numerical integration of (9) we should take into account that the preservation of the invariants of the problem in the numerical integration is associated to a better simulation of the solutions for nonlinear Schrödinger equations (see [5]).

With the purpose of analyzing the importance of preserving the invariants H and N in our problem, we are going to consider several methods for the time integration. Firstly, the simply diagonally implicit Runge–Kutta (SDIRK) method of order three with tableau

$$\begin{array}{c|cc} \frac{3+\sqrt{3}}{6} & \frac{3+\sqrt{3}}{6} & 0 \\ \frac{3-\sqrt{3}}{6} & \frac{-\sqrt{3}}{3} & \frac{3+\sqrt{3}}{6} \\ \hline & \frac{1}{2} & \frac{1}{2} \end{array} \tag{10}$$

This method, although with interesting properties, does not conserve the invariants H and N. That is the reason why it will be modified so that we obtain three different new methods: SDHP that preserves the Hamiltonian H, SDNP that preserves the particle number N, and SD2P preserving both invariants. This preservation is obtained by projecting the numerical approximation onto the manifold levels, defined by the functional H and N. The construction of these methods is explained with more detail in [3]. We also will compare these methods with two symplectic methods: the

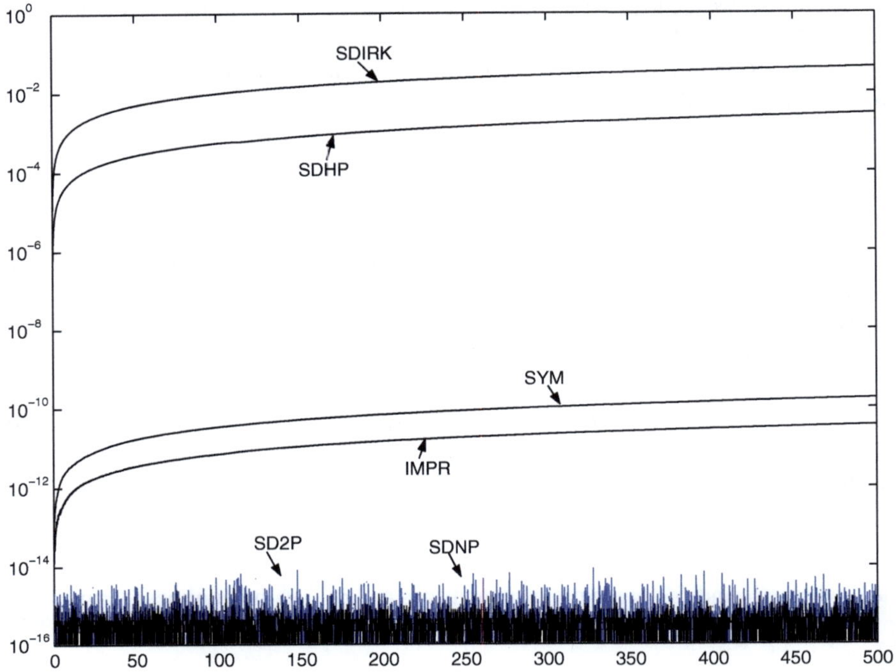

Fig. 1 Experiment with initial condition (12). Time evolution of the error in the invariant N for the different methods used. Time step: $\Delta t = 8.0\text{d}{-}2$ for SD2P, SDNP, SDHP, SYM and SDIRK, $\Delta t = 4.0\text{d}{-}2$ for IMPR

implicit mid-point rule and an order three method consisting on a concatenation of three implicit midpoint steps of length $b_1\Delta t$, $b_2\Delta t$ and $b_3\Delta t$ respectively (see [12]) with

$$b_1 = b_3 = \frac{1}{3}\left(2 + 2^{1/3} + 2^{-1/3}\right), \qquad b_2 = 1 - 2b_3. \tag{11}$$

4 Numerical Experiments

In the numerical experiments we present next, we are going to determine which of the time integrators used is more competitive in each case and we are going to show the advantages of using numerical integrators in time that preserve the invariants of the problem.

In the first experiment we have considered as initial condition

$$u_0(x) = \sqrt{\frac{2\alpha}{\nu}} \exp\left(\frac{iV}{2}x\right) \operatorname{sech}(\sqrt{\alpha}x), \qquad x \in [-30, 30] \tag{12}$$

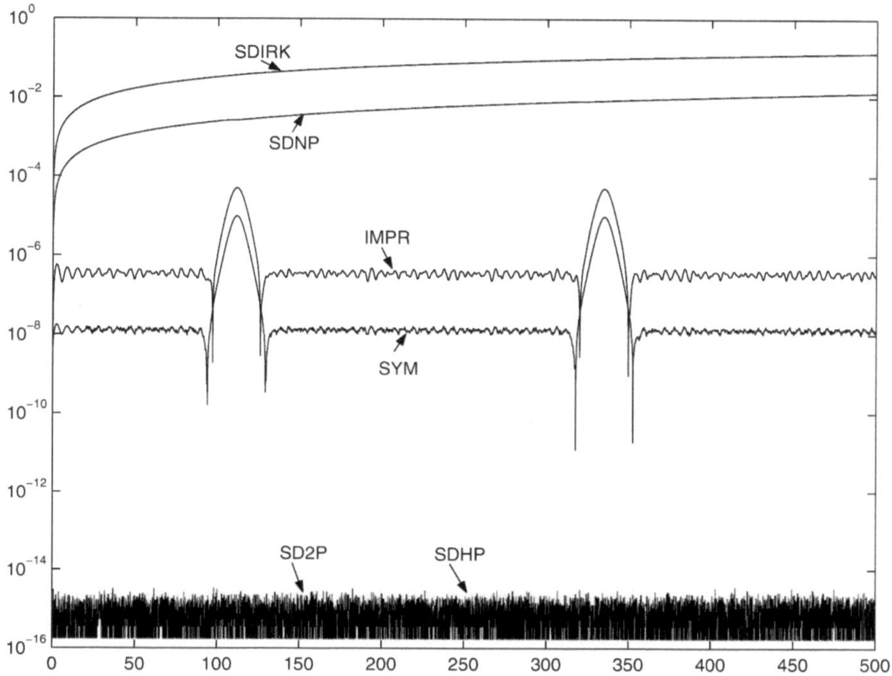

Fig. 2 Experiment with initial condition (12). Time evolution of the error in the invariant H for the different methods used. Time step: $\Delta t = 8.0\text{d}{-}2$ for SD2P, SDNP, SDHP, SYM and SDIRK, $\Delta t = 4.0\text{d}{-}2$ for IMPR

where $\alpha = 1$, $\nu = 1$ and $V = 0.25$. This initial condition would give rise to a soliton traveling with velocity 0.25 in the case of the pure initial value problem. When we use Dirichlet boundary conditions, while the support of the solution is inside the computational window, the numerical solution behaves as such a soliton. Nevertheless, in this case, when the numerical solution "arrives" to the boundary it "bounces" into the computational window again.

We have integrated the problem with the spatial discretization previously explained and, in order to make a selection of the more convient methods, with several integrators in time: the simply diagonally implicit Runge–Kutta method of order three (SDIRK) given by (10), the implicit mid-point rule (IMPR), the symplectic method (SYM) with coefficients given by (11), the method SDHP that preserves the Hamiltonian H, SDNP that preserves the particle number N, and SD2P preserving both invariants.

Figure 1 shows, in logarithmic scale, the time evolution of the relative error in the invariant N for the six methods considered. We can observe three different kind of behaviour. As expected, the worst results are for the non conservative method SDIRK and for the method SDHP that, although preserving the invariant H it does not preserve the invariant N that is being studied in this figure. A better behaviour is

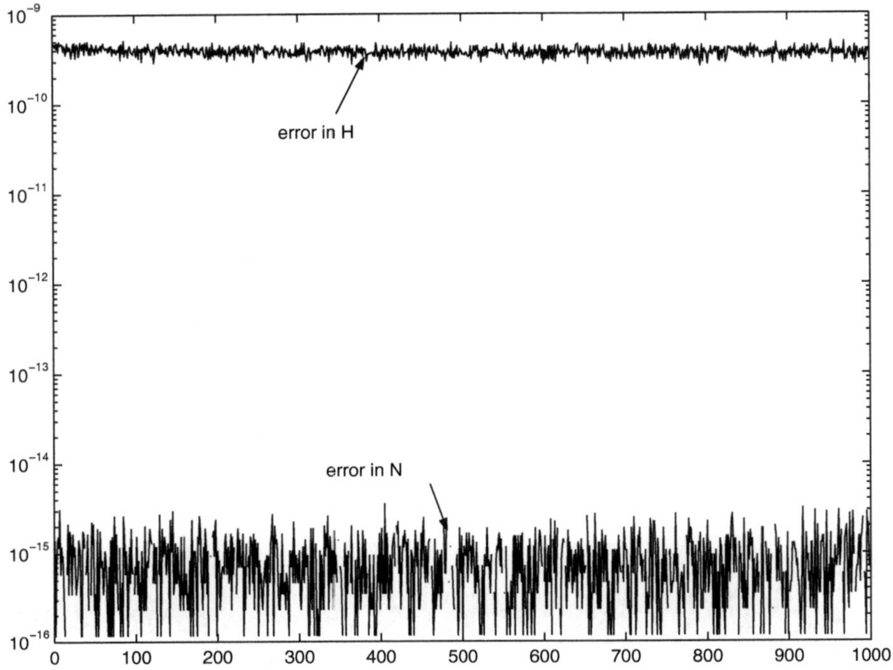

Fig. 3 Experiment for a ground state. Time evolution of the error in the invariants for the method SDNP. Time step: $\Delta t = 1.0\text{d}{-}2$

obtained with the symplectic methods IMPR and SYM. Nevertheless, the results are optimal (the errors are similar to the precision used in the computation) when we use the method SDNP that preserves the invariant N or the method SD2P that preserves both invariants N and H.

In a similar way, Fig. 2 shows, in logarithmic scale, the time evolution of the relative error in the invariant H for the six methods used. The comments for this figure are equivalent as those for Fig. 1. We conclude then, that the best option is to use integrators in time that preserve the invariants of the problem.

In the next experiment we are interested in studying the evolution of a ground state. In order to obtain initial profiles that give rise to ground states of our problem it is enough to consider a discrete version of (6) that after some manipulation can be solved with standard algorithms (more precisely, we have used the NAG library). Taking into account the conclusions of the previous experiment we are going to consider now only time integrators that preserve one of the invariants of the problem. The algorithm SD2P is not possible to use in this case. The reason is that there exists a dependence between the variational derivatives of the two invariants at the ground state which affects the numerical resolution needed to carry out the two projections (see [3] for more details).

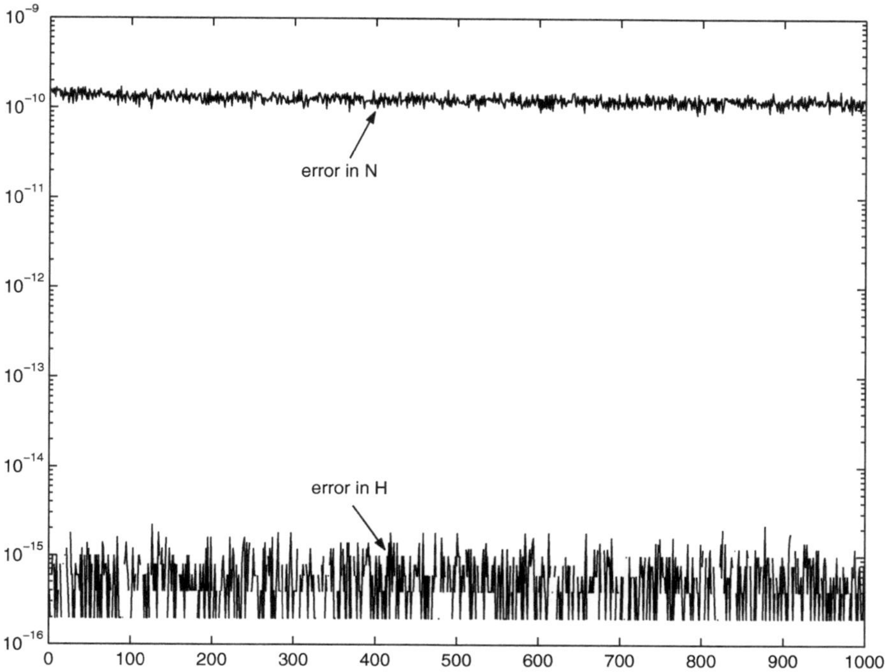

Fig. 4 Experiment for a ground state. Time evolution of the error in the invariants for the method SDHP. Time step: $\Delta\, t = 1.0\text{d}{-}2$

In Fig. 3 we can observe the time evolution of the relative error in both invariants when the method SDNP is used as time integrator. As expected, the errors in the invariant N are similar to the precision used in the computation. Moreover, notice that the results obtained when we measure the errors in the Hamiltonian H are quite good and they do not grow with time. Equivalent results are obtained when we use the integrator SDHP which can be seen in Fig. 4. The results are as expected since now the invariant that is preserved is H.

Finally, we are going to consider an initial condition with a gaussian profile which is similar to a ground state although it is not, and it can be considered as a perturbation of a ground state. More precisely, $u_0(x) = A \exp(-(x-10)^2)$, $x \in [0, 20]$, where A is chose so that the norm of $u_0(x)$ is equal to the norm of the ground state used in the previous experiment.

The purpose of this experiment is to see the behavior of the numerical solution after a long time integration. For the integration in time we have used the two projection method SD2P for which we expect the best behavior. In Fig. 5 we see the modulus of the numerical solution at three different times: $t = 0$, $t = 25000$ and $t = 50000$. In this quite long interval of time the solution has developed to a structure localized at the mid point of the space interval that persists with time along with some turbulence of small size. It is reasonable to suppose that the previous structure is a ground

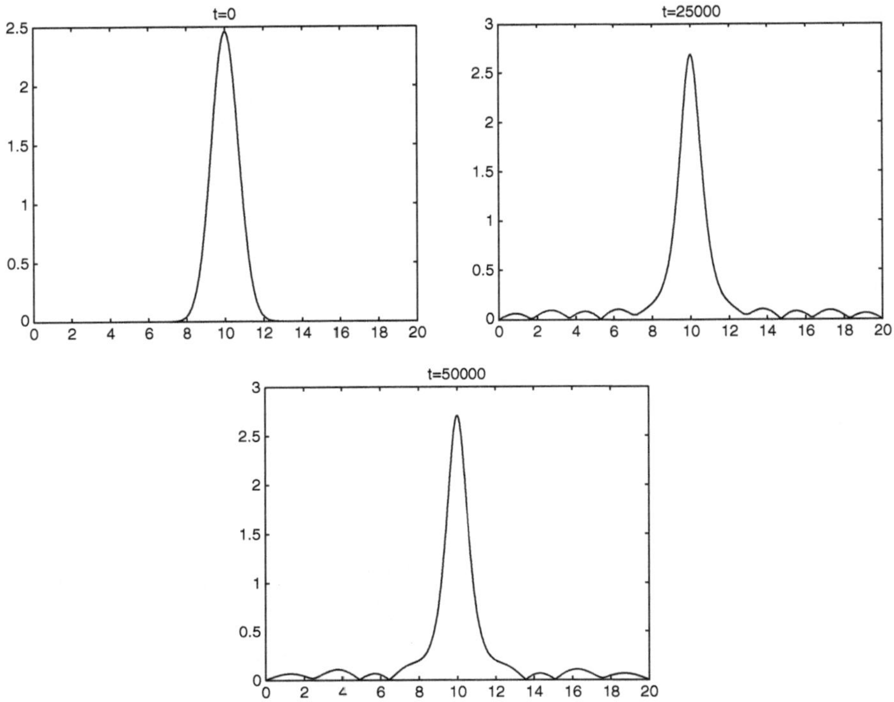

Fig. 5 Modulus of the numerical solution at different times when the initial condition is a gaussian profile. Time step: $\Delta t = 1.0\text{d}{-2}$

state. In order to have more numerical evidence of this, we have done the following experiment. If the numerical solution was a ground state (4), denoting by $u_{j,n}$ this solution at (x_j, t_n), the value of λ at this time can be estimated as

$$\lambda \approx \frac{\arg\left(\frac{u_{j,n-1}}{u_{j,n}}\right)}{\Delta t}.$$

We can see the evolution with time of this estimation for this experiment (at $x = 10$) in Fig. 6. We observe that this estimation has small oscillations around a constant value. This constant in time behavior of the numerical estimate of the velocity λ is associated to the conservative character of the time integrator and it is not obtained in the corresponding simulation with nonconservative methods [3].

Acknowledgment This research has been supported by MICINN projects MTM2011-23417 and MTM2010-19510/MTM.

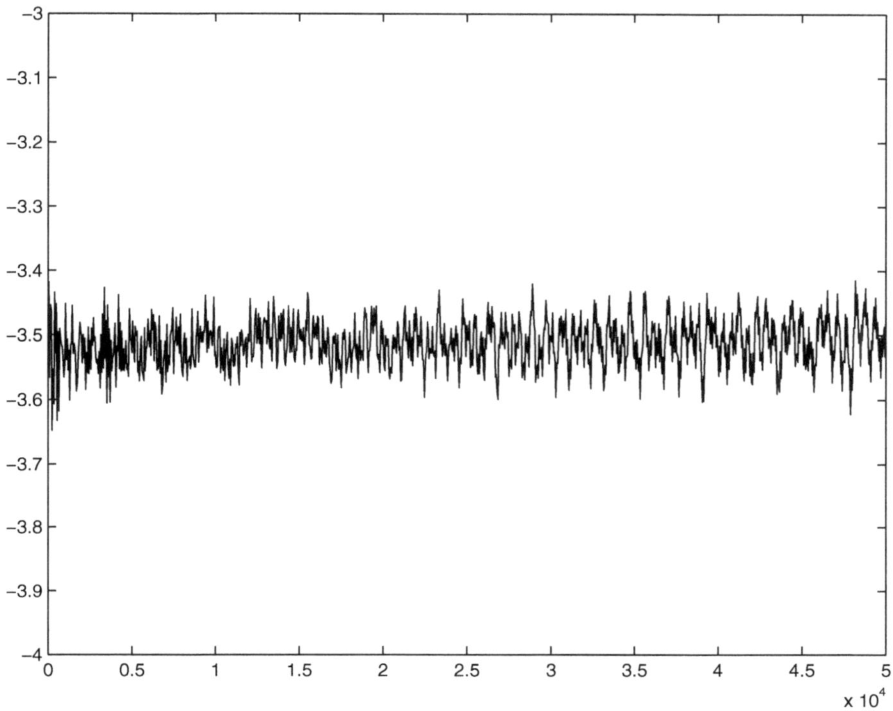

Fig. 6 Experiment with a gaussian profile as initial condition. Time evolution of λ. Time step: $\Delta t = 1.0\text{d}{-2}$

References

1. Ablowitz M, Segur H (1979) On the evolution of packets of water waves. J Fluid Mech 92: 691–715
2. Alonso-Mallo I, Reguera N (2006) A high order finite element discretization with local absorbing boundary conditions of the linear Schrödinger equation. J Comput Phys 220(1):409–421
3. Alonso-Mallo I, Durán A, Reguera N (2010) Simulation of coherent structures in nonlinear Schrödinger-type equations. J Comput Phys 229(21):8180–8198
4. Cai D, McLaughlin DW, McLaughlin K (2002) The nonlinear Schrödinger equation as both a PDE and a dynamical system. Handb Dyn Syst 2:599–675
5. Durán A, Sanz-Serna JM (2000) The numerical integration of relative equilibrium solutions. IMA J Numer Anal 20:235–261
6. Hasegawa A (1985) Self-organization processes in continuous media. Adv Phys 34:1–42
7. Hasegawa A, Kodama Y (1995) Solitons in optical communications. Oxford University Press, Oxford
8. Hasegawa A, Tapper F (1973) Transmission of stationary nonlinear optical pulses in dispersive dielectric fibers. I. anomalous dispersion. Appl Phys Lett 23:142–144
9. Jordan R, Josserand C (2000) Self-organization in nonlinear wave turbulence. Phys Rev E 61:1527–1539

10. Jordan R, Josserand C (2000) Statistical equilibrium states for the nonlinear Schrödinger equation. Math Comput Simul 1897:1–15
11. Jordan R, Turkington B, Zirbe CL (2000) A mean-field statistical theory for the nonlinear Schrödinger equation. Physica D 137:353–378
12. Sanz Serna JM, Calvo MP (1994) Numerical Hamiltonian problems. Chapman and Hall, London
13. Sulem C, Sulem PL (1999) The nonlinear Schrödinger equation, self-focusing and wave collapse. Springer, New York
14. Zakharov VE (1972) Collapse of Langmuir waves. Sov Phys JETP 35:908–922

A Statistical Approach for Tuning the Windowed Fourier Transform

Miguel F. M. Lima and J. A. Tenreiro Machado

Abstract A time frequency analysis is used in many fields for studying signals with a time-varying spectral content. The windowed Fourier transform is one of the most used time-frequency representations. In order to use this technique several parameters must be defined, including the type, the length and the overlap of the windows. For tuning the windowed Fourier transform a new method based on the information theory is presented. Several tests with robotic signals illustrate the appropriateness of the proposed method.

Keywords Windowed Fourier transform · Short time Fourier transform · Robotics · Signal processing · Time-frequency analysis · Mutual information

1 Introduction

Very often real-world processes are non-stationary containing a time-varying frequency content. In many applications we are interested in the frequency content of a signal at a given period of time. In the case of a non-stationary signal, the classical Fourier transform (FT) is not suitable for its analysis. In fact, information localized in time, such as spikes, impacts, seismic events, and high frequency bursts, are not easily detected by the FT. Therefore, a time frequency analysis is used in many fields for studying signals with a time-varying spectral content.

There are several approaches to achieve the time frequency analysis of non stationary signals. Among others, the most popular are the Wigner distribution, the Gabor transform, the windowed Fourier transform (WFT) and the wavelet transform [1]. Textbooks that address the time-frequency representations can be referenced in [2–4]. The comparison between the different approaches, for achieving the time

M. F. M. Lima (✉)
Dept. of Electrical Engineering, Superior School of Technology,
Polytechnic Institute of Viseu, Viseu, Portugal
e-mail: lima@ipv.pt

J. A. Tenreiro Machado
Dept. of Electrical Engineering, Institute of Engineering,
Polytechnic Institute of Porto, Porto, Portugal
e-mail: jtm@isep.ipp.pt

N. M. Fonseca Ferreira, J. A. Tenreiro Machado (eds.), *Mathematical Methods in Engineering,* DOI 10.1007/978-94-007-7183-3_25,

frequency analysis, was developed by several authors [5–7] and it was verified that the choice of the best representation depends on the application [5].

The WFT, also known as short time (or term) Fourier transform (STFT) or time-varying Fourier transform (TVFT), is one of the most widely used time-frequency representations. In fact, this technique is adopted in many fields of engineering, such as in audio (speech and musical) signal processing, vibration signal processing [8] seismic signal processing, electromagnetic radiation [9] and robotics [10]. The WFT is an extension of the classical FT, where the transform is evaluated repeatedly for a running windowed version of the time domain signal. Each FT gives a frequency domain 'slice' associated with the time instant at the window center.

There are several studies for implementing WFT recursive algorithms [11–14]. One important aspect of the WFT is the window length that is related with the time–frequency resolution. The frequency-resolution of the WFT is proportional to the effective bandwidth window. Consequently, for the WFT we have a trade-off between the time and the frequency resolutions: on one hand, a good time resolution requires a short window, while, on the other hand, a good frequency resolution requires a long window. Several authors addressed this issue [5, 6, 15]. In order to adjust the desired resolution, the window length can be adjusted adaptively [16–19] based on an instantaneous quality measurement of the time frequency content.

Another aspect of the WFT is the type of window adopted [20, 21]. Several authors studied the effect of the WFT window [1, 22, 23] and verified that the best choice depends on the type of signal [14].

In summary, there are distinct parameters that must be defined to use the WFT. In this line of thought the need of indices for tuning adequately the WFT motivated the work presented here. In fact the authors developed several experiments and indices that were tested for tuning the WFT. The indices included statistical, entropy and information theory approaches. In this field several authors investigated the connections between the information theory (entropies and mutual information) and the time-frequency representations [24–27]. A method based on the information theory is presented in this work, revealing to be a promising strategy.

To show the behavior of the information theory approach, the WFT is applied to a set of signals captured in a robotic manipulator, which is briefly described in the following section. In the third section are presented some fundamental concepts. The fourth section presents the results based on experimental signals and, finally, the fifth section outlines the main conclusions and points out future work.

2 Apparatus and Experimental Signals

In order to analyze signals that occur in a robotic manipulator an experimental platform was developed. The platform has two main parts: the hardware and the software components [28]. The hardware architecture is shown in Fig. 1. Essentially it is made up of a mechanical manipulator, a computer and an interface electronic system. The interface box is inserted between the arm and the robot controller, in order to acquire

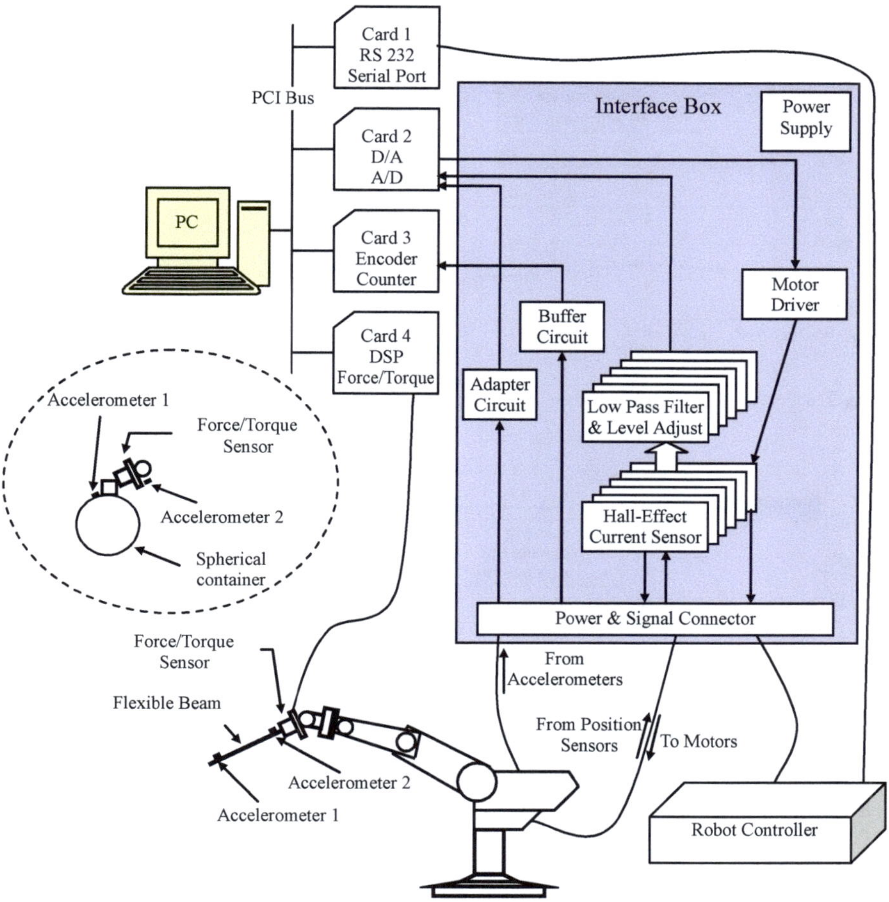

Fig. 1 Block diagram of the hardware architecture

the internal robot signals; nevertheless, the interface captures also external signals, such as those arising from accelerometers and force/torque sensors. The modules are made up of electronic cards specifically designed for this work. The function of the modules is to adapt the signals and to isolate galvanically the robot's electronic equipment from the rest of the hardware required by the experiments.

The software package runs in a Pentium 4, 3.0 GHz PC and, from the user's point of view, consists of two applications: (*i*) the acquisition application is a real time program responsible for acquiring and recording the robot signals; (*ii*) the analysis package runs off-line and handles the recorded data. This program allows several signal processing algorithms such as, FT, WFT, correlation, time synchronization, etc.

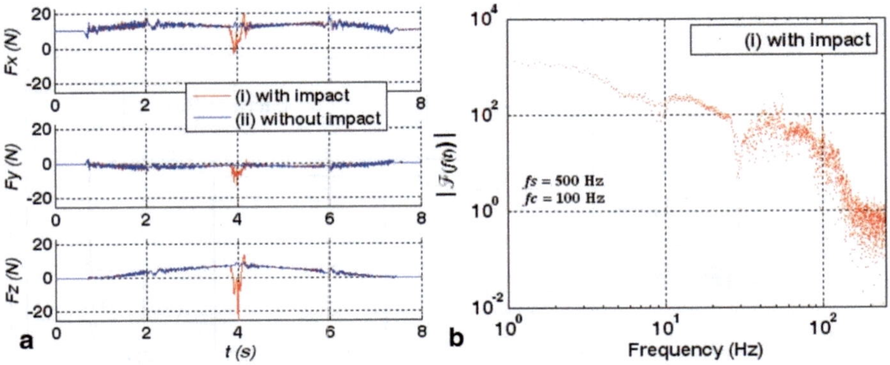

Fig. 2 **a** Forces at the gripper sensor. **b** f_z^{imp} spectrum

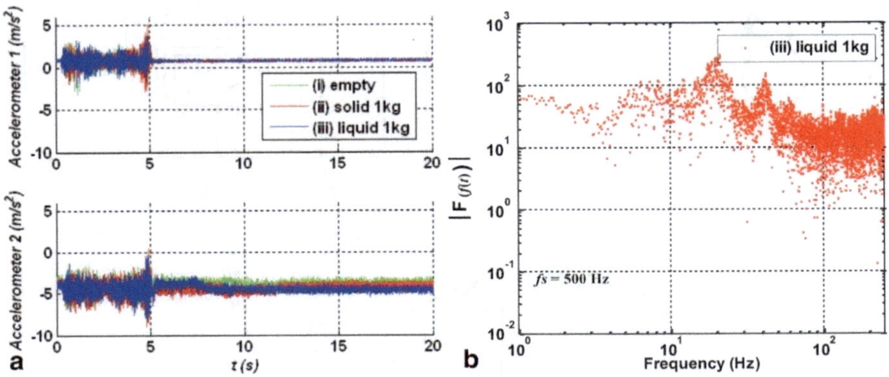

Fig. 3 **a** a_1^{liq} and a_2^{liq} signals. **b** a_1^{liq} signal spectrum

To test the phenomenon of mechanical impacts, in the experimental setup it is used a flexible link that consists of a long, thin, round, flexible steel rod clamped to the end-effector of the manipulator. The robot motion is programmed in a way such that the clamped rod collides with a surface and several signals are recorded with a sampling frequency of $f_s = 500$ Hz. The signals come from different sensors, such as accelerometers, wrist force and torque sensors, position encoders and joint actuator current sensors. Additionally, in another experiment, it is adopted a spherical container carrying a liquid that oscillates during the acceleration/deacceleration transients. To test the behavior of the variables in different situations, the container (Fig. 1) can remain empty or can be filled with a liquid or a solid. The robot motion is programmed in a way that the container moves from an initial to a final position following a linear trajectory.

Figures 2 and 3 depict a typical time evolution of some variables and the corresponding spectrum. Figure 2a shows the forces at the end-effector of the manipulator captured during a total period of $t_T = 8$ s for the impact analysis. These signals present clearly a strong variation at the instant of the impact, that occurs approximately at $t = 4$ s. The Fourier spectrum of f_z^{imp} (force z component for the case of impact) is shown in Fig. 2b.

Figure 3a shows the accelerations a_1^{liq} (accelerometer 1 at the clamped end of the container) and a_2^{liq} (accelerometer 2 at the terminal link of the robot) when the robot carries the liquid container. The signals are captured during a total period of $t_T = 20$ s. The a_1^{liq} signal spectrum is shown in Fig. 3b.

Figures 2b and 3b show the spectrum of signals that contains information which is localized in time, due to the rod impact and the liquid vibration, respectively. Occasionally the signal spectra are scattered. In order to deal with these issues a multiwindow algorithm is used in the next sections.

3 Main Concepts

3.1 The Windowed Fourier Transform

One way of obtaining the time-dependent frequency content of a signal is to take the FT of a function over an interval around an instant τ. The WFT transform accomplishes this by using a general window function. The concept of this mathematical tool is straightforward. We multiply the signal to be analyzed $x(t)$ by a moving window $g(t - \tau)$ and, then, we compute the Fourier transform of the windowed signal $x(t)g(t - \tau)$. Each FT gives a frequency domain 'slice' associated with the time value at the window centre. Actually, windowing the signal improves local spectral estimates [1]. The WFT for a window function centered at time τ, is represented analytically by:

$$F(\omega, \tau) = \int_{-\infty}^{+\infty} x(t)g(t - \tau)e^{-j\omega t} dt \tag{1}$$

where $\omega = 2\pi f$ is the frequency.

Each window has a width t_w and the distance between two consecutive windows can be defined in a way so that they become overlapped during a percentage of time β in relation to t_w. Therefore, the frequencies of the analyzing signal $f < 1/t_w$ are rejected by the WFT. Diminishing t_w reduces the frequency resolution and increases the time resolution. Augmenting t_w has the opposite effect. Therefore, the choice of the WFT window entails a well-known duration-bandwidth trade-off.

The rectangular window can introduce an unwanted side effect in the frequency domain. As a result of having abrupt truncations at the ends caused by the window, the spectrum of the FT will include unwanted "side lobes". This gives rise to an oscillatory behavior in the time domain called the Gibbs phenomenon [22]. In order to reduce this unwanted effect, usually is used a weighting window function that attenuates the signals at their discontinuities. For this reason there are several popular windows normally adopted in the WFT as, for example, the Hanning, Hamming,

Gaussian and Blackman [22]. Harris [20] and Nuttall [21] present several windows with its spectrum characteristics.

If the windows do not overlap, then it is clear that some data are lost. Additionally, if the windows overlap in a short period of time a significant part of the time signal is ignored due to the fact that most windows exhibit small values near the boundaries. To avoid this loss of data, overlap analysis must be performed.

In resume, in order to apply the WFT there are several parameters that must be defined, namely the window type, the window's width t_w and the overlapped time β. Some windows have also a parameter α that affects its shape. In this study are adopted three types of windows: the Gaussian, the fractional, and the Hanning window.

The Gaussian window has the following expression:

$$g(t) = e^{-\frac{1}{2}\left(\alpha \frac{t}{t_w/2}\right)^2}, \quad t \in \left[-\frac{1}{2}t_w; \frac{1}{2}t_w\right] \tag{2}$$

where $\alpha, t_w \in \Re^+$ are parameters.

Expression (3) represents a window that we call fractional due to the fact that the parameter $\alpha \in \Re$ can present any real value in the interval $0 < \alpha < \alpha_{max}$. The window is centered at time τ and the parameters (α, t_w) affect its shape and width.

$$g(t) = 1 - \left|\frac{t-\tau}{t_w}\right|^\alpha, \quad t \in \left[-\frac{1}{2}t_w; \frac{1}{2}t_w\right] \tag{3}$$

This window is interesting due to the fact that the variation of α modifies significantly its shape. If $\alpha = 1$ it yields the well known Bartlett (or triangular) window.

The Hanning window [22] is represented by:

$$g(t) = 0.5\left[1 - \cos\left(\frac{2\pi t}{t_w}\right)\right], \quad t \in \left[-\frac{1}{2}t_w; \frac{1}{2}t_w\right] \tag{4}$$

where t_w is the width of the window. In this case there is no shaping parameter.

Many authors studied the windows applied to the WFT in the perspective of their own characteristics. As referred previously, the choice of the window for a particular signal depends of the signal itself. Therefore, the automatic tuning of the window parameters is also dependent from the signal. Bearing these facts in mind, this article considers the window together with the signal.

3.2 Mutual Information

The WFT denoted by $F(\omega, \tau)$ can be interpreted as a bidimensional probability density function with two variables ω and τ as long as we normalize it according with the expression:

$$F_1(\omega, \tau) = \frac{\int_{t\min}^{t\max} x(t)g(t-\tau)e^{-j\omega t}dt}{\int_\tau \int_\omega \left|\int_{t\min}^{t\max} x(t)g(t-\tau)e^{-j\omega t}dt\right| d\omega\, d\tau} \tag{5a}$$

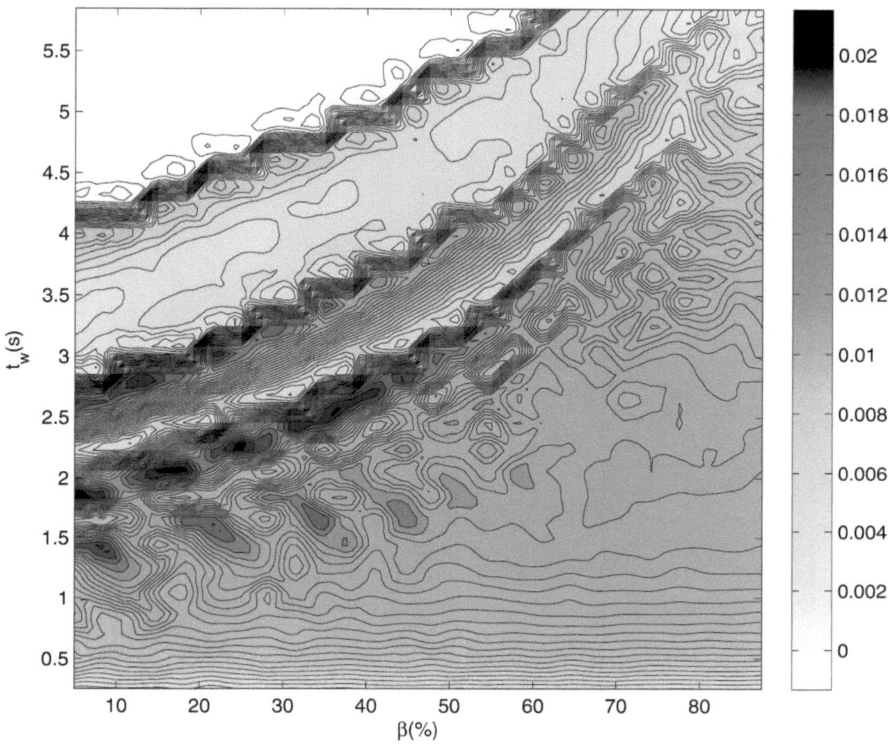

Fig. 4 The index $I_{av}(\omega, \tau)$ vs (β, t_w) of f_x^{imp} signal for the Gaussian window with $\alpha = 2.5$, $t_T = 8$ s

The marginal probability distributions of the variables ω and τ are $F_2(\omega)$ and $F_3(\tau)$, respectively, according with the expressions:

$$F_2(\omega) = \int_{\tau} F(\omega, \tau) d\tau \tag{5b}$$

$$F_3(\tau) = \int_{\omega} F(\omega, \tau) d\omega \tag{5c}$$

The mutual information [29, 30], or transinformation [31], is the index that measures the dependence of two variables in the viewpoint of the information theory. The mutual information for the two values of variables ω and τ is:

$$I(\omega, \tau) = \log_2 \frac{F_1(\omega, \tau)}{F_2(\omega) F_3(\tau)} \tag{6}$$

The average mutual information $I_{av} \in \Re$ between the two variables is given by:

$$I_{av}(\omega, \tau) = \int_{\tau} \int_{\omega} F_1(\omega, \tau) \log_2 \frac{F_1(\omega, \tau)}{F_2(\omega) F_3(\tau)} \, d\omega \, d\tau \tag{7}$$

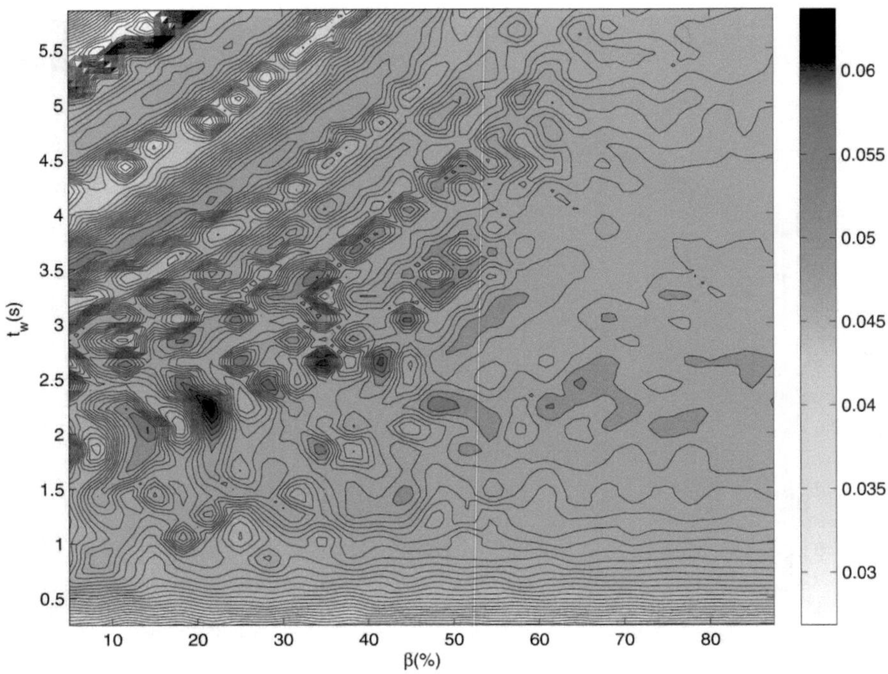

Fig. 5 The index $I_{av}(\omega, \tau)$ vs $(\beta,\ t_w)$ of the a_2^{liq} signal for the Gaussian window with $\alpha = 2.5$, $t_T = 20$ s

One application of I_{av} is to obtain the time lag required to construct the pseudo phase space. The I_{av} connects two sets of measurements with each other and establishes a criterion for their mutual dependence based on the idea of information connection. Additionally, I_{av} recognizes the non-linear properties of the variables [32]. By other words, the mutual information presents good results both for linear and nonlinear relationships between the variables. In this line of thought, the mutual information will be tested for tuning the WFT.

4 Results

To evaluate the average mutual information for WFT tuning, a set of signals captured in a robotic manipulator is used. Due to space limitations we depict only the most relevant features.

Figure 4 depicts the average mutual information $I_{av}(\omega, \tau)$ for the f_x^{imp} signal (force x component at the gripper of the robot for the rod impact) for the Gaussian window acquired during $t_T = 8$ s. The Gaussian window's width t_w and the overlapping time β vary in the ranges $0.25 < t_w < 6$ s and $5 <\leftrightarrow \beta < 90\,\%$, respectively, while adopting $\alpha = 2.5$. There are three locus of peaks and several experiments demonstrated

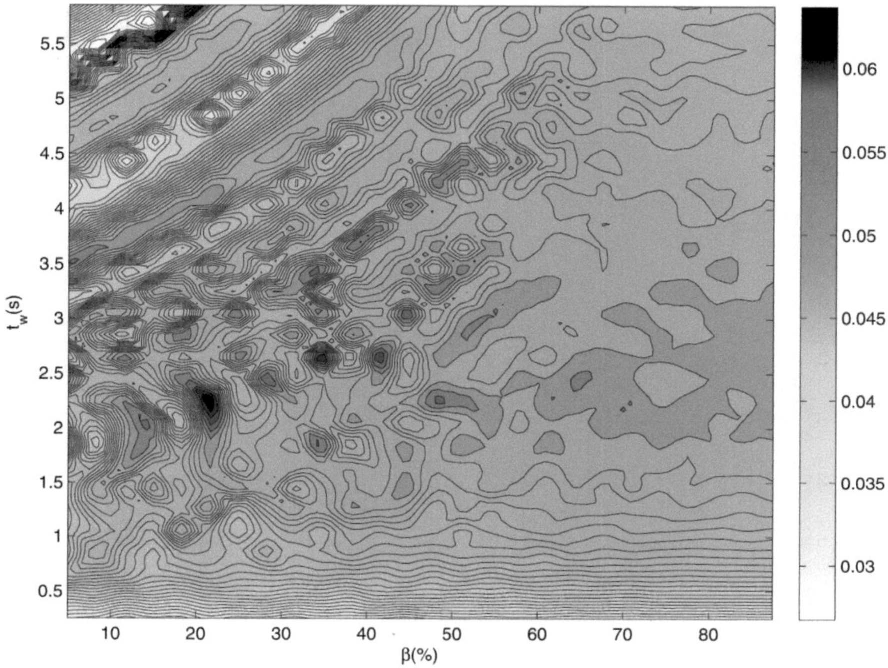

Fig. 6 The index $I_{av}(\omega, \tau)$ vs (β, t_w) of the a_2^{liq} signal for the Hanning window with $t_T = 20$ s

that the best tuning is found in the first curve that occurs in the increasing direction of t_w. Therefore, the best tuning parameters corresponds to the higher peak at $(\beta, t_w) = (36.7, 2.6)$.

Figure 5 depicts the average mutual information $I_{av}(\omega, \tau)$ of the a_2^{liq} signal (accelerometer 2 at the terminal link of the robot) when the robot carries the liquid container for the Gaussian window, acquired during $t_T = 20$ s. The range values of t_w, β and α is identical to those adopted in the previous example. Again, we choose the higher peak, located at the first curve in the increasing direction of t_w. In this case, the higher peak occurs at $(\beta, t_w) = (20.83, 2.29)$ which is the higher absolute peak of $I_{av}(\omega, \tau)$.

In the previous examples was adopted the Gaussian window and now we test the Hanning window. Figure 6 shows $I_{av}(\omega, \tau)$ for the signal analyzed in Fig. 5 (a_2^{liq}). The higher peak occurs at $(\beta, t_w) = (20.83, 2.29)$ corresponding to the best WFT tuning that, in fact, is the one obtained for the Gaussian window. The tests proved that the results for the Hanning window are very close to those obtained for the Gaussian window with $\alpha = 2.5$. For instance, analyzing the same signal, for the Gaussian window with $\alpha = 2.0$, the higher peak occurs at $(\beta, t_w) = (21.67, 2.29)$ which is different from that obtained for the Hanning window.

We can test also the fractional window (3). Figure 7 depicts the average mutual information $I_{av}(\omega, \tau)$ of the f_x^{imp} signal (force x component of the robot gripper

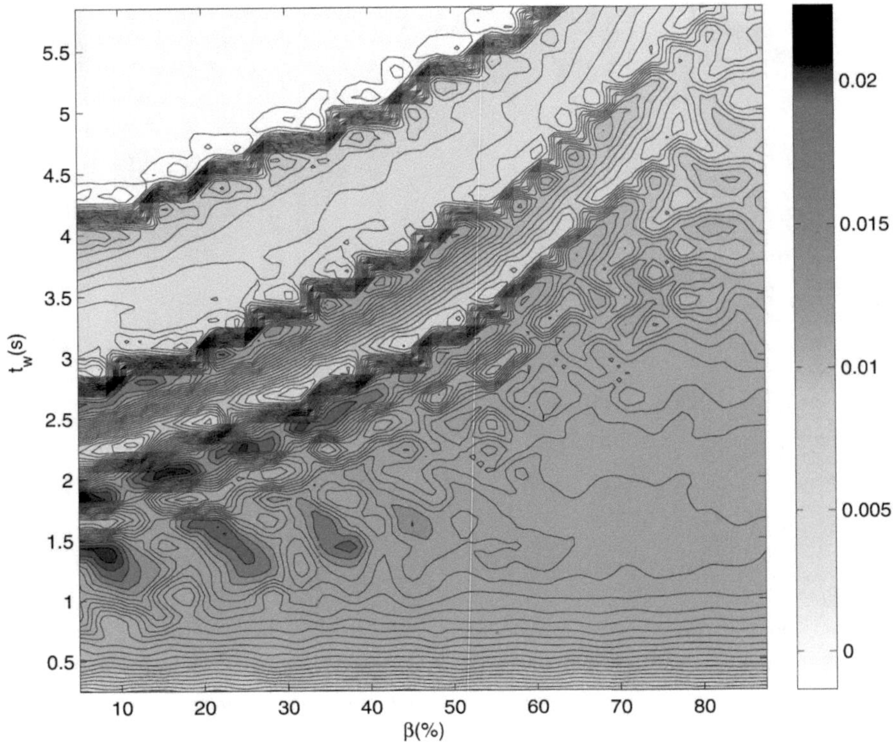

Fig. 7 The index $I_{av}(\omega, \tau)$ vs (β, t_w) of the f_x^{imp} signal for the fractional window with $\alpha = 1$, $t_T = 8$ s

for the rod impact) for the fractional window, acquired during $t_T = 8$ s. The range values of t_w and β are those used in the previous examples. If we choose the higher peak, located at the first curve in the increasing direction of t_w, we get the tuning parameters $(\beta, t_w) = (31.7, 2.3)$.

Previous examples show the applicability of the proposed method. Nevertheless, the practice reveals for some signals that it is difficult to choose the adequate tuning parameters (β, t_w). Figure 8 shows $I_{av}(\omega, \tau)$ vs (β, t_w) of the i_2^{liq} signal. There are several curves of peaks with identical values, and consequently it is difficult to select the most appropriate. Therefore, a deeper insight into the nature of this feature must be envisaged to better understand the behavior of $I_{av}(\omega, \tau)$.

5 Conclusions

The WFT is one of the most widely used time-frequency representations that is adopted in many fields of engineering. In order to use this technique several parameters must be defined according to the signal analyzed.

This work presents the average mutual information as an index that can be used for tuning the WFT. The window settings obtained with the proposed index revealed

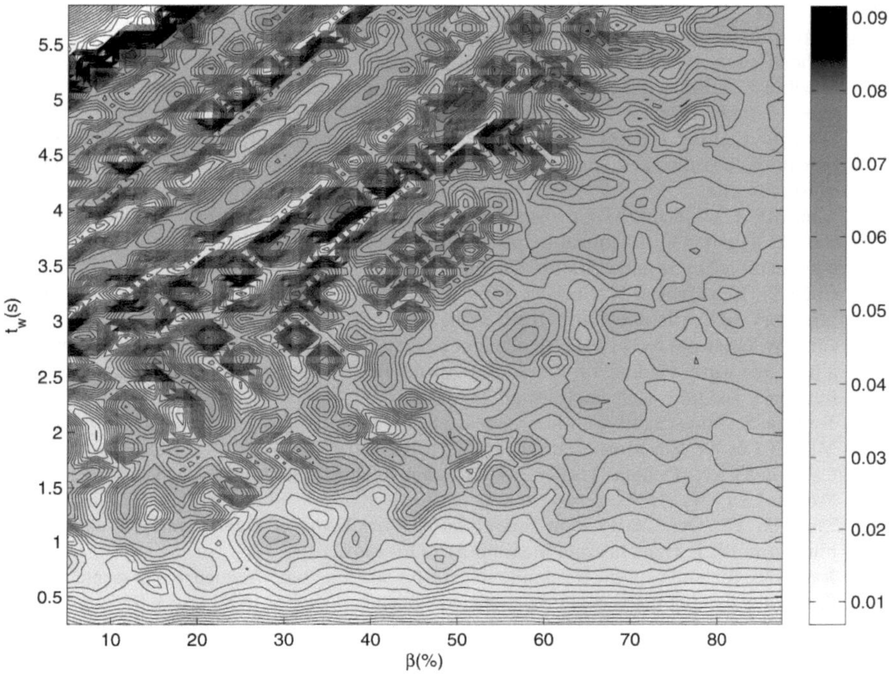

Fig. 8 The index $I_{av}(\omega, \tau)$ vs (β, t_w) of the i_2^{liq} signal for the Gaussian window with $t_T = 20$ s

to constitute a good compromise between the time and the frequency resolutions for the signals under analysis. The results based on experimental signals are promising and demonstrate the applicability and the effectiveness of the new approach. Nevertheless, the practice reveals for some signals it is difficult to choose the adequate tuning parameters based on the proposed method. Therefore, a deeper insight into the nature of this feature must be envisaged to overcome this limitation.

References

1. Allen RL, Duncan M (2004) Signal analysis, 1st edn. IEEE Press, Wiley-Interscience, New York. ISBN: 978-0-471-23441-8
2. Cohen L (1995) Time-frequency analysis: theory and applications, 1st edn. Prentice Hall, Wiley-Interscience, New York. ISBN: 978-0-471-23441-8
3. Flandrin P (1999) Time-frequency/time-scale analysis, wavelet analysis and its applications, 1st edn, vol 10. Academic, New Jersey. ISBN: 9780135945322
4. Mallat S (1999) A wavelet tour of signal processing, 2nd edn. Academic, San Diego. ISBN: 0080543030
5. Jones DL, Thomas W (1989) A resolution comparison of several time-frequency representations. In: Proceedings IEEE 1989 international conference on acoustics, speech, and signal processing, May 23–26, Academic, Glasgow, pp 2222–2225. ISBN 9780123743701

6. Jones DL, Thomas W (2 Feb. 1992) A resolution comparison of several time-frequency representation. IEEE Trans Signal Proc 40:413–420
7. Cohen L (July 1989) Time-frequency distribution: a review. Proc IEEE 77:941–981
8. Scheffer C, Girdhar P (2004) Practical machinery vibration analysis and predictive maintenance, 1st edn. Elsevier. ISBN 0750662751
9. Ozdemir C, Ling H (1997) Joint time-frequency interpretation of scattering phenomenology in dielectric-coated wires. IEEE Trans Antennas Propagat 45:1259–1264. Burlington, Massachusetts
10. Lima MFM, Machado JAT, Crisóstomo M (2006) Windowed Fourier transform of experimental robotic signals with fractional behavior. In: Proceedings IEEE international conference on computational cybernetics, Tallin, Estonia, pp 21–26
11. Chen W, Kehtarnavaz N, Spencer TW (7, July 1993) An efficient recursive algorithm for time-varying Fourier transform. IEEE Trans Signal Proc 41:2488–2490
12. Chen W, Griswold NC (1994) An efficient recursive time-varying Fourier transform by using a half-sine wave window. In: Proceedings of the IEEE-SP international symposium on time-frequency and time-scale analysis, pp 284–286
13. Tomazic S, Znidar S (1996) A fast recursive STFT algorithm. In: Electrotechnical conference, MELECON '96. 8th Mediterranean, vol 2, pp 1025–1028
14. Czerwinski RN, Douglas L (Feb. 1997) Adaptive short-time fourier analysis. IEEE Signal Process Lett 4:42–45
15. Zielinski TP (Oct. 2001) Joint time-frequency resolution of signals analysis using gabor transform. IEEE Trans Instrum Meas 50(5):1436–1444
16. Jones DL, Richard G (1992) A simple scheme for adapting time-frequency representations. In: Proceedings of the IEEE-SP international symposium, October 1992, pp 83–86
17. Jones G, Boashash B (5, May 1997) Generalized instantaneous parameters and window matching in the time-frequency plane. IEEE Trans Signal Proc 45(5):1264–1275
18. Djurovic I, Stankovic L (2003) Adaptive windowed fourier transform. Signal Process 83: 91–100
19. Stankovic L (2001) A measure of some time-frequency distributions concentration. Signal Process 81:621–631
20. Harris FJ (1978) On the use of windows for harmonic analysis with the discrete Fourier transform. Proc IEEE 66(1):51–83
21. Nuttall AH (1 Feb. 1981) Some windows with very good sidelobe behavior. IEEE Trans Acoust SpeechSignal ProcessASSP-29(1):84–91
22. Oppenheim AV, Schafer RW,Buck JR (1999) Discrete-time signal processing, 2nd ed. Prentice-Hall, Inc., Upper Saddle River
23. Ha YH, Pearc JA (2 Feb. 1989) A new window and comparison to standard windows. IEEE Trans Acoust Speech Signal Process 37(2):298–301
24. Aviyente S (2005) A measure of mutual information on the time-frequency plane. In: IEEE International Conference on acoustic speech and signal processing ICASSP2005, Philadelphia, vol 4, pp iv/481–iv/484
25. Aviyente S, Williams WJ (2005) Minimum entropy time-frequency distributions. IEEE Signal Process Lett 12(1):37–40
26. Baraniuk RG et al (2001) Measuring time-frequency information content using the Rényi entropies. IEEE Trans Inf Theory 47(4):1391–1409
27. Loughlin PJ, Cohen L (2004) The uncertainty principle: global local or both? IEEE Trans Signal Process 52(5):1218–1227
28. Lima MFM, Tenreiro Machado JA, Crisóstomo M (5, May 2005) Experimental set-up for vibration and impact analysis in robotics. WSEAS Trans Syst 4:569–576
29. Shannon CE (July, Oct. 1948) A mathematical theory of communication. Bell Syst Tech J 27:379–423; 623–656
30. Cover TM, Thomas JA (2006) Elements of information theory, Wiley series in telecommunications and signal processing. Wiley-Interscience

31. Spataru Al (1970) Theorie de la Transmission de l'Information—Signaux et Bruits. Editura, Bucharest
32. Trendafilova I, Brussel H Van (6, Nov 2001) Non-linear dynamics tools for the motion analysis and condition monitoring of robot joints. Mech Syst Signal Process 15:1141–1164

Can People With High Physical Movement Restrictions Access to Any Computer? The CaNWII Tool

N. Rodrigues, N. Martins and J. Barbosa

Abstract The potential of the common webcam, allied to the technology of the command of the well known Nintendo's game console, the WII, enlarge the possibilities of new ways to interact with computers. The presented work describe one of those ways, an accessibility tool to people with very restrict physical movements. The CaNWII tool allows an easy and robust way to interact with any computer.

Keywords Accessibility · Face tracking · WII controller · Mouse controller through head movements

1 Introduction

This paper describes a first phase of a developed work, with the main purpose settled in a research related with the creation of new interfaces human-machine, involved with the accessibility concepts. In this phase, the work made sure only in the use of movements capture techniques based in the use of vision sensors and infrared sensors, both available as low cost equipments.

The vision sensor chosen was the common webcam due its lower cost and portability. Although its reduced resolution, when used to perform simple tasks related to computer vision, the intended results could be achieved.

In December of 2006, Nintendo launch a new game machine, the Nintendo WII. The great innovation of this new console was its controller. It contains an accelerometer and an infrared light sensible camera. These functionalities allowed millions of users to interact with computers in new ways and where the main reason we chose to use that controller, named WII Remote or Wiimote [1].

N. Rodrigues (✉) · N. Martins · J. Barbosa
Institute of Engineering of Coimbra,
Polytechnic Institute of Coimbra, Coimbra, Portugal
e-mail: nunorod@isec.pt

N. Martins
e-mail: ncmartin@isec.pt

J. Barbosa
e-mail: jorbar@isec.pt

N. M. Fonseca Ferreira, J. A. Tenreiro Machado (eds.), *Mathematical Methods in Engineering,* DOI 10.1007/978-94-007-7183-3_26,

In this context, our first experiences done focused in the creation of an application to control the mouse cursor, capturing the head movements, through the combined use of an webcam and a Wiimote. This application is being used successfully, in the Association of Cerebral Paralysis of Coimbra (APPCC—Associação de Paralisia Cerebral de Coimbra). This association helps people of different age levels and different disability levels, both intellectual level and physical level. Physical disabilities imply, many times, the lack of motor coordination translated in the impossibility of these users using the interaction traditional equipments like the mouse or the keyboard.

Due the low cost, ease of installation and use and solution portability here proposes, the users of this center now have the ability to use the computer in their homes besides their stay in the association.

The name for the proposed system, CANWII, derives from the names of the two types of sensors used: the CAmera aNd the WII command.

The article is organized as follows. In the second section, it is described how to control the cursor through head movements, using a simple webcam. In the third section, it is described how to control the cursor through head movements, using the Wiimote. In the fourth section, we present some of the problems that each of the previous approaches has, and how to overcame those problems, joining both of the applications. In the end it was made some conclusions are made and some ideas are presented to improve the created tool.

2 Cursor Controller Using a Webcam

In this section we describe a system that tries to control the traditional mouse cursor of graphical user interfaces tracking the user's head movements, relying on a regular consumer webcam.

2.1 Introduction

The main goal of this system is to allow users to control the mouse cursor, used as a primary device of human-computer interaction in most modern computer systems, without requiring the use of the hands. This way, be it by convenience of by physical disabilities, users can still use computer programs, even if they can't use a hand to handle the mouse.

Choosing a low cost webcam allows the system to be considered for use in a wide range of situations where a high budget is not an option. Furthermore, the tracking of user's movements is based on computer vision techniques that do not require the usage of additional devices like colored markers or similar accessories.

In order to control the mouse cursor on screen, this system tries to track head movements analyzing the positions of the user's nose within the images that the webcam is continuously capturing when the service is running.

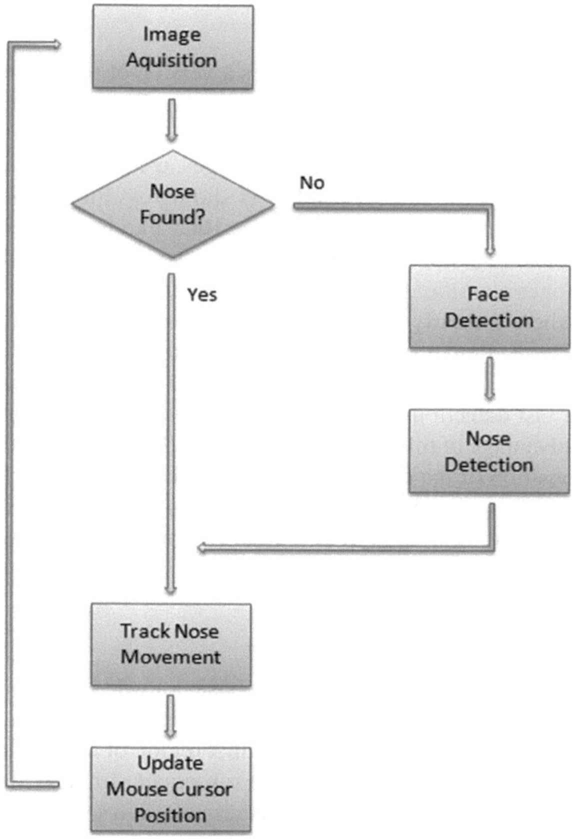

Fig. 1 Mouse through webcam subsystem architecture

2.2 *Implementation*

The overall software architecture for the mouse through webcam subsystem is illustrated in Fig. 1.

The system runs in the background, performing an endless loop and, for each iteration, the first step consists in capturing a low-res image (320 × 240 pixels) easily achievable by any low-end consumer webcam. Once the image is acquired, and if the system is just starting (or the user's nose position has been lost), control is handed to a face detection module which tries to find within the image the location of the user's face. The next step is to isolate the nose region and choose the most viable features to track the user's head movements. If the nose position was already determined in a previous iteration, the system tries to track the new position in order to find the movement that occurred and, using that value, update the mouse cursor position within the screen.

2.2.1 Face Detection

The face detection module uses a version of the technique developed by Paul Viola and Michael Jones [2], later extended by Rainer Lienhart and Jochen Maydt [3]. This technique is based on a supervised classifier, which was previously trained to recognize frontal human faces in images, using a boosting machine learning algorithm. To detect the face, a window is slid across the image and the pre-trained classifier tries to detect a series of simple Haar-like features (mostly based on rectangles or rotated versions of rectangles identifying lines and edges in the image). The Viola-Jones classifier has been implemented as a cascade of increasing complexity nodes, where the early nodes quickly reject areas of the image where a face is not present. These simpler nodes still identify a lot of false positives which are later discarded as more complex nodes are applied to the image. This kind of approach allows for a fast detection of the face to the degree where it can be used in real time. Taking advantage of this, in our system, to further enhance the correct detection of the initial position of the head in front of the webcam, we require that the user is sufficiently still for a specified number of frames such that the detected positions of the head within the captured images in those frames are approximately the same.

2.2.2 Nose Detection

Detection of the nose relies on basic knowledge of anthropometrical characteristics of human face—the approximate position of the nose is around the second third of the face. With this knowledge, that region of the image is scanned for those features which will be easier to track in subsequent frames. Intuitively, we easily understand that the best features to find recurrently in a sequence of images of the same subject are those that are nearly unique. So a good approach would be to look for points with significant change in them, for example points with a strong derivative. Of course, a point with a strong derivative can be a point belonging to an edge of some kind and this point will probably be very similar to other points in the same edge and, as such, hard to identify in later images. So a better approach would be to choose points where strong derivatives are present in two orthogonal directions. These features are called corners and are those with the more information to be more easily tracked from one image to the next.

A commonly definition of corner is based on a matrix of second-derivatives of the image intensities and the autocorrelation matrix of the image formed by the second-derivative values, over a small window around each point [4]. According to this definition, the corners are identified by points where the smallest of the eigenvalues of the autocorrelation matrix for that point is above a given minimum threshold [5]. In our system, we try to find a specified number of these features around the nose region (second third of the detected face rectangle) and define the nose position to be the mean of the selected features.

2.2.3 Track Nose Movement

To obtain nose movement between subsequent captured user images we have to track the new positions of the selected features for each new image we capture. The tracking is made relying in a sparse optical flow algorithm, based in the Lucas-Kanade algorithm [6]. This algorithm estimates the "velocity" each of the features moved from one frame to the next, thus allowing to obtain their new position. Obtaining the new center of mass of these points we get the new location of the nose. Sometimes, because of lighting inconsistencies or the angle the user is facing the camera the tracked position of some of the features is too distant from that center of mass. In those situations, those features are no longer tracked, and the position is determined by the remaining points.

The difference of position of the nose within the image is then smoothed by means of applying a low pass filter to the movement history so that excessive jittering of the cursor is avoided.

2.2.4 Update Mouse Cursor Position

The final step consists in sending a message to the operating system to simulate that the mouse moved by an amount proportional to the nose movement and this way update the cursor position accordingly.

3 Cursor Controller Using the WII Command

Another part of this work was the attempt to do what was presented in the previous section replacing the camera by the Wiimote and a pair of infrared lights. For this situation, it's necessary that the computer used by this system have the capacity of communication via Bluetooth, since is the technology used by the Wiimote command.

3.1 Related Work

Regarding the use of the Wiimote command, it was the HCI, Johnny Lee, the first to show the word what could be done with this new command and that no one knew it was possible. From his projects stand the Wiimote whiteboard and the head tracking system. The first application allows the creation of an interactive board which transforms any surface (wall, upper table, computer screen, etc.) in an interactive screen. To surfaces that aren't a computer screen it's necessary to use a projector. The second updates the vision angle accordingly with the user position. This causes the creation of the effect that the screen is a visualization window to a three dimensional world, that is beyond it. The application presented here follows the same concept of this last application.

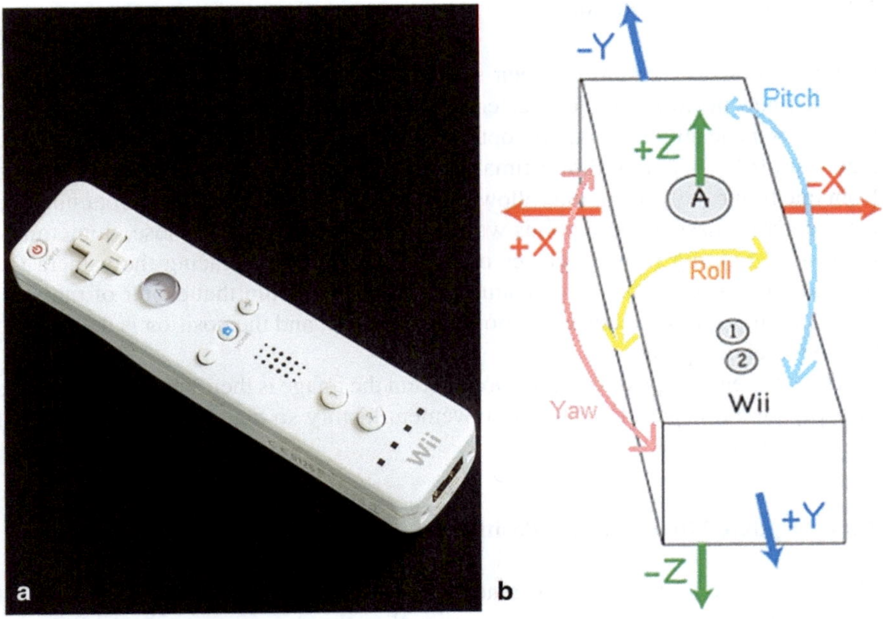

Fig. 2 **a** Wiimote controller. **b** Wiimote rotation angles

Other works also very important were those that resulted on libraries' creation to use the Wiimote command, specifically, control and data obtaining that Wiimote sends to the computer. From these libraries stands out the Wiiuse library, the WiimoteLib library and the Wiiyourself! library. The Wiiuse library [7] was chosen to the development of the application.

3.2 *Introduction to the Wiimote*

Wiimote command appearance can be seen in Fig. 2a. This command uses a combination of movements' sensor technology and infrared lights detection [8], so to create on games a certain level of interaction, considered impossible before.

Wiimote command uses an infrared camera to monitor light points that will be interpreted by a receiving station. These points can be any infrared light sources. However, camera isn't limited to the infrared bandwidth, because it uses a filter, placed ahead, in order to eliminate any strange light to that frequency range. The command technology associated to the camera can detect, at most, four points of infrared light. The resulting image, with the captured points, is used to estimate command position in the real world. Like this, when light points are in the lower part of the image, we know that in the command front part is raises in relation to the rear part. On the other hand, when Wiimote is bended down, the referred points are in the upper part of the image captured by the camera. The same concept is applied with the left and right movement, using now the lateral parts of that image.

The command has, also, an accelerometer that keeps the direction and gravity forces applied to the command. Its function is detecting the directions where the command is moving regarding the three axes of the world referential system/ZZ axis, XX axis and YY axis). These forces are used to calculate, in a discerning way, the direction of the applied force. The accelerometer can also detect the rotations regarding to the ZZ axis and XX axis. For a better perception, see Fig. 2b.

Finally, a Wiimote characteristic, useful in any application, is the wireless connectivity to a base station. Wiimote has a Bluetooth chip responsible for the information transmission and reception. The fact of not being applied any of the Bluetooth authentication or encryption functionality, makes possible, that Wiimote sent data can be recognized by any device.

3.3 Implementation

The system of using user movements of any computer like mouse movements works fixing Wiimote next to the monitor or to the surface where its content is being projected. The controlled user, like this, the mouse with infrared lights placed, for example, on glasses or helmet. Like this, the command is going to capture the infrared light points, under the user application control and translate them to the position where the lights point to.

Be able to translate the light point's movements into mouse cursor movements it was necessary to obtain the position of the light point in a certain moment, and immediately in the following moment to obtain the position again, so to be able to calculate the light direction and shift. After it, to introduce lights position in the mouse controllers, it was enough to follow what was done and presented in the previous section, this is, initialize the INPUT structure, used by Windows function, ant that synthesizes mouse clicks and movements. Mouse click emulation is done by the time the cursor stays in the same place.

The number of lights that can be used in this situation depends on the user difficulty getting moves them, but with a single light the movement won't be too smooth. To smooth the movement to make calculations more robust it were used two infrared lights to control user movements.

3.4 Results

Figure 3, shows the system being used by a user. Notice that Wiimote is placed on the computer monitor and the lights are on the users glasses. In this case the person only has to be able to move the head to use the computer. The writing can be done through any accessibility software since controlled by the mouse. Lights don't make the user uncomfortable, because they are placed in an accessory already belonging to the user. Aren't, because of that, non invasive.

Fig. 3 Mouse control system in operation

The fact that the user can control the cursor position through the lights, allows accessibility to people with several mobility difficulties, because those lights can be placed in any part of the user body. Despite the option taken here, the choice of the lights position to be placed in a user must be based of course on his comfort.

4 The CaNWII Tool

By themselves, each of the two subsystems has some limitations.

It is common sense that any light could affect the images captured by the camera. In fact, the cursor controller through a webcam is somewhat sensitive to light changes (for example, when a window is opened or a light is switched on). Sometimes, when this happens, the nose tracker is unable to identify the features that it was working with and the system must reset the tracking procedure, meaning the user has to adopt his initial position, centered in front of the webcam.

On the other hand, as described in Sect. 3, the Wiimote has an infrared camera that captures the position of the infrared light emitters. It was also said that the Wiimote camera captures, at most, four lights. Specifically, the controller captures the first four points of light it finds, ignoring the rest, if they exist. Knowing that the sun emits infrared light, we can understand that sun rays could affect the performance of the cursor controller through the Wiimote.

To overcome those problems, we combine both approaches presented in Sects. 2 and 3, creating the CaNWII tool. It is a redundant system, but it is this redundancy that makes the system more robust. In general terms, the main loop of the CaNWII program, each os the subsystems is running and obtaining its own cursor position. Then, it is calculated the mean of the two obtained positions. The final cursor position is saved.

When a new position, detected by any of the subsystems, is different from the saved position by more than a specified threshold, this means that one of the subsystems failed. In that case, only the position of the other subsystem is used to update the actual mouse cursor, while the failing subsystem resets its tracking procedures to restart providing correct positions. If both approaches fail, the whole system informs the user that it must assume his initial position so both tracking procedures can be restarted.

5 Conclusion

The described work in this paper is the initial result of a broader and deeper study of new human-machine interfaces. The results obtained till now are pretty promising, because the CaNWII tool is a robust system of low cost, easy to install, configure and to use, having regard to similar systems owners. It is a tool of high potential to APCC users, because besides using it in the time they are in the association, they can install it at their homes (now here are two users to use it effectively at their homes).

Was thus reached the main purpose of this project, this is, to facilitate the access to the new informatics technologies to people with special needs due to their physical limitations, that otherwise would be almost impossible.

Wiimote has also a vibration motor, a speaker and a sound amplification chip, which can give feedbacks. These functionalities are of extreme importance in the struggle of the lack of accessibilities, because they can help people with a low degree of vision or hearing. These were not used yet, but are being studied in order to gain the CaNWII tool.

Acknowledgments We would like to thank the APCC contributors, as well as the users, which helped us to gather the useful information to its development.

References

1. http://en.wikipedia.org/wiki/Wii_remote. Accessed 15 April 2010
2. Viola P, Jones MJ (2001) Rapid object detection using a boosted cascade of simple features. IEEE CVPR, Cambridge
3. Lienhart R, Maydt J (2002) An extended set of haar-like features for rapid object detection. IEEE ICIP, Santa Clara, pp 900–903
4. Harris C, Stephens M (1988) A combined corner and edge detector. In: Proceedings of the 4th Alvey Vision Conference, 147–151
5. Shi J, Tomasi C (1994, June) Good features to track. In: 9th IEEE conference on computer vision and pattern recognition, Seattle, WA
6. Lucas BD, Kanade T (1981) An iterative image registration technique with an application to stereo vision. In: Proceedings of the 1981 DARPA Imaging Understanding Workshop, 121–130
7. http://www.wiiuse.net. Accessed 15 April 2010
8. http://www.youtube.com/watch?v=ETAKfSkec6A. Accessed 15 April 2010

A Proposal for Detection and Estimation of Golf Putting

Gonçalo Dias, J. Miguel A. Luz, Micael S. Couceiro, Carlos M. Figueiredo, Nuno Ferreira, Pedro Iglésias, Rui Mendes, Maria Castro and Orlando Fernandes

Abstract This study presents an experimental research design of a PhD work, studying the effects of the variability in the performance of the Golf putting. The instruments used to analyze the putting were two digital cameras to detect the relevant dynamic objects (i.e., ball and putter) and a biaxial accelerometer to confirm the exact moment at which the putter hits the ball. To synchronize the instruments, it was used a trigger. The ball's trajectory and the putting movement were automatically analyzed based on visual detection and parameter estimation. The kinematic analysis

G. Dias (✉)
Faculty of Sport Sciences and Physical Education,
University of Coimbra, Portugal
e-mail: goncalodias@fcdef.uc.pt

J. M. A. Luz · M. S. Couceiro · C. M. Figueiredo · N. Ferreira
Department of Electrotechnics Engineering, Coimbra Institute of Engineering, Portugal
e-mail: miguel.luz@isec.pt

M. S. Couceiro
e-mail: micael@isec.pt

C. M. Figueiredo
e-mail: cfigueiredo@isec.pt

N. Ferreira
e-mail: nunomig@isec.pt

P. Iglésias · R. Mendes
Coimbra College of Education, Polytechnic Institute of Coimbra, Portugal
e-mail: piglesias@esec.pt

R. Mendes
e-mail:rmendes@esec.pt

M. Castro
Coimbra College of Health Technology, Polytechnic Institute of Coimbra, Portugal
e-mail: mac@estescoimbra.pt

O. Fernandes
Proto-Department of Sport and Health, University of Evora, Portugal
e-mail: orlandoj@uevora.pt

N. M. Fonseca Ferreira, J. A. Tenreiro Machado (eds.), *Mathematical Methods in Engineering,* DOI 10.1007/978-94-007-7183-3_27,

of the putting was performed using the Matlab software, to determine the amplitude, velocity and acceleration of the players' gestures. We concluded that the Golf putting action parameters are divided into different stages (i.e., backswing, downswing and follow-through) and that those can be useful to investigate the effects of variability in this movement.

Keywords Golf putting · Performance · Kinematic analysis · *Matlab* · Process variables

1 Introduction

The literature presents several studies on the Golf putting analysis, mainly focusing around its biomechanical properties [1–3]. However, both followers of cognitive theories of Motor Control [1] and Dynamic Systems Theories [4] analyze the same biomechanical variables in their studies, its use is quite different in movement analysis [2]. As "traditional" cognitive theories use experimental designs that privilege the analysis of product variables, the Dynamic Systems Theories approach drives its research design to studies that privilege process variables, being closer to the real and ecological situation [5, 6].Few studies have been made analyzing process variables, such as position, velocity/speed or acceleration in Golf putting execution (linear or angular) [2, 7]. Pelz work [3], as a reference for the study of the putting, suggests the possibility of stability and variability aspects in this movement execution, analyzed in expert and inexperienced players, presents considerable differences in amplitude, velocity and acceleration of execution.Accordingly to empiric knowledge as well as the considerations already made, it is concluded that the Golf putting need further analysis not only from a quantitative point of view, but also in a qualitative way privileging the study of process variables.Attending to the exposed information, this work presents the experimental design and methodological aspects that support a PhD thesis, in the analysis of the effects of variability in the Golf putting performance in expert subjects.

2 Experimental Design

It is presented the experimental design and methodological aspects that support this research, which analyzes the effects of variability on Golf putting performance. The adopted task was the Golf putting, implying the strike of a ball (*Titleist*; *model Pro V1*) with a putter (*Putter Jumbo Black Beauty*; size 35; standard) on a horizontal and still surface, placed on the ground over a ramp.

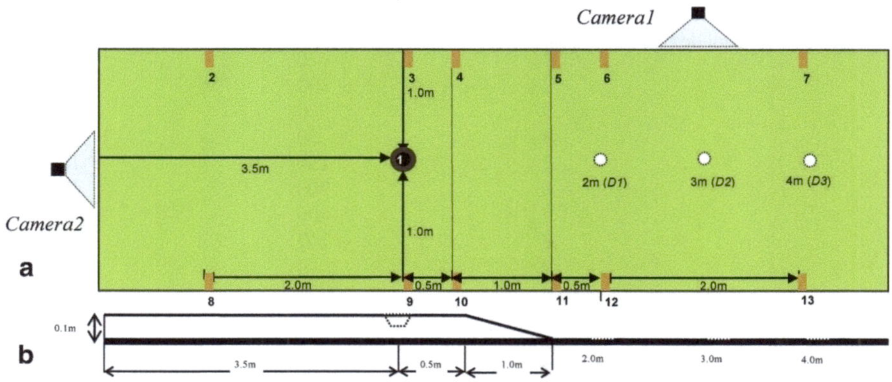

Fig. 1 Apparatus **a** Top view—reference points and cameras **b** Side view

2.1 *Apparatus and Procedure*

It was used an apparatus which included an artificial plain green carpet, used by *Minigolf* professionals, rectangular with no flaws, quite similar to the green's natural surface texture, with 10 m long, 2 m wide and 4 mm thick [8].

The ball's rolling speed on the carpet was measured with a stimpmeter, corresponding to 10 m, which is an acceptable value accordingly to the green's validation criteria of the Professional Golf Association (*PGA Tour*).

A real Golf hole was placed at 3.5 m of carpet's ending and at 1 m of each lateral extremity. Three black dots marked the putting places at 2 m (*D1*), 3 m (*D2*) and 4 m (*D3*). The dots were in the same direction of the hole, at 1 m of each lateral extremity of the carpet as well (Fig. 1).

Under the carpet it was placed a ramp with 1 m long, leveling up the carpet's surface in 10 cm high. Next to the ramp, it was placed a platform with 4 m long to keep that height (Fig. 1). The ramp allows the ball to get to the level of the hole.

2.2 *Coordinate System*

Since the putting for this study was recorded with digital cameras, in order to aid the data analysis, there were marked 13 references points on the carpet, according to Table 1 and Fig. 1, corresponding to the real coordinates of the experimental device. The centre of the hole is the reference for the adopted coordinate system, with the point (0.0). This coordinate system allows knowing the exact location of the ball in the apparatus, just by reading the result tables. These points were determined based on the work of [9], as it clearly characterizes the 3 plans of the experimental device.

The reference points were digitalized in a file, in the same format and order of the file used to store the real coordinates of the apparatus. These 13 reference points

Table 1 Real coordinates of the experimental device

Reference	x-axis (cm)	y-axis (cm)
1	0	0
2	– 200	100
3	0	100
4	50	100
5	150	100
6	200	100
7	400	100
8	– 200	– 100
9	0	– 100
10	50	– 100
11	150	– 100
12	200	– 100
13	400	– 100

(Table 1 and Fig. 1), are needed to make a match with the virtual coordinates (pixel points) of the video recordings, so that a correspondence between real and virtual coordinates could be made [9]. It was used the high level computational tool *Matlab*, to deal and analyze all the data [10].

2.3 Data Recording and Synchronization

To perform this study, two similar digital cameras were used, *Casio Exilim/High Speed EX-FH25*. The autonomy of the digital cameras was also considered, and in order to smoothly record the entire session without further problems, it were used rechargeable 2700 mA batteries and 16 GB memory cards.

One camera (*Camera1*) was placed at 4 m from the experimental device, in front of the subject while the other (*Camera2*) was placed 2 m after the apparatus ending capturing the entire device, in order to retrieve ball's trajectory and eventual error to the hole. These procedures were based on Knudson and Morrison's work [11], suggesting shooting distances of 2 to 10 m in studies of this kind. Both cameras were working still in their tripods and all the positioning and calibration features mentioned were used the same way for the entire study, guaranteeing reliability for later data analysis.

Camera1 was placed at 55 cm from the ground, pointing forward. It was shooting at 210fps at a resolution of 480×360 pixels and its lens with a focal length of 26 mm. *Camera2* was placed on a tripod as well, at 1 m and 55 cm high, with an inclination of 22 pointing down. It was shooting at 30fps at a resolution of 1280×720 pixels and its lens with a focal length of 26 mm. Some previous studies about putting performance analysis used 25fps to 50fps [12] which leads to the conclusion that the 210fps considered is an adequate procedure to study a gesture as precise as the Golf putting.

The apparatus and digital cameras were always in the exact same place, so that everything was recorded in the same conditions.

The digital cameras recordings allowed retrieving the following information:

1. Ball's trajectory through the apparatus;
2. Golf putting action parameters in distinct stages, backswing, downswing, ball impact and follow-through;
3. Position, velocity and acceleration of the putter during the movement;
4. Error distances in vertical length (VE), horizontal width (HW) and radial error (RE) to the hole.

Putter's movement monitoring was performed with a *Biovision* 2003 accelerometer, biaxial movement sensor with two orthogonal axis. This accelerometer dimensions are: 9.0 mm × 11.5 mm, with 50 g and it is sensitive to acceleration changes of 2G (gravitical force). The accelerometer was placed in the upper side of the putters head. The accelerometer sensor cables were connected to an input box (*Inputbox Biovision*) through two independent channels. A software called *DASYLab* v9.0 was used to retrieve and storage all data provided by the accelerometer. *DASYLab* was configured so that all signals would be recorded at 840 Hz (which is a multiple of both 30 Hz and 210 Hz) allowing the synchronization of all the information of both the accelerometer and the two camera footages. Additionally, red LEDs (light-emitting diodes) were placed in the frontal side of the putters head allowing a better understanding of the putter movement on *Camera1*.

The data synchronization of both digital cameras and the accelerometer was performed using a trigger. This instrument had a pressure button connected to the *Biovision Inputbox* and to two independent boxes with blue LEDs on them. These boxes were placed in a visible place in the corner of each camera field of view. In practical terms, every time this pressure button was manually triggered, the blue LED would turn on in both boxes (and therefore recorded in the videos), and the signal was simultaneously received in the *Biovision Inputbox* (and recorded through *DASYLab*).

2.4 Data Storage and Processing

The accelerometer simultaneously with the digital cameras recordings allowed retrieving the following information: i) putting action parameters in backswing, downswing, ball impact and follow-through stages (Fig. 2); and ii) position, velocity and acceleration of putter movement.

Although the accelerometer has retrieved putting action parameter simultaneously with the digital cameras, its main purpose consisted in giving the exact moment at which the putter hits the ball (it worked as some kind of an auxiliary trigger). This procedure turned out to be more reliable and precise then the video recordings processing for this specific moment. The retrieved data was saved through *DASYLab* in *ASCII* format files. The information on these files is easily imported in other programs, such as *Microsoft Excel* or *Matlab*. Every file obtained with accelerometer data or from the digital cameras was renamed according to a pre-established codification, allowing organizing all the information by the order of execution of

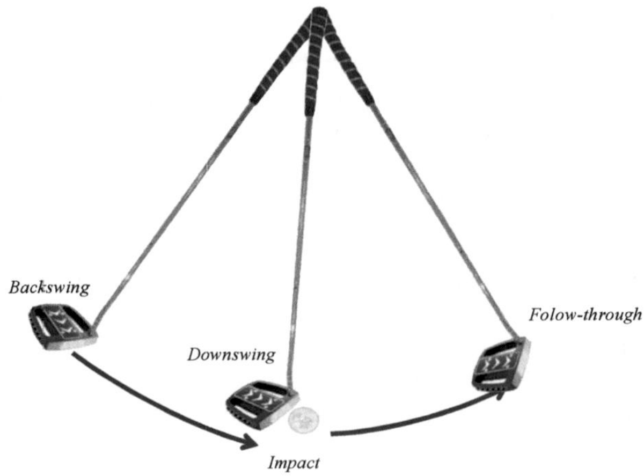

Fig. 2 Putter movement action parameter analysis (adapted from [12])

the subject, making it easier to access and deal with. All the information from the accelerometer and both digital camera recordings were processed using *Matlab*. A program was developed in order to determine biomechanical parameters, such as: position, velocity and acceleration of the putter, allowing the qualitative analysis of the putting [12].

Another program was developed in order to track ball's trajectory and perform quantitative analysis, by determining the distance error of the ball to the hole (if any existed). Both putter and ball's movement analysis were performed using automatic tracking of these objects. All data retrieval was performed by the same researcher and the adopted procedures were the same for all subjects. A trial study was made in order to check all instruments performance and validate the adopted procedures.

To confirm if the obtained information was reliable, all data of this previous study was compared with other researches concerning putt movement [12]. For data quality control, were also taken into account the researches of [12, 15].

3 Visual Data Analysis

Algorithms for object detection are one of the fundamental issues in several fields such as robotics [16]. The vision-based techniques can be classified into two categories, the stereo vision approach and the motion-based approach [17]. In the latter one, the motion field is computed from consecutive images obtained from the same camera, and other static or dynamic objects are then detected when their motions are dominant in the scene. One of the most reliable methods for object recognition is the color-based recognition algorithms (e.g., color histogram intersection, the color region adjacency graph and methods which use the statistics of color space components). Visual data analysis can be divided in two distinct steps: i) detect the several objects in the scene identifying the ball and the putter; and ii) estimate the ball and putter's trajectory.

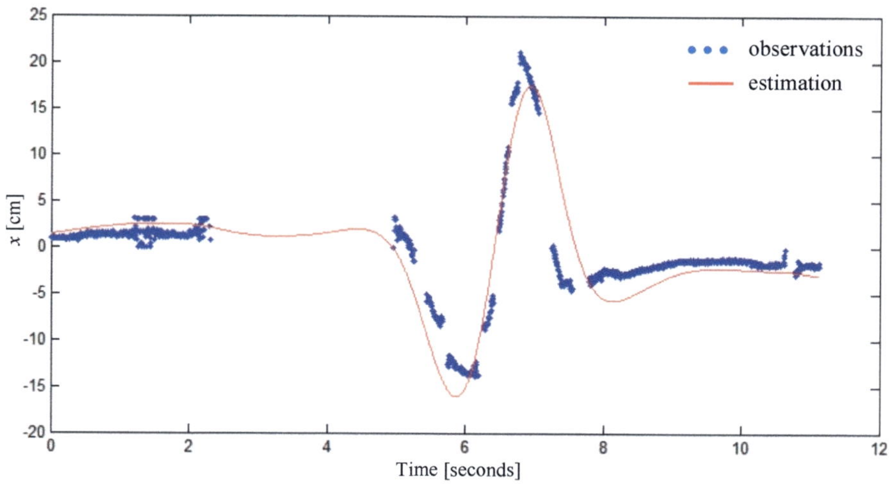

Fig. 3 Observations of the putter's trajectory retreived from Camera1 and estimation using the cubic smoothing spline method

Our detection algorithm is based on Schettini [18] that perform a search for an object with a similar shape, using color histogram intersection for object color match verification afterwards. One of the major challenges in this design is to estimate the ball and putter's trajectory, since the obtained videos (image quality) is affected by the ambient light. The problem of tracking dynamic objects and estimating their time-varying position has been studied extensively in robotics, engineering, computer vision, and several other areas [19]. In that sense, the appearance of objects is ambiguous, partly occluded, may vary quickly over time, and is perceived via a high dimensional measurement space.

The estimation of the ball and putter's trajectory was based on the cubic smoothing spline method. Considerable effort has been devoted over several decades to developing the mathematics of spline functions. In statistics, smoothing splines have been used in fitting curves to data ever since workable algorithms first became available in the late sixties [20]. The cubic smoothing spline f minimizes:

$$\sum_{j=1}^{n}\left(y_j - f\left(x_j\right)\right)^2 + (1-p)\int \ddot{f}(x)^2 dx \tag{1}$$

where y and x are the values of the observations over time respectively, n is the number of entries of observations, and the integral is over the smallest interval containing all the entries of x. The default value for the smoothing parameter, p, is chosen in dependence on the given data sites x defined between 0 and 1. We chose a smoothing parameter $p = 0{,}4$ based on the average spacing of the data sites.

Figure 3 depicts the retrieved data relative to the putter's trajectory obtained through image analysis using a simple color and area-based recognition algorithm and the respective estimation of the trajectory based on the cubic smoothing spline method.

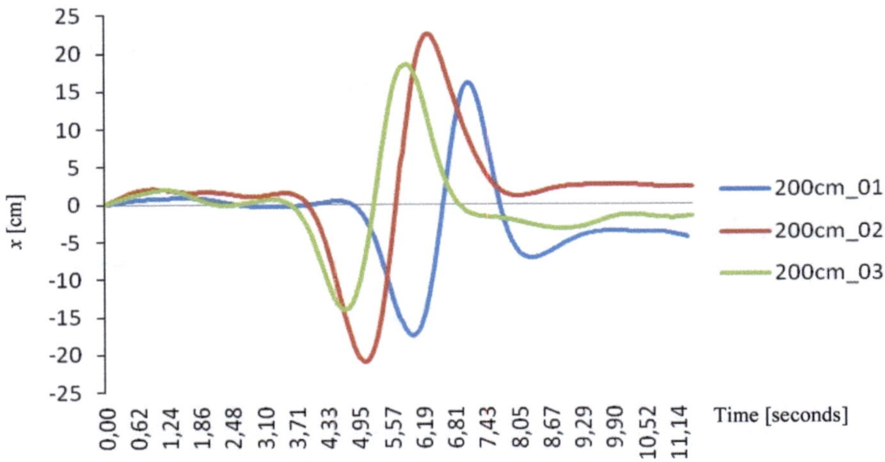

Fig. 4 Putter's position in the x-axis for the first subject at 2 m of the hole (distance D1) in his first three attempts

These results seam satisfactory and close to what is seen in the respective literature [2, 7].

4 Results

In order to get validation of the procedures described, it was performed a previous study including 3 inexperienced subjects (results of this study in Fig. 4, Table 2 and Fig. 5). The obtained results show that putter's trajectory during the movement is similar to a sinusoidal function (Fig. 4). It was also verified that after the impact on the ball (when the sinusoidal function passes through the origin), end the negative semi-cycle composed by both backswing and downswing, the subject tends to perform a positive semi-cycle (follow-through) with a maximum amplitude in module similar to the amplitude of the negative semi-cycle, getting near the origin in the end.

Table 2 allow analysing the subjects' performance based on position, velocity and acceleration amplitudes of each attempt. With the obtained data, two distinct analyses can be performed: i) intra-subject—analyzing the motor behavior and cinematic of the same subject in different attempts at the same distance and at distinct distances; and ii) inter-subject—analyzing the motor behavior and cinematic of each subject in different attempts at the same distance and distinct distances.

The most relevant intra-subject feature reports to ones performance in the three trials at the same distance. It is possible to see reduced putter amplitude in position, velocity and acceleration in the first strike at each distance.

The data shows that when a subject has a radial error different than zero (Fig. 5), the tendency is to raise the amplitude in position, velocity and acceleration of the putter in the second trial, in order to correct that error. In the third and last trial,

Table 2 Intra and inter-subject performance analysis of putter's maximum amplitude position, velocity and acceleration at 2, 3 and 4 m to the hole

	Distance (cm)	S1			S2			S3		
		E1	E2	E3	E1	E2	E3	E1	E2	E3
Position (cm)	200	33.53	43.43	32.52	63.66	64.61	54.42	31.58	36.92	28.93
	300	41.21	28.02	26.54	31.66	60.52	37.39	26.17	40.03	41.87
	400	21.18	24.60	29.09	66.93	81.11	55.74	37.00	27.04	48.25
Velocity (cm/s)	200	51.64	61.97	47.35	79.11	86.11	73.12	52.62	57.14	45.78
	300	63.91	45.68	46.65	53.41	81.59	56.82	44.10	58.80	59.08
	400	38.33	42.72	40.71	80.49	102.19	75.68	51.47	37.26	65.34
Acceleration (cm/s)	200	116.26	130.82	96.38	129.87	172.96	163.72	139.49	145.47	116.33
	300	130.42	101.95	116.53	136.06	168.29	129.95	109.54	131.34	132.11
	400	102.41	103.69	64.04	155.25	208.04	156.35	104.99	59.42	141.09

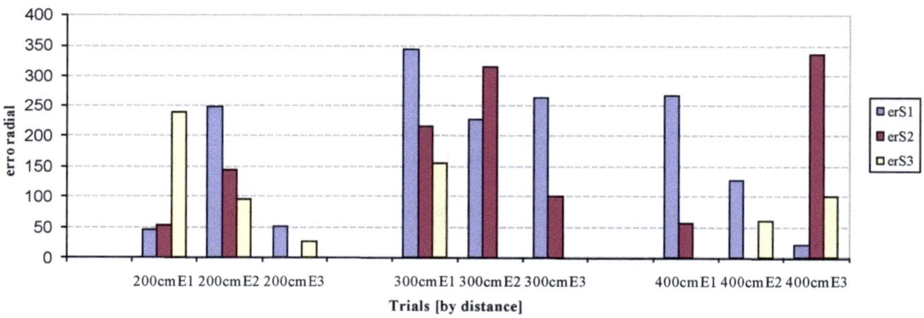

Fig. 5 Intra and inter-subject performance analysis of radial error of the ball to the hole (Legend: *E1/E2/E3* Trials, *cm* centimeters)

the subject has tendency to correct once again the amplitude in position, velocity and acceleration of the putter. Anyway, exceptions to this situation occur, when the subject can reduce or even eliminate the radial error in the second attempt, tends to maintain or raise the amplitude in position, velocity and acceleration of the putter.

About the inter-subject analysis, one feature stands out, and it's related with amplitude's maximum value in position, velocity and acceleration of the putter. For instance, subject 2 (S2) tends to have higher values in any of the attempts at the same distance, and at different distances (Table 2). The data available confirm that Golf putting is a complex gesture and differently executed from subject to subject.

5 Conclusion

According to the experimental design and methodological aspects mentioned, it can be concluded that the Golf putting action parameters are divided into different stages (i.e., backswing, downswing and follow-through) and that those can be useful to investigate the effects of variability in this movement.

The Golf putting action parameters can be accurately determined by processing the video information using detection and estimation techniques. With the implementation of those techniques, the study benefits by using automatic tracking to analyse the putter movement as well as ball's trajectory.

Specific lighting techniques must be studied and applied, since the exclusive use of ambient light should be avoided in the recordings, because its unpredictability can cause significant color changes in the obtained videos.

It is recommended that the instruments and the adopted methodological aspects in this work may be validated in other studies in order to consolidate and go even deeper in the known data about it.

References

1. Delay D, Nougier V, Orliaguet JP, Coelho Y (1997) Movement control in golf putting. Hum Mov Sci 16:597–619
2. Hume PA, Keogh J, Reid D (2005) The role of biomechanics in maximising distance and accuracy of golf shots. Sports Med 35(5):429–449
3. Pelz D (2000) Putting Bible: the complete guide to mastering the green. Publication Doubleday, New York
4. Kelso JAS (1995) Dynamics patterns: the self-organization of brain and behavior. MIT, Cambridge
5. Araújo D, Ripoll H, Raab M (2009) Perspectives on cognition and action in sport. Nova Science Publishers Inc., London, UK
6. Davids K, Button C, Bennett S (2008) Dynamics of skill acquisition—A constraints-led approach. Human Kinetics, Illinois
7. Karlsen J, Smith G, Nilsson J (2008) The stroke has only a minor influence on direc-tion consistency in golf putting among elite players. J Sports Sci 26(3):243–250
8. Mendes R, Dias G, Chiviacowsky S (2010) Golfe e Aprendizagem Motora: o Efeito da Interferê ncia Contextual na Aprendizagem do Putt. Braz J Motor Behav 5(Supplement):21–22
9. Fernandes O (2008) Tool for applied and contextual time-series observation (TACTO). 2nd Internacional Congress of Complex Systems in Sport. Madeira
10. Couceiro MS, Luz JMA, Figueiredo CM, Ferreira NMF (2011) Modeling and control of biologically inspired flying robots. Robotica 30(1):107–121 (Cambridge University Press)
11. Knudson DV, Morrison CS (2002) Qualitative analysis of human movement. Human Kinetics Publishers, Illinois
12. Paradisis G, Rees J (2000) Kinematic analysis of golf putting for expert and novice golfers. In: Proceedings the 18th International Symposium on Biomechanics in Sports. Hong Kong, China
13. Wholesalegolf. Golf Putters (2009) Jumbo black beauty. http://www.wholesalegolf.co.uk.htm. Accessed 21 Dec 2009
14. Fernandes O, Caixinha P, Malta P (2007) Techno-tactics and running distance analy-sis using one camera. J Sports Sci Med 6:204–205
15. Barros RML, Brenzikofer R, Leite NJ, Figueiroa PJ (1999) Desenvolvimento e avaliação de um sistema para análise cinemática tridimensional de movimentos humanos. Revista Brasileira de Engenharia Biomédica 15(2):79–86
16. Ohya A, Kosaka A, Kak A (1997) Vision-based navigation of mobile robot with obstacle avoidance by single camera vision and ultrasonic sensing. In: Proceedings of the 1997 IEEE/RSJ international conference on intelligent robots and systems, IROS '97, Ibaraki, Japan
17. Kim YG, Kim H (2004) Layered ground floor detection for vision-based mobile robot navigation. In: Proceedings IEEE international conference on robotics and automation, ICRA'04, Incheon, South Korea

18. Schettini R (1994) Multicolored object recognition and location. Pattern Recogn Lett 15: 1089–1097
19. Luber M, Arras KO, Plagemann C, Burgard W (2009) Classifying dynamic objects. Autonomous Robots 26(2–3):141–151
20. Reinsch CH (1967) Smoothing by spline functions. Numer Math 10:177–183

Analysis of Electricity Market Prices Using Multidimensional Scaling

Filipe Azevedo and J. Tenreiro Machado

Abstract This paper studies the impact of the energy upon electricity markets using Multidimensional Scaling (MDS). Data from major energy and electricity markets is considered. Several maps produced by MDS are presented and discussed revealing that this method is useful for understanding the correlation between them. Furthermore, the results help electricity markets agents hedging against Market Clearing Price (MCP) volatility.

Keywords Multidimensional scaling · Electricity markets · Energy markets · Hedging · Econometric models · Electricity price volatility

1 Introduction

Due to the specific nature of the electricity commodity, namely its non-storability, and due to the necessity of maintaining the electrical system constantly in balance, wide fluctuations on spot market prices occur. This effect, when associated to heat or cold climate waves, can stimulate the spot price to climb up to 1,000 % for short periods of time [1]. Therefore, the volatility is unusually high even when compared with other energy markets such as oil or gas. Another implication of the electricity non-storability is the impossibility of transferring a certain amount of energy from one part of the world to another one, without considering transmission restrictions. However, besides the instantaneous nature of the product electrical energy, factors like the uncertainty associated to fuel prices, energy demand, generation availability or, even, social and political events have also a high impact on price volatility [2–5].

Facing this state of affairs, electricity market agents have to deal with the necessity of understanding phenomena that are at the basis of market price evolution. The knowledge of those factors allows decision makers to develop the most adequate set

F. Azevedo (✉)
INESC TEC (formerly INESC Porto) and School of Engineering,
Polytechnic Institute of Porto, Porto, Portugal
e-mail: filipe.azevedo@inescporto.pt; fta@isep.ipp.pt

J. T. Machado
School of Engineering, Polytechnic Institute of Porto, Porto, Portugal
e-mail: jtm@isep.ipp.pt

N. M. Fonseca Ferreira, J. A. Tenreiro Machado (eds.), *Mathematical Methods in Engineering,* DOI 10.1007/978-94-007-7183-3_28,

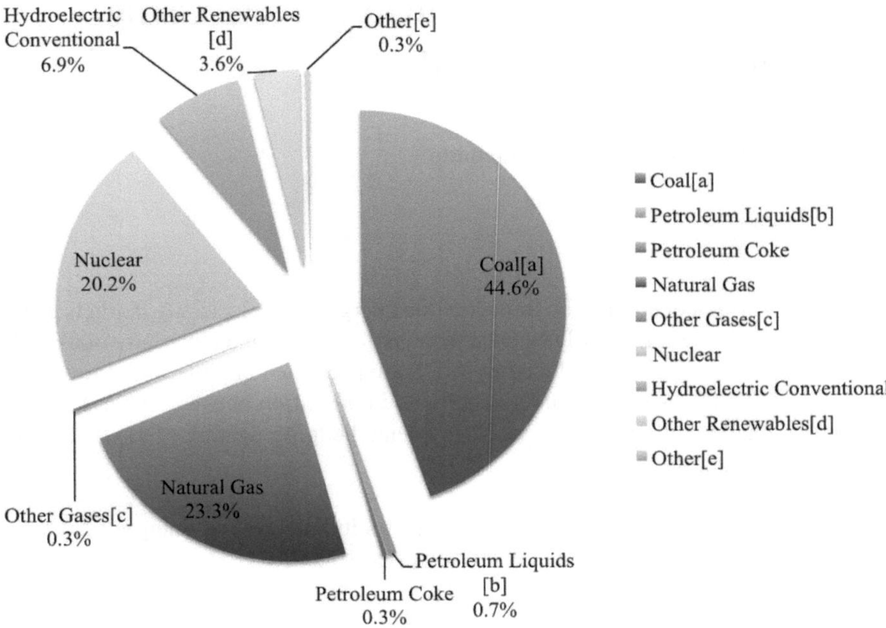

Fig. 1 Sources of electricity in the U. S. during 2009 *a* Anthracite, bituminous, subbituminous, lignite, waste coal, and coal synfuel. *b* Distillate fuel oil, residual fuel oil, jet fuel, kerosene, and waste oil. *c* Blast furnace gas, propane gas, and other manufactured and waste gases derived from fossil fuels. *d* Wood, black liquor, other wood waste, biogenic municipal solid waste, landfill gas, sludge waste, agriculture byproducts, other biomass, geothermal, solar thermal, photovoltaic energy, and wind. *e* Non-biogenic municipal solid waste, batteries, chemicals, hydrogen, pitch, purchased steam, sulfur, tire-derived fuel, and miscellaneous technologies

of strategies to sell, or to buy, electric energy in the spot, forward and futures market. In addition, those strategies are important to practice the hedge against electricity market price volatility and, simultaneously, to increase the profits.

Derivatives markets were introduced in electricity markets to allow their agents to eliminate the risk of credit and to turn the market more liquid. This effect is mainly due to the appearance of new agents that operate in traditional markets, that see in electricity markets an opportunity to withdraw dividends and to increase the efficiency in risk management. In addition, some of the new agents described above are also active participants on energy markets, like oil and natural gas.

The first power plants were driven by waterpower or by coal, but today we rely on a larger variety namely, coal, nuclear, natural gas, hydroelectric and petroleum, with a small contribution from solar energy, wind generators and geothermal sources. Figure 1 illustrates the production of electricity in the U. S. by source for the year 2009.

From Fig. 1 it is clear that, in the U. S. and for the year 2009, the main sources for the production of electricity are coal, natural gas and nuclear.

Table 1 Energy markets

Energy market	Abbreviation	Country
West Texas intermediate	WTI	USA
BRENT crude	BRENT	UK
Natural gas	NG	USA

Table 2 Electricity markets

Electricity market	Abbreviation	Country
OMEL electricity market	OMEL-PT	Portugal
OMEL electricity market	OMEL-ES	Spain
Energy exchange Austria	EXAA	Austria Germany
Gestore Mercati Energetici	GME	Italy
PJM interconnection	PJM	USA

For better understanding electricity markets, price behavior and their correlation with the evolution of energy prices, the Multidimensional Scaling (MDS) technique is used in this paper [6–10].

MDS is adopted in distinct scientific areas such as visualizing time-varying correlations across stock markets [11, 12], signal processing [13, 14], digital communications [15], adaptive controllers [16] and music [17]. However, presently there are no studies about applying MDS for analyzing electricity market prices and their correlation with the energy price evolution.

Monthly historical data, from July 2007 up to August 2010, for energy and electricity markets is used. It is considered data from July 2007, because OMEL defined prices for Portugal and Spain separately due to market splitting, from that date. In Tables 1 and 2 are presented the energy and the electricity markets used in this study. For PJM Interconnection electricity market is used Locational Marginal Price (LMP) Load Weighted Mean Price.

Bearing these ideas in mind, the paper is organized as follows: Sect. 2 introduces the MDS method. Section 3 presents a case study. Section 4 discusses the results out coming from the MDS processing. Finally Sect. 5 outlines the main conclusions.

2 Multidimensional Scaling

MDS is a technique for the analysis of similarity or dissimilarity data on a set of objects [18]. Its main purpose is to find a configuration of the data points in a low n-dimensional space, such that the original distance between objects in the full-dimensional space is represented with some degree of fidelity by the distances between points in the low-dimensional space. This means that observations that are close together in a high-dimensional space should be close in the low-dimensional space and vice-versa. Many aspects of MDS were originally developed by researchers in the social science community and the method is now widely available in some statistical packages [19].

2.1 Classical Multidimensional Scaling

Classical scaling is also known under the names Torgerson scaling and Torgerson-Gower scaling, because the first practical method available was a technique presented in [6–9], and it is based on theorems developed in [7, 10]. The fundamental idea of classical multidimensional scaling is to transform the distance matrix into a cross-product matrix and, then, to find its eigen-decomposition, which gives a Principal Component Analysis (PCA). Due to this reason in some literature classical multidimensional scaling is also referred as PCA. Like PCA, MDS can be used with supplementary or illustrative elements, which are projected into the dimensions after they have been computed.

2.2 Nonclassical Multidimensional Scaling

Nonclassical multidimensional scaling creates a configuration of points whose inter-point distances approximate the given dissimilarities. This is sometimes a too strict requirement and non-metric scaling is designed to relax it a bit. Instead of trying to approximate the dissimilarities themselves, non-metric scaling approximates a nonlinear, but monotonic, transformation of them. Because of the monotonicity, larger or smaller distances on a plot of the output will correspond to larger or smaller dissimilarities, respectively. However, the nonlinearity makes only an attempt to preserve the ordering of dissimilarities. Therefore, there may be contractions or expansions of distances at different scales.

There are two forms of nonclassical multidimensional scaling namely, metric scaling and nonmetric scaling. In metric MDS it is created a configuration of points such that their inter-point distances approximate the original dissimilarities. One measure of the goodness of fit of that approximation is known as the "stress". Nonmetric MDS has a slightly less ambitious goal than metric scaling. Instead of attempting to create a configuration of points, for which the pairwise distances approximate the original dissimilarities, it attempts only to approximate the ranks of the dissimilarities. Another way of saying this is that nonmetric MDS creates a configuration of points whose inter-point distances approximate a monotonic transformation of the original dissimilarities [8, 9,18–21].

3 Case Study

This study aims to study the correlation between energy and electricity markets prices using nonmetric scaling. To achieve this goal, historical data from major energy, stock and electricity markets is used.

3.1 Energy and Electricity Markets

Tables 1 and 2 present the energy market prices, and the electricity markets, respectively.

3.2 Nonmetric Scaling Stress Function

In this case study is adopted the Nonmetric Scaling form of Nonclassical MDS. Moreover, two measures, namely the City Block and Standardized Euclidean distance metric function defined in (1) and (2), respectively, are compared and used to measure the distance (d_{st}) between each pair of observations:

$$d_{st}^{CB} = \sum_{j=1}^{m} \left| x_{sj} - x_{tj} \right| \tag{1}$$

$$\left(d_{st}^{SE}\right)^2 = (x_s - x_t)\, V^{-1}(x_s - x_t)' \tag{2}$$

where m represents the number of observations, x the variables, V the n $\times$ n diagonal matrix whose j^{th} diagonal element is $S(j)^2$ and S the vector of standard deviations.

The effect of the two alternatives is compared in the sequel.

3.3 Number of Dimensions in MDS

The variation on the "stress" value with the number of dimensions to use is presented in Fig. 2. The goodness-of-fit criterion used, also known as the "stress", is the sum of squares of the inter-point distances.

From Fig. 2 we conclude that the required number of dimensions to use is $n = 3$ for the Cityblock and Standard Euclidean distances. However, we can verify that Cityblock metric function has, for all dimensions n of the MDS plot, lower "stress" values; therefore, in this work will be adopted the metric function Cityblock.

4 Results

The nonmetric MDS solution plot of the configuration for $n = 3$ is represented in Fig. 3. It is clear the emergence of three major clusters: U. S. PJM electricity market and energy group {PJM, NG, WTI, BRENT}, Iberian electricity market {OMEL-PT, OMEL-ES} and electricity market group {EXAA, GME}.

In U. S. the main sources for the production of electricity are coal, natural gas and nuclear. This is the reason why PJM electricity market is closer to natural gas

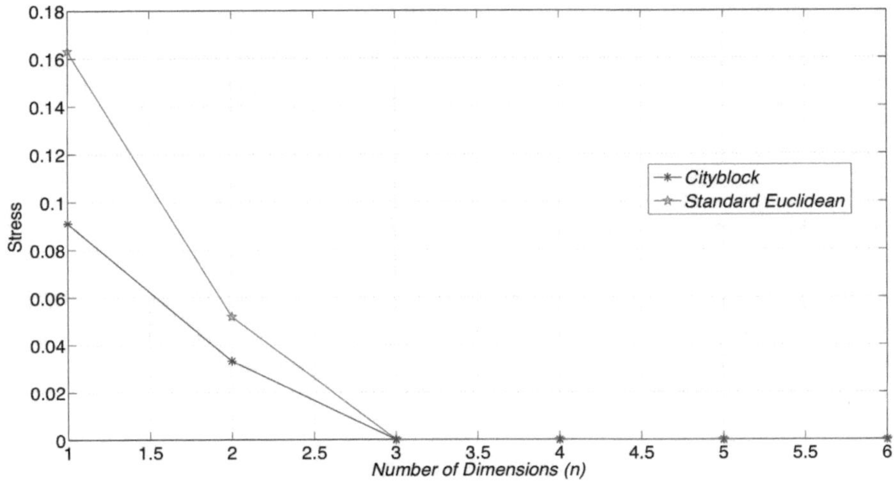

Fig. 2 Stress variation versus the number n of dimensions of the MDS plot

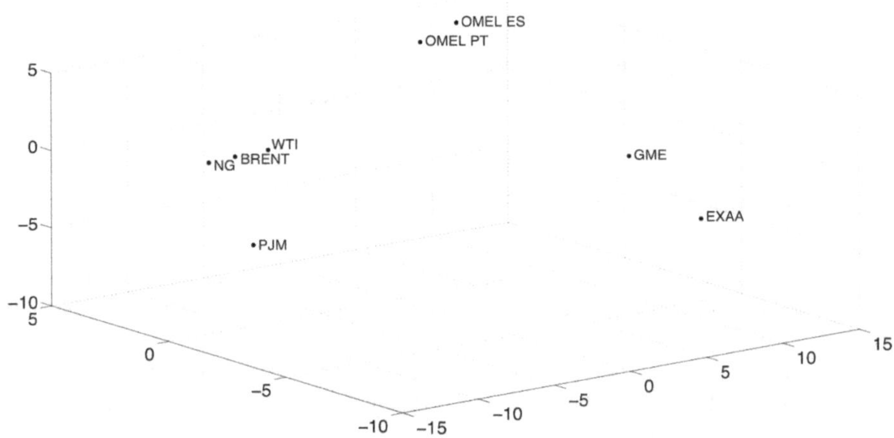

Fig. 3 Nonmetric MDS solution

energy markets (NG) than oil markets (WTI and BRENT). Moreover, the natural gas market (NG) used in this case study is for U. S., which reinforces its proximity to the PJM electricity market.

To check the fitting of the output MDS configuration and to analyze the disparities, it is useful to analyze the Shepard chart depicted in Fig. 4.

Figure 4 reveals that MDS has found a configuration of points in three dimensions whose inter-point distances approximates the disparities, which, in turn, are a nonlinear transformation of the original dissimilarities. The concave shape of the disparities as a function of the dissimilarities indicates that fitting tends to contract

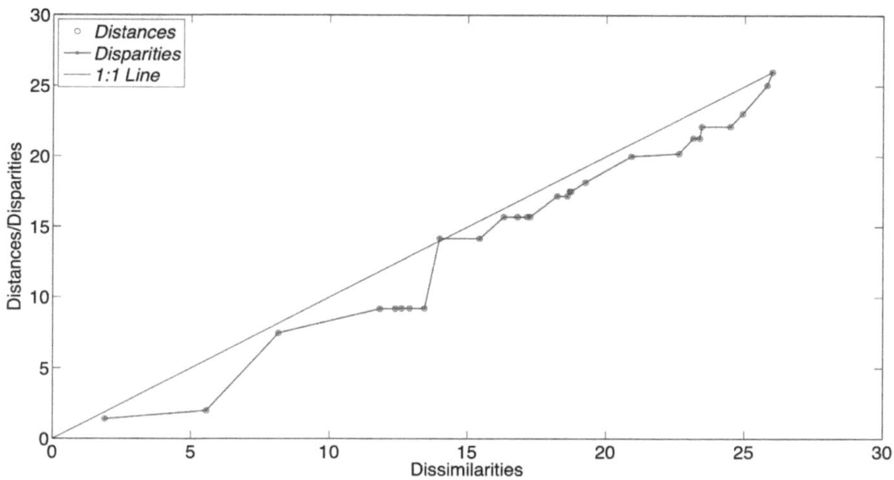

Fig. 4 Shepard plot for $n = 3$

small distances relative to the corresponding dissimilarities. This result is perfectly acceptable in practice and demonstrates that MDS can be easily adopted for the visual analysis of energy and electricity market prices.

5 Conclusions

In this paper we proposed a statistical graphical method for visualizing time-varying correlations between energy market and electricity market behavior. We illustrated the MDS-based method on the basis of monthly price average for three energy markets and five electricity markets.

The results show, clearly, the emergence of three major groups: U. S. PJM electricity market and energy group {PJM, NG, WTI, BRENT}, Iberian electricity market {OMEL-PT, OMEL-ES} and electricity market group {EXAA, GME}. From the described groups, natural gas is closer to PJM electricity market then oil group. This is due to the importance of combined cycle power plants upon the electricity production. The natural gas market (NG) used in this case study is for U. S., which reinforces its proximity to the PJM electricity market. In European electricity markets this effect is not so strong.

There are several issues relevant for further research. A first issue concerns applying the proposed method to alternative data sets, to see how informative the method can be in these cases. A second issue concerns incorporating the graphical evidence in an econometric time series model for improving empirical specification strategies.

Acknowledgments The author thanks the following organizations for allowing access to the data:

- U. S. Energy and Information, http://www.eia.doe.gov/
- Yahoo! Finance, http://finance.yahoo.com/
- OMEL Mercado de Electricidad, http://www.omel.es/inicio
- PJM: Interconnection, http://www.pjm.com/
- EXAA Energy Exchange Austria, http://www.exaa.at/
- GME Gestore Mercati Energetici, http://www.mercatoelettrico.org/En/Default.aspx

The authors acknowledge the help of João Soares to obtain the historical data.

References

1. Hull JC (2002) Fundamentals of futures and options markets, 4th edn. Prentice-Hall, Upper Saddle River
2. Shahidehpour M, Alomoush M (2001) Restructured electrical power systems—operation, trading, and volatility. Marcel Dekker, New York
3. Azevedo F, Vale ZA, Oliveira PBM, Khodr HM (2010) A long-term risk management tool For electricity markets using swarm intelligence. Electr Pow Syst Res 80:424–433 (Elsevier)
4. Azevedo F, Vale ZA, Oliveira PBM (2007) A decision-support system based on particle swarm optimization for multi-period hedging in electricity markets. IEEE Trans Power Syst 22(3): 995–1003
5. Skantze PL, Ilic MD (2001) Valuation, hedging and speculation in competitive electricity markets: a fundamental approach. Springer, Berlin (Kluwer Academic)
6. Gower JC (1966) Some distance properties of latent root and vector methods used in multivariate analysis. Biometrika 53:325–338
7. Eckart C, Young G (1936) Approximation of one matrix by another of lower rank. Psychometrika 1:211–218
8. Torgerson WS (1952) Multidimensional scaling: theory and method. Psychometrika 17(4): 401–419
9. Torgerson WS (1958) Theory and methods of scaling. Wiley, New York
10. Young G, Householder AS (1938) Discussion of a set of point in terms of their mutual distances. Psychometrika 3:9–22
11. Groenen PJ, Franses PH (2000) Visualizing time-varying correlations across stock markets. J Empir Financ 7:155–172
12. Tenreiro Machado JA, Duarte GM, Fernando B (2011) Identifying economic periods and crisis with the multidimensional scaling. Nonlinear Dyn 64:611–622
13. Martínez-Torres MR, García FJ, Marín SL, Vázquez SG (2005) A digital signal processing teaching methodology using concept-mapping techniques. IEEE Trans Educ 48(3):422–429
14. Rosario D, Romano J (2010) Multidimensional image processing for remote sensing anomaly detection. In: Proceedings of 2nd international conference on image processing theory tools and applications (IPTA), Paris, pp 471–476
15. Chen S, Mulgrew B, Grant PM (1993) A clustering technique for digital communications channel equalization using radial basis function networks. IEEE Trans Neural Netw 4:570–578
16. Bingulac SP (1994) On the compatibility of adaptive controllers (published conference proceedings style). In: Proceedings of 4th annual Allerton conference circuits and systems theory, New York, pp 8–16
17. Tenreiro Machado JA, Costa AC, Lima MFM (2010) Dynamical analysis of compositions. Nonlinear Dyn 65:399–412. doi:10.1007/s11071-010-9900-6
18. Borg I, Groenen P (2005) Modern multidimensional scaling—theory and applications, 2nd edn. Springer, Berlin
19. Martinez W, Martinez A (2005) Exploratory data analysis with MATLAB. Chapman & Hall/CRC, Boca Raton

20. Takane Y, Young FW, Leeuw J (1977) Nonmetric individual differences multidimensional scaling: an alternating least squares method with optimal scaling features. Psychometrika 42:7–67
21. Cox TF, Cox MA (2001) Multidimensional scaling, 2nd edn. Chapman & Hall/CRC, Boca Raton
22. Smith G, Pisecki W (September 2006) The time-triggered sensor paradigm. In: Proceedings of the 10th IEEE international conference on intelligent engineering systems, Honolulu, Hawaii, pp 297–300
23. Collina PL, Berben I (February 1997) Computer-based plant surveillance. IEEE Trans Comput 6(1):33–40

Mathematical and Statistical Concepts Applied to Health and Motor Control

Filipe Melo and Catarina Godinho

Abstract Variability and complexity are characteristic of human motor behavior. Research concerning movement patterns generation is a subject of interest shared by different areas like sports, health, or neurosciences. Among other motor abilities, postural control or gait, are abilities studied in normal and disabled subjects of different ages, using different types of methodologies and analytical approaches, including linear and nonlinear models. Nevertheless, depending on what we are looking for, these approaches can be more or less accurate for our purposes. Humans as biological systems, must be analyzed in a dynamical way, employing specific tools. The knowledge of the information given by these tools can be very helpful in medical research allowing the clinicians to identify and differentiate specific motor manifestations, like tremor, or postural instability, that are common to different pathologies, or even different levels of severity, like in Parkinson's Disease.

Keywords Motor Control · Movement dynamics · Variability and complexity · Parkinson's Disease

The research related with movement patterns generation, concerning sports and health, psychology, or neuroscience, uses concepts and methodologies related to the analysis of variability and complexity in human motor behavior. This type of approach includes mathematical models, as well as non linear tools, to explain movement dynamics. Several studies have been conducted in order to analyze motor behavior of different populations (normal subjects, PD patients, athletes, etc.), performing different tasks, such as postural control or gait tasks, in different conditions (with or without vision, with stable or unstable area of support, etc.).

Many clinicians specialized in medical research and clinical evaluation use linear models for prediction and intervention. However, it is very clear that linear models are considered limited in many cases and, in some specific cases, they are not the optimal approach [4].

F. Melo (✉)
Fac Motricidade Humana, Univ Tecn Lisboa, CIPER,
1499002 Cruz Quebrada Dafundo, Portugal
e-mail: fmelo@fmh.ulisboa

C. Godinho
Cooperativa de Ensino Superior Egas Moniz, Almada, Portugal

N. M. Fonseca Ferreira, J. A. Tenreiro Machado (eds.), *Mathematical Methods in Engineering,* DOI 10.1007/978-94-007-7183-3_29,

The term linear can be associated to only one dimension. Linear tools measuring variability give information about the quality, but not about the time-evolving dimension of the signal. These tools, including the statistics of range, mean, standard deviation, and coefficient of variation, while providing correct descriptions of inherent variability, are not helpful in explaining what is actually happening or varying within a system. They are incomplete in their justification about human movement variability. Mean values eliminate movement temporal variation and cover the correct composition of variability present in the movement action. The observed variations between the repetitions of a task are, usually, considered random and independent of past and future repetitions, which have been shown to be false. Perturbations to a dynamic system may lead to different patterns of macroscopic order that are not predictable by traditional methods [3].

The term nonlinear, in association with the term dynamics – *nonlinear dynamics*, can be associated to a system relating multiple dimensions, whose output is not proportional to its input. Nonlinear systems are related with the production of unpredictable responses revealing chaotic characteristics. Humans, as biological systems, are, generally, good examples of complex nonlinear systems, showing a great amount of inherent variability, in space and time, in their behaviors.

This variability, attested by differences in the observed behavior, when performing multiple repetitions of a task, reflects the numerous solutions available, traduced by the different adopted strategies. This plasticity, contrasting with the idea of an inflexible programming process, is guaranteed by the multiple complex synergies related to the neuromuscular system.

The idea of an optimal variability, associated to a characteristic movement behavior, is essential in a nonlinear perspective. For the Dynamical Systems Theory (DST) the increased variability in a system is related to an increasing instability which may indicate a possible change to another behavior. A biological system presenting no variability corresponds to an inert organism associated to a non dynamic condition. This invariance in movement behavior, must conduct to an atypical mapping of the sensory-motor homunculus, resulting in a disturbed motor function, usually related with more primitive behaviors, characterizing less complex systems.

The concept of optimal movement variability can be associated to a system whose dynamics lie between great variability and complete repeatability. Considering the great inventory of human motor actions, and more specifically gait as an example of a cyclic task, the different steps produced when walking cannot be consider either random or totally repeatable, showing instead a normal and healthy variability that can be considered optimal (within certain limits).

The problem of quantifying exactly the amount of optimal (normal) variability that a system should present, can be related to the identification and understanding of movement dynamic patterns.

1 Postural Tasks

Many studies tried to understand the strategies used by the postural control system to maintain the complex multi-degrees-of-freedom process controlled by the musculoskeletal system, in equilibrium with external forces, during quiet standing, or during the execution of an action.

The competence of maintaining an upright posture implies the use of a complex sensory-motor control system. We cannot adopt an upright position without producing a sway associated to an oscillation of the Center of Gravity (COG) or of the Centre of Pressure (COP). The analysis of the time-varying coordinates of the centre of pressure, known as a stabilogram, can show two types of control: a) a more reflexive closed-loop control in response to external perturbations; b) a more stabilizing open-loop control during longer periods of undisturbed stance. Earlier studies limited the analysis of these time series plots to statistics concerning the calculation of the length of the sway path, concerning both antero-posterior and medio-lateral directions, the average sway amplitude and radial area, ignoring the dynamic characteristics of the stabilograms (magnitude and direction of the COP displacements, the temporal ordering of COP time series coordinates, etc.). Mathematical techniques, employing non linear analysis, like stabilogram-diffusion analysis, recurrence quantitative analysis (RQA), can be applied to the study and interpretation of stabilograms, conducting to the extraction of repeatable, physiologically meaningful parameters.

The **Lyapunov Exponent(LyE)**, is a measure that quantifies the level of separation, or divergence with time, of nearby trajectories, in the state space. This separation of nearby trajectories is usually associated to instability, which can be characterized by Lyapunov Exponent, meaning that the higher the instability (divergence) of a system, the larger the value of the LyE.

The **Entropy** is a primary mathematical concept, firstly presented in information theory, representing a measure of the variability of a system. On the other hand, approximate entropy (ApEn) is a specific process to determine complexity, which quantifies the regularity or predictability of a time-series [8, 10]. Approximate entropy quantifies the probability that a series of data points, a certain distance apart, within a state space, will show comparable characteristics on the next incremental comparison [8, 9]. Time series presenting a greater probability of lasting the same distance apart upon comparison, will correspond to lower ApEn values, while time series presenting large differences in distances between data points will correspond to higher values of ApEn. In other words a more regular and predictable time-series is less complex than a less regular and predictable one.

2 Application in Medical Sciences and Research

The ApEn has been used in several medical settings during the last decade. It has been used to study different aspects like the effect of aging on cardiovascular dynamics [5], differences in heart rate control in normal and sudden infant death syndrome

Table 1 Analysis of the joint angular time series data from one PD patient and one healthy age-matched control subject during treadmill walking, using different non linear parameters. (Adapted from [[1], p. 39–43])

	LyE		ApEn		CoD	
	Control	Parkinson	Control	Parkinson	Control	Parkinson
Hip	0,117	0,102	0,314	0,278	2,012	2,015
Knee	0,166	0,191	0,353	0,505	3,151	3,441
Ankle	0,188	0,195	0,337	0,446	3,246	3,673

[11], or the effect of gender in growth hormone secretion [12]. Studies on the effect of aging on cardiovascular dynamics generally showed a good correlation between sickness and aging, and decreased ApEn values. These findings are in agreement with the general hypothesis advanced in the medical sciences that atypical physiological behavior can be related with more regularity, while normal physiological behavior is related with less regularity (great complexity) [9].

Some researchers have also used ApEn to characterize human movement. In such studies, learning and behavior reorganization are associated with changes in complexity [7, 15]. Morrison and Newell [6] used ApEn to analyze the level of active control during limb motion. In particular, they observed that the lower the ApEn value, the more active the control at the particular segment analyzed.

Different human movement studies have also used the ApEn measure with pathological populations. Vaillancourt and Newell [15] examined the complexity of resting and postural tremor in Parkinson's patients using finger accelerometer signals.

Buzzi [1] in his laboratory attempted to understand differences in locomotor variability of Parkinson's disease patients. The author examined the angular displacements of the lower limb joint for regularity changes in PD, during treadmill walking, at the off cycle of their dopamine treatment (Fig. 1).

The results reported in Table 1, concerning LyE and ApEn, showed that the knee and the ankle of both subjects presented more complexity than the hip, attested by higher values for these two joints, and also that the Parkinson's subject presented even more complexity in the time series than the control subject. An interesting observation was that the Parkinson's subject presented lower values for the hip joint. This finding possibly demonstrates an adaptation at the hip for the Parkinson's subject to compensate for the increased complexity and local stability at the more distal joints. The author suggested that a possible explanation for these results is a loss of independent sources of control due to the pathology.

However, statistical analysis didn't show significant differences in the LyE and ApEn values between PD patient and the control (Fig. 1). Nevertheless, the results showed a decreasing regularity from distal to proximal joints. More studies are needed in order to understand Parkinson's disease motor behavior.

Schmit et al. [13], in a study concerning sport activities, compared the spatiotemporal profile of postural sway, of trained ballet dancers and track athletes, during four different balance conditions (standing on a stable, or unstable surface, with the eyes open, or closed). Linear analysis of the results did not present significant differences between both groups during the normal vision condition, but presented increased variability in both groups during closed eyes condition and on a foam surface.

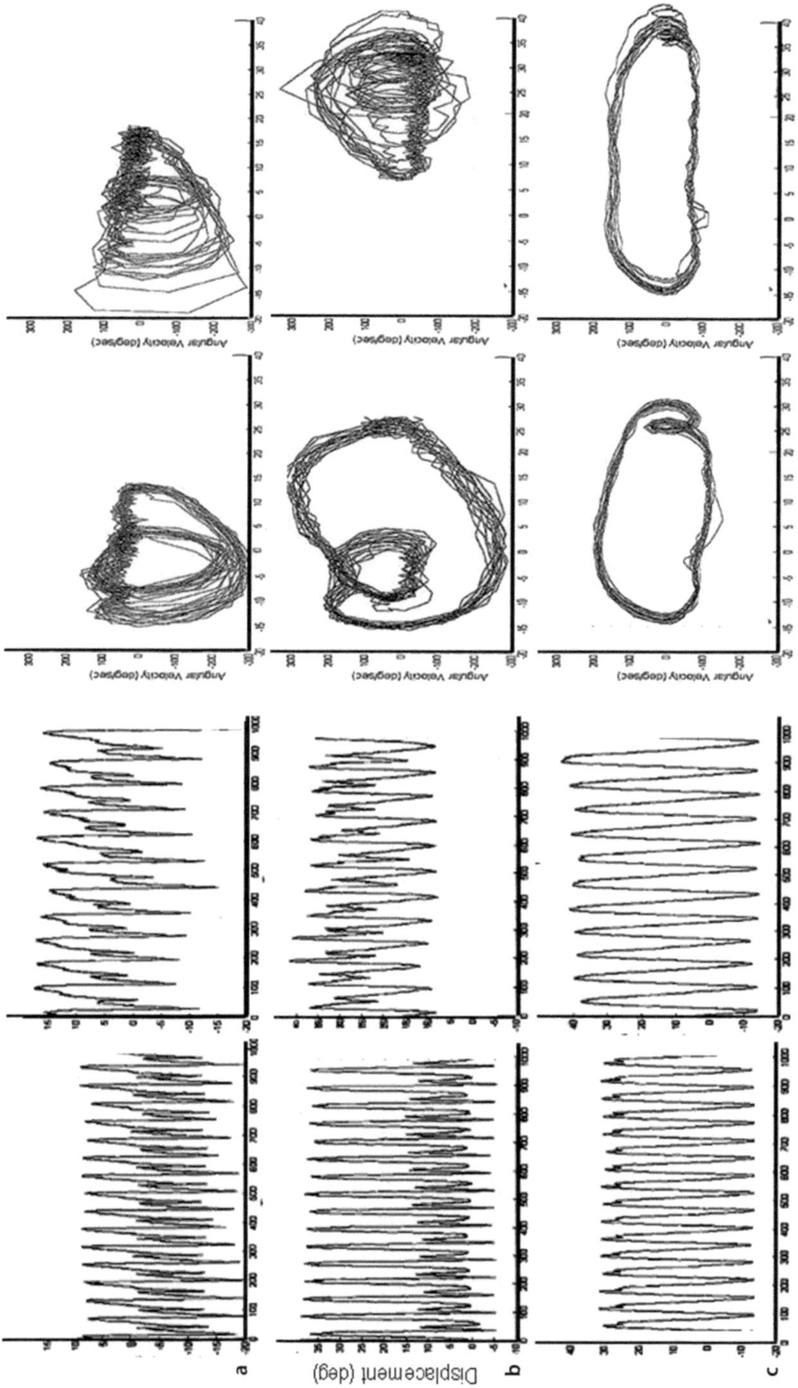

Fig. 1 Time series data (*left*) and phase plane plots (*right*) from different joints **a** ankle, **b** knee, and **c** hip) of one control (*right*) and one Parkinson's disease subject (*left*) when walking on a treadmill. (Adapted from [14])

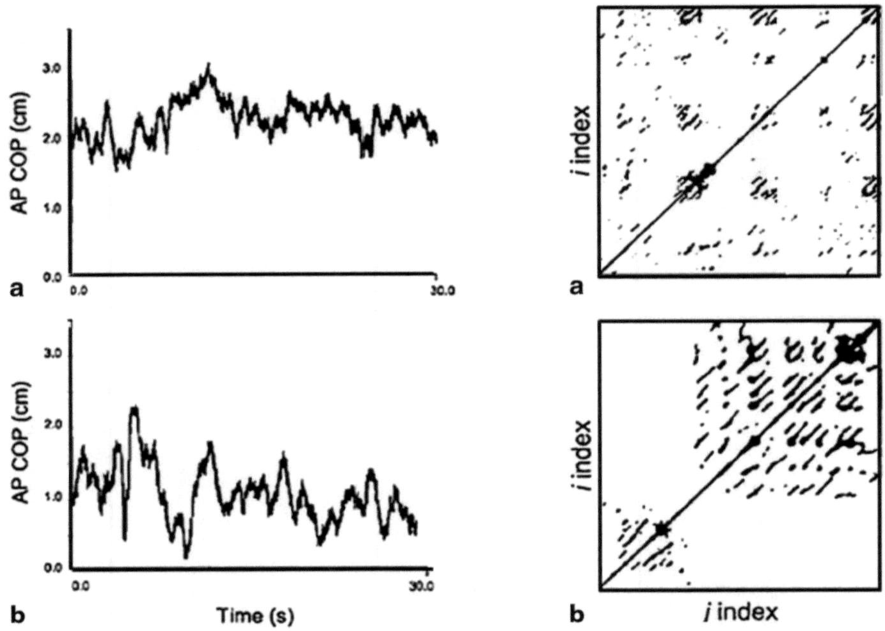

Fig. 2 Centre of Pressure (*COP*) time series in the anterior/posterior (*AP*) axis (*left*) and Recurrence Plots (*right*) of the data of *COP* time series of a dancer (**a**), and a track athlete (**b**). (Adapted from [13])

Non linear analysis (Fig. 2), and more specifically recurrence quantification analysis of the data, of both groups, revealed significant differences in postural sway. The postural sway of the dancers was less regular (lower recurrence), less sTable(lower maxline), less complex (lower entropy), and more stationary (lower absolute trend) than that of track athletes. Dancers, possibly as a result of focused balance training, exhibited different dynamic patterns of postural sway.

There are numerous oscillatory phenomena in motor control that occur regularly or irregularly, both in health and disease processes. These behaviors are clinically evaluated, during specific sessions. The observation of motor function time series presenting an irregular behavior, like tremor (Fig. 3), does not allow a clinician to infer, by visual inspection, whether the underlying process should be characterized as a deterministic (regular) or a stochastic (irregular) process.

Outcome assessment has become important in evaluating upper limb extremities in patients suffering from movement disorders. Nevertheless, some of the instruments used in clinical evaluation are quite generic, measuring grip strength, and range of motion, but are not able to evaluate daily life activities.

More instruments and methodologies of analysis are needed in order to accurately and objectively characterize patients' data during clinic evaluation sessions. Nonlinear analysis seems to provide promising methods to help diagnose and intervention in clinical settings.

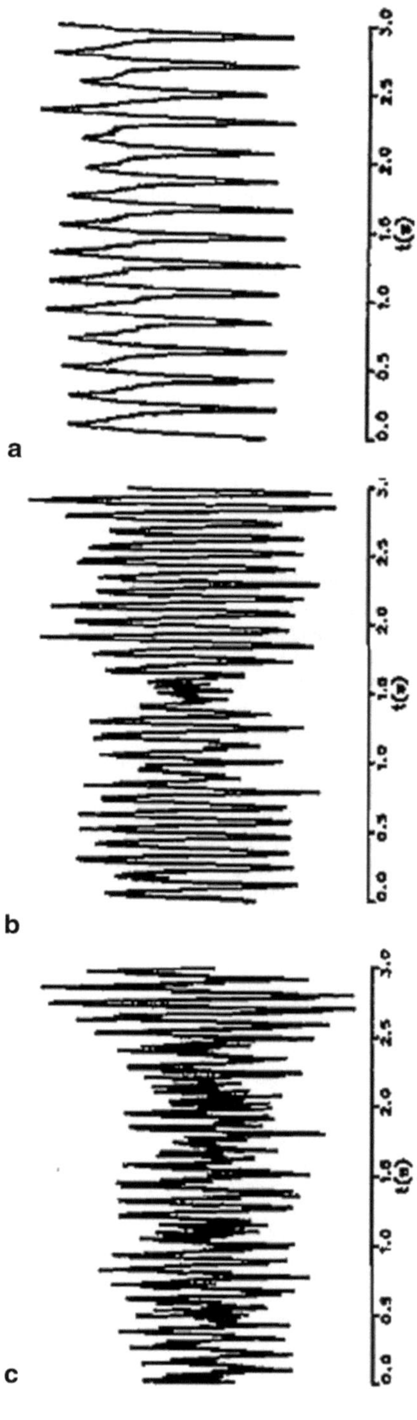

Fig. 3 Data concerning physiological (**a**), essential (**b**), and Parkinsonian (**c**) hand tremors. (Adapted from [2])

References

1. Buzzi UH (2001) An investigation into the dynamics of Parkinsonian gait. MS thesis, University of Nebraska at Omaha
2. Gantert C, Honerkamp J, Timmer J (1992) Analyzing the dynamics of hand tremor time series. Biol Cybern 66:479–484
3. Goldberger AL, West BJ (1987) Applications of nonlinear dynamics in clinical cardiology. Ann N Y Acad Sci 504:195–213
4. Habourne R, Stergiou N (2009) Movement variability and the use of nonlinear tools: principles to guide physical therapist practice. Phys Ther 89(3):267–282
5. Kaplan DT, Furman MI, Pincus SM, Ryan SM, Lipsitz LA, Goldberger AL (1991) Aging and the complexity of cardiovascular dynamics. Biophys J 59:945–949
6. Morrison S, Newell KM (1996) Inter- and intra-limb coordination in arm tremor. Exp Brain Res 110(3):455–464
7. Newell KM (1997) Degrees of freedom and the development of posture center of pressure profiles. In: Newell KM, Molenaar PCM (eds) Applications of nonlinear dynamics to developmental process modeling. Erlbaum, Mahwah
8. Pincus SM (1991) Approximate entropy as a measure of system complexity. Proc Natl Acad Sci USA 88:2297–2301
9. Pincus SM, Goldberger AL (1994) Physiological time-series analysis: what does regularity quantify? Am J Physiol 266:H1643–1656
10. Pincus SM, Gladstone IM, Ehrenkranz RA (1991) A regularity statistic for medical data analysis. J Clin Monit 7(4):335–345
11. Pincus SM, Cummins TR, Haddad GG (1993) Heart rate control in normal and aborted-SIDS Infants. Am J Physiol 264:R638–646
12. Pincus SM, Gevers EF, Robinson ICAF, Van der Berg G, Roelfsema F, Hartman ML, Veldhuis JD (1996) Females secrete growth hormone with more process irregularity than males in both humans and rats. Am J Physiol 270:E107–115
13. Schmit JM, Regis DI, Rilley MA (2005) Dynamic patterns of postural sway in ballet dancers and track athletes. Exp Brain Res 163(3):370–378
14. Stergiou N (2004) Innovative analysis of human movement. Analytical tools for human movement research. Human Kinetics, Champaign
15. Vaillancourt DE, Newell KM (2000) The dynamics of resting and postural tremor in Parkinson's disease. Clin Neurophysiol 111(11):2046–2056

MIX
Papier aus verantwortungsvollen Quellen
Paper from responsible sources
FSC® C105338

If you have any concerns about our products,
you can contact us on
ProductSafety@springernature.com

In case Publisher is established outside the EU,
the EU authorized representative is:
Springer Nature Customer Service Center GmbH
Europaplatz 3, 69115 Heidelberg, Germany

Printed by Libri Plureos GmbH
in Hamburg, Germany